VOLUME 11

AERATION:
Principles
and Practice

WATER
QUALITY
MANAGEMENT
LIBRARY

LIBRARY EDITORS
W. W. ECKENFELDER
J. F. MALINA, JR.
J. W. PATTERSON

WATER QUALITY MANAGEMENT LIBRARY

The immense environmental challenges facing the world now and in coming years can only be met through marshalling the talents of the best environmental engineers and scientists and through the use of innovative, cost-effective solutions.

The **Water Quality Management Library** addresses these challenges and reflects the organized efforts of leading international experts. Collectively, the eleven volumes in this library are a pertinent and timely compendium of water pollution control and water quality management. They form a unique reference source of international expertise and practice in key aspects of modern water pollution science and technology. With such valuable communication of knowledge using these and other books, we can hope to overcome the critical environmental issues challenging us today.

VOLUME 11

AERATION: Principles and Practice

James A. Mueller, Ph.D., P.E.
William C. Boyle, Ph.D., P.E.
H. Johannes Pöpel, Dr.-Ing

with significant contributions from:
Martin Wagner
David E. Gibson
Yeong-Kwan Kim

WATER
QUALITY
MANAGEMENT
LIBRARY

LIBRARY EDITORS
W. W. ECKENFELDER
J. F. MALINA, JR.
J. W. PATTERSON

CRC PRESS

Boca Raton   London   New York   Washington, D.C.

Library of Congress Cataloging-in-Publication Data

Mueller, James A.
    Aeration : principles and practice / James A. Mueller, William C. Boyle, H. Johannes
Pöpel ; with significant contributions from Martin Wagner, David E. Gibson,
Yeong-Kwan Kim.
        p.   cm. — (Water quality management library)
    Includes bibliographical references and index.
    ISBN 1-56676-948-5 (alk. paper)
    1. Sewage—Purification—Aeration. I. Boyle, William C. (William Charles), 1936– II.
Pöpel, H. Johannes. III. Title. IV. Series.

TD758 .M84 2002
628.3′5—dc21   2001052466
                                                                     CIP

Visit the CRC Press Web site at www.crcpress.com

No claim to original U.S. Government works
International Standard Book Number 1-56676-948-5
Library of Congress Card Number 2001052466
Printed in the United States of America 1 2 3 4 5 6 7 8 9 0
Printed on acid-free paper

Dedication

To our wives:
MaryBeth, Nancy, and Ursula

Preface

The use of aeration in the wastewater treatment field has been in existence for over a century. Each of the authors has been involved with the theory and application of aeration systems for a little less than half a century. It was a daunting task to put together what we considered the important principles underlying the mechanisms involved in aeration and show how they are applied in practical applications. The objective was to not only provide the basic theory, but also the current practice and latest applications, so the book would be useful to today's professional engineers as well as to future engineers now studying the field.

The task was conceived in the early 1990s by Wes Eckenfelder, who recognized a gap in the field. After a number of false starts, and with Bill and I soliciting the assistance of Johannes at the WEF convention in Chicago in 1997, it was begun in earnest in 1998—taking several years to complete. Johannes supplied an in-depth theoretical background as well as the European experience, especially in deep tank aeration. Bill supplied his experience in the diffused aeration area, and his desire to continually find the state of the art and how it is—and should be—practiced today. I enjoyed tying the theory and practice together to attain a good understanding of the most recent applications.

We received much assistance from our colleagues in the field. Especially noted on the title page are those who spent a great deal of time and effort providing critical input. They provided a needed jolt for each author to finish the endeavor by their knowledge of the field, review of concepts, and critical editing when required.

I would especially like to mention the assistance of a number of former students at Manhattan College. Richard Carbonaro scanned critical pictures while Rosanne Schirtzer, Clayton Conklin, Kevin Clarke and Sue Hildreth dug into the economics data from various agencies, a daunting task in itself. John Gormley, Engineering Librarian at Manhattan, continually obtained needed references and ran critical interference allowing me to ignore due dates.

The assistance of large municipal agencies in supplying critical information is acknowledged. The New York City Department of Environmental Protection, NYCDEP (especially Robert Adamski, John Leonforte, James G. Mueller (son), Hilary Einsohn, and Siobahn Rohan), coordinated efforts to obtain cost information on the New York City plants. The Metropolitan Water Reclamation District of Greater Chicago, MWRDGC (especially Hugh McMillan), provided the latest developments on the Chicago side channel aeration systems. The Middlesex County Utilities Authority, MCUA (especially Victor Santamarina), supplied insights into their high purity oxygen system upgrade.

Most of all I would like to thank God for giving us the energy and insights to complete this book. I look forward to it continuing to shed light on the profession and leading to the design and development of better aeration systems.

The poem that follows was composed by Jim McKeown, a member of our original oxygen transfer standards committee, who died of cancer in the winter of 1990–1991. It gives a bit of the history of the standards work, supported by the USEPA and ASCE, that Bill and I were involved with since 1976. It is a reminder that our work should never get the best of us—not above our relationships with each other, and with our God.

James A. Mueller

To the Study of the Drop and the Bubble

James J. McKeown

This is a poor story about
the dirty water band
who took to the field
when standardization was at hand.

After all, wasn't it clear,
although the data wasn't "purty,"
what was named the clean water test
was really very dirty.

The next step was upon us
it took only a spark of inspiration
for our band to begin the search
for the transfer of mass during respiration.

So we left the mainstream,
unfortunately, to no one's real sorrow
to pursue our fair dream
in a breach where Whittier did Narrow.

The first results were so startling,
every possible relationship linear,
we had to move east—to avoid the critique—
our findings were true, but only in Califor-ni-a.

Where we could test
to avoid bias oracle;
where wastewater was
by all standards, categorical.

Who could argue with respiration, although lazy
extracted from sewage
undergoing renovation
in New Jersey?

Convinced by such rationale
supported by those seeking to prove
that if things aren't quite right once
then they are always right when dual.

We joined the band within site
of sometime energetic Indian Point
where sometime aeration interfered
with our living in an otherwise elegant joint.

Although we did proudly stand,
our bloom soon lost its peak
when K_La escaped us
through an insidious leak.

Suitably humbled, we moved on
to further learn that
the non-steady test couldn't be rushed
when for nearly 20 minutes all in Ridgewood
town, everyone, refused to flush.

Let's not forget good can come from bad
for here in course bubbly, we examined off-gas.
And also, it can now be reported to superman's value,
we corralled fair krypton here by switching from
plastic to glass.

Undaunted we moved on to finale grand
all planned to succeed where Miller had fallen
now was the time to again make our stand.

We would continue to search to stoop to
lower ourselves to the depths where oxygen did lurk
barely dissolved in such dirty water
that we even enlisted one we called daughter—uh
clerk.

But success was to come
from more than mere traces.
Rather, from working together
with methods as different
as different as the looks on our faces.

Now, you think we were done,
but an epilogue beckons.
Because this band, as a group
learned of martinis Cajun and riverboat soup,
not to mention, the proper way to eat grapefruit.

But most important, to leave some work undone
so we could meet once more
to march to the cadence and the lure
in search of a sponsor to help us continue to work
toward making dirty water—pure.

March 23, 1984
ASCE Oxygen Transfer Standards Committee
Coronado, California

Table of Contents

1 Introduction

1.1 PURPOSE

1.1.1 Need and Growth in Field

At the beginning of the 20th century, activated sludge systems were developed into an economically viable secondary treatment method. Aeration, used to transfer oxygen to the biologically active masses of organisms within these systems, has been an important part of wastewater treatment as the use of activated sludge proliferated in the field. Significant changes have occurred in these systems as a result of not only advances in technology but also variations in the cost of energy required to operate them. The driving force of economics in some instances has brought the technology used in older systems back to the forefront. Due to the efficiency of power utilization, fine pore diffused aeration systems with full floor coverage have been rediscovered as an outstanding example of this technology.

Different types of aeration systems have been employed in the field, depending on location and specific treatment requirements. Large urban areas, where land is at a premium, have tended to use high rate systems. In contrast, areas that are more rural have used lower rate systems, generally requiring less operator involvement. The requirements for increased nutrient removal and better effluent quality have fostered the growth of systems that now incorporate not only the typical aerobic regions in aeration tanks, but the anaerobic and anoxic regions as well. Thus, numerous types of activated sludge systems have been developed to incorporate these different demands. These include deep tank aeration, high-purity oxygen, carousel or racetrack systems, anaerobic selector, and biological nutrient removal systems that attain nitrification and denitrification in different sections of the same tank. The basic principles governing the transfer of oxygen into the aerobic portion of these aeration systems are similar for all applications.

The impact of aeration systems on plant capital and operating costs is one measure of the importance of this unit operation to wastewater treatment. Table 1.1 summarizes the capital and operating costs of the aeration systems as a fraction of total plant costs. These costs were obtained for a number of plants in the New York metropolitan area, as well as a plant in Seattle, Washington, and one in Darmstadt, Germany. The date of the plant capital costs is given at substantial plant completion when secondary treatment is begun. Many of the contracts are written on a multiyear basis, sometimes spanning 10 to 20 years, especially for the large New York plants being upgraded. Construction of the Red Hook plant, a new facility, was begun in 1982 and completed in 1989 with secondary treatment on line in 1988.

Based on Table 1.1, the capital costs for aeration systems are typically between 15 and 25 percent of the construction costs for the total treatment plant. The exception to this statistic is the relatively low 5.57 percent aeration capital costs for

TABLE 1.1
Impact of Aeration Systems on Activated Sludge Treatment Plant Costs

Plant Name	Location	Design Flow, m³/s (MGD)	Type Aeration System	Capital Costs		Yearly Operating Costs		Reference
				Total Plant 10⁶ $ (year)	% Due to Aeration	Total Plant 10⁶ $/yr (year)	% Due to Aeration	
Coney Island	Brooklyn, NY	4.4 (100)	Diffused, fine pore	650 (1990)	20	4.43 (1998) 4.05 (1999)	20.1–25.5* 20.3–25.2*	(Conklin, 2001)
North River	Manhattan, NY	7.5 (170)	Diffused, fine pore	968 (1986)	5.57	7.12 (1998) 7.43 (1999)	15.7 16.8	(Conklin, 2001; Leonforte, 1998)
Red Hook	Brooklyn, NY	2.6 (60)	Diffused	232 (1988)	16.8	2.49 (1998) 2.29 (1999)	25 24	(Conklin, 2001; Leonforte, 1998)
Owls Head	Brooklyn, NY	5.3 (120)	Diffused	380 (1995)	27	7.15 (2000)	17	(Clarke, 2001)
West Point	Seattle, WA	5.8 (133)	High purity O₂ •surface •4 stage	229 (1995)	19.3			(Hildreth, 1999; Hildreth, et al. 1997)
MCUA	Sayreville, NJ	6.5 (147)	HPO •turbine •surface	95.5 (1974) +8.9 (1995)	19.3 100 Upgrade	16.4 (1997) 15.2 (1999)	19.5 before 13 after upgrade	(Schirtzer, 2000)
Darmstadt Central	Germany	0.46 (10)	Diffused, fine tubes with propellers •racetrack	95 (1995)	15	3.4 (1997)	11.4	(Poepel, 2001; Wacker, 1998)

* Including air scrubbers.

the North River plant in New York City. This plant, located in upper Manhattan, has two additional major construction costs associated with it. One is construction of the plant on piles over the Hudson River, and the other is the park constructed on top of the plant for use by local residents. The costs of the Coney Island and Owls Head plants include a complete plant upgrade, during which the facility maintained operations. This scenario is typically more costly than new plant construction. Due to the proximity of the local population, as in many New York plants, the Coney Island costs include covered tanks for all but the secondary clarifiers and a scrubber system to capture and treat air emissions before discharge.

FIGURE 1.1 Original submerged turbine system for MCUA plant showing aeration tank turbine drives (A), gear reducer (B), high purity oxygen delivery piping (C) and compressor room (D). (Photos courtesy of Middlesex County Utilities Authority, Sayreville, New Jersey.)

Operation costs for aeration in treatment plants typically account for 15 to 25 percent of the total plant operational costs including labor and chemical use. The energy consumed at the Coney Island plant by the blowers is 40 percent of the total energy, the remainder due to the numerous pumping systems and air scrubbers at the plant. For the high purity oxygen system in the Middlesex County Utility Authority (MCUA) plant in New Jersey, operational costs for aeration were reduced significantly from 19.5 percent of total costs to 13 percent after upgrading from turbine to surface aeration. A significant reduction in power demand occurred with the elimination of the large recirculating compressors and the cryogenic oxygen generation facility. A pipeline oxygen source was economically feasible and allowed simpler operation and maintenance with lower labor requirements for the treatment plant. Total operational costs for this facility are high due to the significant costs for sludge disposal after cessation of ocean dumping. Figures 1.1 and 1.2* illustrate the

---

* Figures 1.1 and 1.2 also appear in the color insert following page 84.

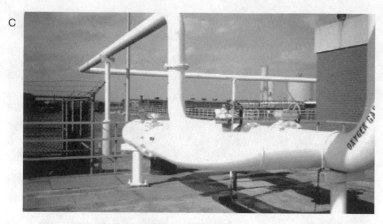

FIGURE 1.1 (continued)

differences in equipment requirements of the MCUA plant before and after upgrade to surface aeration (Schirtzer, 2000).

Costs due to aeration at the relatively simple racetrack system used in Darmstadt, Germany are only 11.4 percent of the operational costs. The capital and operating costs are high for such a small plant compared with the larger facilities in the U.S. This is due in part to economy of scale and to the higher degree of treatment obtained by the plant, which discharges into a small creek. The per-cubic-meter sewer charge for the contributing population is the second highest in Germany.

In addition to the wastewater treatment plants, where aeration systems have been employed historically, new applications of aeration systems are being used in the natural environment. Typically, these are used to improve dissolved oxygen concentrations to desired levels in natural waters where the demand for oxygen is greater than can be supplied by natural reaeration. These applications have the same basic principles governing the transfer of oxygen as those used in plant aeration systems.

In order to effectively incorporate the principles governing the design and analysis of aeration systems into this myriad of applications, an understanding of the basic principles involved in oxygen transfer is required. However, along with the principles,

FIGURE 1.2 (A) New surface aeration system for MCUA plant showing (B) compact surface aeration drives, (C) with elimination of most overhead piping, and (D) elimination of most equipment from compressor room. (Photos courtesy of Middlesex County Utilities Authority, Sayreville, New Jersey.)

the actual practice in the different applications is desirable to provide the field with a useful product.

This book incorporates the approach of presenting the basic theory behind aeration processes and then providing specific applications to several processes and types of systems used in the field.

1.1.2 LONG-TERM INVOLVEMENT OF ASCE COMMITTEE

A significant portion of the material and work conducted for this book was developed during the authors' involvement with the American Society of Civil Engineers (ASCE) committee on Oxygen Transfer Standards. This committee, composed of numerous practitioners in the aeration field from around the world, was started in 1976 with the initial purpose of developing a standard for the testing of aeration equipment in clean water. A number of conferences were held and reports generated not only to develop the state of the art in clean water testing but also to extend the testing techniques to process (dirty) water. With the financial assistance of the

FIGURE 1.2 (continued)

USEPA, the work of the committee was extended to include design applications as well as full-scale testing at various sites throughout the U.S. The many reports already developed by this committee, as well as the ongoing work to continually reevaluate and upgrade the state of the art in aeration testing, have supplied a significant portion of the background material for this endeavor.

1.1.3 SUMMARIZE STATE OF THE ART IN ONE LOCATION

This book is intended to summarize, in one location, the state of the art in aeration principles and practice. The numerous reports available from the above committee as well as the ever-changing body of technical literature in the field are incorporated into this work to show present practice.

Diffused air systems are considered in detail due to their present predominance in the field, with mechanical aeration systems providing the breadth of use. To minimize land area requirements in industries and metropolitan areas, experiences with deep tank aeration are presented along with their impacts on the equipment required for air supply. Design applications with both U.S. practice and European experience are included along with testing techniques to evaluate performance. For high rate systems, the oxygen transfer principles to describe high purity oxygen aeration are developed along with the current application. Finally, use of constructed aeration systems in natural waters is evaluated due to recent full-scale applications in rivers.

1.2 INTENDED AUDIENCE

Professionals involved in the design and analysis of aeration systems should find this book a primary resource to understand and effectively evaluate various alternatives based on a consistent set of principles. It is also aimed at the academic profession, both students and professors, since the principles involved in aeration are fully developed to allow application to practice. Various examples applying the principles to design will be useful to both groups.

1.3 BIBLIOGRAPHY

Clarke, K. (2001). "Treatment Plant Costs for Owls Head NYC Water Pollution Control Facility." Masters Degree Special Project, Department of Environmental Engineering, Manhattan College, NY.

Conklin, C. (2001). "Development of Capital and Operating Costs for Three NYC Water Pollution Control Plants—Coney Island, North River and Red Hook." Masters Degree Thesis, Department of Environmental Engineering, Manhattan College, NY.

Hildreth, S. B. (1999). "Aeration Capital Costs for West Point, Seattle WWTP." Personal communication, 13 Jan., 1999.

Hildreth, S. B., Finger, R. E., Hammond, R. R., and Daigger, G. T., (1997). "Full Scale High Purity Oxygen Activated Sludge Performance at the West Point WWTP, Seattle, Washington." *WEFTEC '97, 70th Annual Conference of the Water Environment Federation*, Chicago, IL, 617–628.

Leonforte, J. P. (1998). Letter on NYC Wastewater Plant capital costs—4 Nov., 1998. Chief, Division of Intergovernmental Coordination, Bureau of Environmental Engineering, NYCDEP.

Pöpel, H. J. (2001). Personal communication breaking down costs of Darmstadt plant. Emails, 3–5 Feb., 2001.

Schirtzer, R. (2000). "Submerged Turbine Aeration Conversion to Surface Aeration—Middlesex County Utility Authority (MCUA) Cost Data." Masters Degree Special Topic, Department of Environmental Engineering, Manhattan College, NY.

Wacker, J. (1998). Fax to H. Johannes Pöpel with costs information on Darmstadt Central Treatment Plant, Germany on 17 Mar., 1998.

2 Principles

2.1 MASS TRANSFER PRINCIPLES

2.1.1 PHYSICAL MECHANISMS INVOLVED IN TRANSFER

Mass transfer refers to the movement of molecules or mass from one location to another due to a driving force. This movement can occur within one fluid phase or among a number of fluid phases. Of particular concern to mass transfer in aeration is the transfer between two phases. This chapter specifically addresses the transfer between a gas and a liquid, which can be considered to occur in three stages. Oxygen molecules are initially transferred from a gas phase to the surface of a liquid. Equilibrium is quickly established at the gas–liquid interface. The oxygen molecules then move from the interface into the main body of the liquid.

The diffusion process in the liquid phase is initially considered with emphasis on the speed of diffusive transport and the factors influencing it. Interphase transport between the gas and the liquid is then addressed to establish the relationship between the oxygen saturation concentration in the liquid and the oxygen concentration in the gas phase. The basic equation describing the transfer of oxygen from the gas to the liquid phase is developed with the factors affecting the important parameters. Finally, the basic equations used for design are presented along with the relationship between process water conditions and the clean water conditions used in manufacturers' specifications for their equipment.

2.1.2 FICK'S LAW–QUIESCENT CONDITIONS

The principles defining the movement of oxygen molecules are similar to those defined in Newton's law, which governs the transfer of momentum in fluid flow, and Fourier's law, which defines the transfer of heat when a temperature gradient is present (Bird et al., 1960). The following equation, Fick's law, describes the transfer process when a concentration gradient is present in the fluid and no convection occurs. In this process, Brownian motion of the molecules in the fluid provides the transport.

$$J = -D \frac{dC}{dy} \tag{2.1}$$

The left-hand side of the equation provides the rate of mass transfer per unit interfacial area or mass flux. The negative sign indicates that transfer occurs in the direction of a decreasing gradient from a higher concentration to a lower value, similar to sliding down hill. The proportionality factor in the equation, D, represents the diffusion coefficient or diffusivity and is used to define the linear dependency of the flux on the associated gradient.

Figure 2.1 shows a schematic of the diffusive transport of oxygen molecules into a quiescent tank. The upper liquid layer is kept saturated by input of oxygen from the outside. The lower liquid layer initially is devoid of oxygen. Brownian motion causes both water and oxygen molecules to be transported across the interface between the two layers. Due to this random motion of molecules, oxygen begins to penetrate to the lower layers of the liquid in the "y" direction. Figure 2.2 shows the lower liquid layer when one-half of the total volume has attained saturation. It should be noted that penetration is not to the same depth in all locations due to the random nature of the diffusive process. Finally, at an infinite time, as shown in Figure 2.3, the total volume of the lower layer is saturated.

By conducting a mass balance on an elemental slice within the liquid layer, the differential equation describing the change in concentration with time is given by Fick's second law of diffusion (Bird et al., 1960) as:

$$\frac{\partial C}{\partial t} = D \frac{\partial^2 C}{\partial y^2}$$

The equation describing the time-space distribution of the oxygen penetration into the above tank is given by (Sherwood et al., 1975).

or
$$
\begin{aligned}
C(t,y) &= C_0 + (C_s - C_0)\text{erfc}\left[\frac{y}{2\sqrt{Dt}}\right] \\
C(t,y) &= C_0 + 2(C_s - C_0)\phi\left[-\frac{y}{2\sqrt{Dt}}\right]
\end{aligned}
\qquad (2.2)
$$

The complementary error function, erfc, and the cumulative Gaussian error function, ϕ, are available on spreadsheet programs and tabulated in statistics and engineering texts (Blank, 1982; Carslaw and Jaeger, 1959).

An example of the rate of molecular diffusion into the upper 5 mm of the tank in Figure 2.1 is given below using the following parameters at 20°C after one hour:

oxygen saturation concentration, C_s = 9.09 mg/L, $D = 1.83 \cdot 10^{-9}$ m²/s, $C_0 = 0$ mg/L, initial oxygen free water.

$$C(t,y) = 0 + 2(9.09 - 0)\phi\left[-\frac{5 \cdot 10^{-3}}{\sqrt{2 \cdot 1.83 \cdot 10^{-9} \cdot 3600}}\right]$$

$$= 2(9.09)\phi[-1.377] = 2(9.09) \cdot 0.0844$$

$$= 1.53 \text{ mg/L or } 16.8\% \text{ of saturation.}$$

This process is slow as demonstrated further for a 0.5 m tank using Equation (2.2). Figure 2.4 illustrates that oxygen penetrates only to a depth of 10 mm after one hour, increasing to about 50 mm after one day. After 100 days, significant oxygen penetration occurs to mid-depth, taking almost one year to reach the bottom of the tank and over 10 years to come close to saturation.

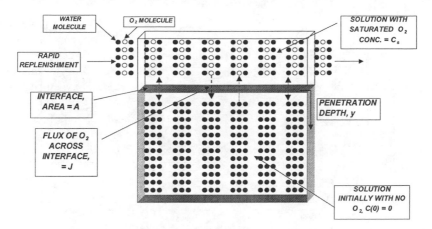

FIGURE 2.1 Oxygen diffusion schematic for quiescent solutions, $t = 0$.

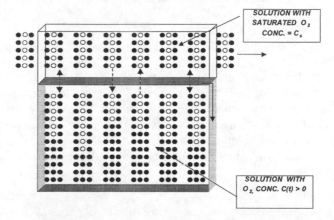

FIGURE 2.2 Oxygen diffusion schematic for quiescent solutions, $t = 1/2\ t$ infinity.

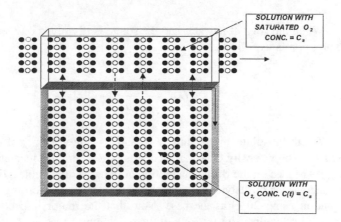

FIGURE 2.3 Oxygen diffusion schematic for quiescent solutions, $t = t$ infinity.

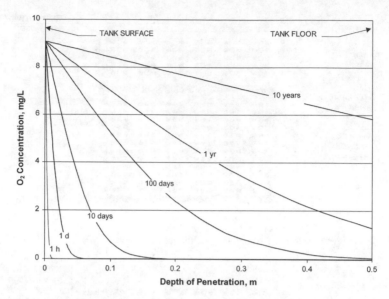

FIGURE 2.4 O_2 Profiles for molecular diffusion into a 0.5-m-deep tank.

Both the saturation and diffusivity values in Equation (2.2) are affected by temperature. Saturation decreases with increasing temperature (as discussed later), while diffusivity increases with temperature. The Wilke-Chang relationship (Reid et al., 1987) is an empirical correlation commonly used to describe the diffusivity, D_{AB}, of a dilute solution of A in solvent B as a function of molecular weight, M_B, and viscosity, μ_B, of the solvent, total volume, V_A, of the solute and absolute temperature, T.

$$D_{AB} = \frac{7.4 \times 10^{-12} T \sqrt{\phi M_B}}{\mu_B V_A^{0.6}} [=] \frac{m^2}{s} \qquad (2.3)$$

When the solvent is water and the solute is dissolved oxygen, the Wilke-Chang expression is as follows.

$$D = \frac{6.85 \times 10^{-12} T}{\mu} [=] \frac{m^2}{s} \qquad (2.4)$$

T is the absolute temperature in K, and μ is the viscosity of water in centipoises (g/m-s). The viscosity of water decreases as temperature increases, and fluid exerts less resistance on the Brownian motion of the water molecules. Figure 2.5 illustrates the increase in diffusivity with increasing temperature according to the Wilke-Chang equation using 20°C as the base. Note that the major impact of the temperature change on the diffusivity is due to the reduction in viscosity.

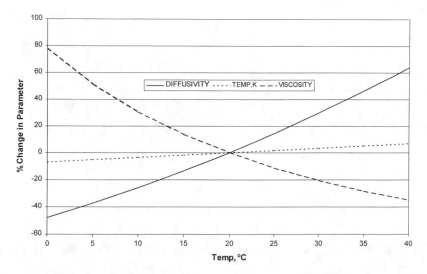

FIGURE 2.5 Relative effects of changes in temperature and viscosity on oxygen diffusivity using Wilke–Chang equation.

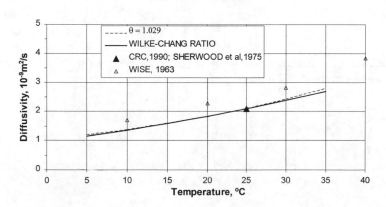

FIGURE 2.6 Effect of temperature on oxygen diffusivity.

An overall expression to relate the effect of temperature on the diffusivity value can be expressed as follows:

$$D_{t,°C} = D_{20°C}\theta^{t-20} \tag{2.5}$$

Figure 2.6 shows that a θ value of 1.029 fits the Wilke-Chang expression using the typical handbook value (Weast, 1989) for oxygen diffusivity at 25°C of 2.1×10^{-9} m²/s. The data provided by Wise (1963) is somewhat higher but fits the general profile.

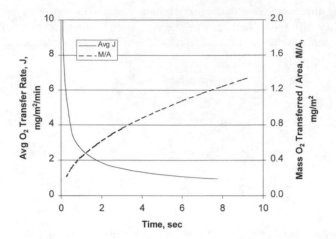

FIGURE 2.7 Initial rate and mass of oxygen transferred to water by Fick's diffusion at 20°C.

The total mass of oxygen transferred by diffusion, M, per unit interfacial area, A, into an infinitely deep tank (Sherwood et al., 1975), similar to the situation in Figure 2.1, is given as:

$$\frac{M}{A} = 2(C_s - C_0)\sqrt{\frac{Dt}{\pi}} \qquad (2.6)$$

The average concentration, $\overline{C}$, attained over the depth of the tank, represented by d, can be obtained as follows:

$$\overline{C} = \frac{M}{Ad} = \frac{M}{V} = \frac{2(C_s - C_0)}{d}\sqrt{\frac{Dt}{\pi}} \qquad (2.7)$$

The average flux of oxygen during the above time is obtained by dividing Equation (2.6) by the time of transfer to attain:

$$\overline{J} = \frac{M}{At} = 2(C_s - C_0)\sqrt{\frac{D}{\pi t}} \qquad (2.8)$$

Figure 2.7 provides the average transfer rate, $\overline{J}$ and total mass per unit area, M/A, during the first seconds of transfer. The initially high rates of transfer are quickly reduced as oxygen begins to build up in the layers adjacent to the interface. This outcome highlights the desirability of removing these upper layers by mixing them into the bulk solution (convective transport) to allow transfer to proceed more rapidly.

2.1.3 COMPARISON OF DIFFUSIVE TO CONVECTIVE TRANSPORT

Mixing and turbulence in the bulk solution destroy any concentration gradients in the major portion of the liquid with molecular diffusion occurring only in a thin layer at the interface. The mass flux is then defined in terms of the measured concentration difference and an empirically determined transfer coefficient, k_L, which represents the liquid film coefficient. This definition is expressed as follows.

$$J = k_L\left(C_s - C\right) \tag{2.9}$$

The mass flux can be expressed in terms of the change in the bulk liquid concentration by multiplying by the interfacial area per unit liquid volume, $a = \dfrac{A}{V}$.

$$J\frac{A}{V} = \frac{dC}{dt} = k_L a\left(C_s - C\right) \tag{2.10}$$

Integrating between the initial conditions and those at time, t, yields the following:

$$\int_{C_0}^{C} \frac{dC}{C_s - C} = k_L a \int_{0}^{t} dt$$

$$\frac{C_s - C}{C_s - C_0} = e^{-k_L a t} \tag{2.11}$$

When the initial concentration is zero, then the fraction saturation attained with time is given as follows.

$$\frac{C}{C_s} = 1 - e^{-k_L a t}; \qquad C_0 = 0 \tag{2.12}$$

The fraction saturation obtained by molecular diffusion as a function of tank depth can be obtained by expressing Equation (2.7) as follows:

$$\frac{\bar{C}}{C_s} = \frac{2}{d}\sqrt{\frac{Dt}{\pi}}; \qquad C_0 = 0 \tag{2.13}$$

Figure 2.8 shows the above two equations for a range of $k_L a$ values, from the high rates encountered in aeration tanks to the lower rates in natural water systems. To approximate the results from the field, it is obvious that molecular diffusion must occur in the thin, centimeters to microns surface layers of these systems. Turbulent or convective transport occurs over the bulk of the depth.

2.1.4 GAS–LIQUID TRANSFER

The mass transfer principles discussed above have not yet addressed the relationship between the gas and liquid phases. Figure 2.9 is a schematic of the two phases

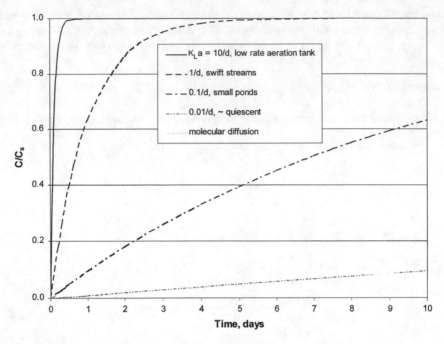

FIGURE 2.8 O_2 Transfer rates for field conditions compared to molecular diffusion at 20°C and 0.5 m depth.

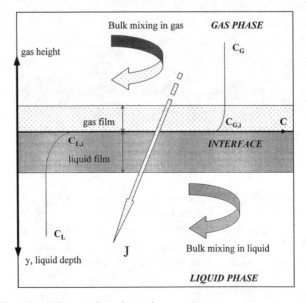

FIGURE 2.9 Two phase O_2 transfer schematic.

showing two resistances to transfer, one in the gas phase and one in the liquid phase. The schematic also reveals a discontinuity occurring between the two phases.

2.1.4.1 Gas and Liquid Films

The oxygen flux is expressed using both liquid, k_L, and gas, k_G, film coefficients, similar to Equation (2.9), but with the concentration difference expressed in each phase from the bulk values, C_G and C_L, to the interface values, $C_{G,i}$ and $C_{L,i}$.

$$J = k_G \left(C_G - C_{G,i} \right) \qquad \text{gas layer} \qquad (2.14)$$

$$J = k_L \left(C_{L,i} - C_L \right) \qquad \text{liquid layer} \qquad (2.15)$$

Note that the oxygen flux through each layer is equal with no buildup of oxygen at the interface.

2.1.4.2 Henry's Law

The relationship between the concentrations at the interface is expressed by Henry's law as follows.

$$C_{G,i} = H C_{L,i} \qquad (2.16)$$

This equation is an equilibrium relationship where the concentrations at the interface have the same activity or chemical potential (fugacity). Both concentrations are expressed in similar units, so H, the Henry's constant, is considered to be dimensionless, although actual units are $(mg/L)_{gas}/(mg/L)_{liquid}$. One must be careful when using handbook values for Henry's constant since it is also expressed as the inverse of the above and called a solubility or absorption coefficient.

2.1.4.3 Overall Driving Force

Combining the above three equations yields the following.

$$J = \left[\frac{1}{k_L} + \frac{1}{Hk_G} \right]^{-1} \left(\frac{C_G}{H} - C_L \right) \qquad (2.17)$$

The first term in the above equation contains the resistances to transfer in both liquid, R_L, and gas, R_G, layers, while the driving force or concentration difference is expressed in terms of measurable concentrations in bulk gas and bulk liquid phases. The first term in brackets is the inverse of the total resistance to transfer (R_T) and can be expressed as follows.

$$\left. \begin{array}{l} R_T = R_L + R_G \quad \text{or} \\[2mm] \dfrac{1}{K_L} = \dfrac{1}{k_L} + \dfrac{1}{Hk_G} \end{array} \right\} \qquad (2.18)$$

K_L is the overall liquid film coefficient taking into account both gas and liquid phase resistances. The relative importance of both resistances can be evaluated using the following expression for the resistance due to the liquid film.

$$\frac{\%R_L}{100} = \frac{R_L}{R_T} = \frac{K_L}{k_L} = \frac{1}{1 + \dfrac{1}{H \dfrac{k_G}{k_L}}} \qquad (2.19)$$

Using typical values of the gas to liquid film coefficient ratio, $\dfrac{k_G}{k_L}$, of 20 to 100, with a Henry's constant for oxygen of 29 at 20°C, shows that the liquid film resistance comprises more than 99.8 percent of the total resistance. The gas phase resistance is insignificant, typical of low solubility compounds such as oxygen and nitrogen. For oxygen transfer, $K_L \cong k_L$ and the gas side resistance can be ignored. Thus, turbulence and mixing has to be applied only to the liquid. The only impact of gas phase turbulence would be shear stress at the interface causing liquid phase turbulence.

2.1.4.4 Liquid Film Coefficient

There are a number of theories to describe the liquid film coefficient. Summaries of the earlier work, given in Sherwood et al. (1975), Aiba et al. (1965), and Eckenfelder and O'Connor (1961) are briefly reviewed here.

First proposed by Nernst in 1904, an equation for the two-film theory using stagnant gas and liquid films was derived by Lewis and Whitman in the 1920's to allow both gas and liquid resistances to be added in series. Through a gross simplification, linear concentration profiles were used in each of the films with sharp discontinuities between film and bulk phase concentration gradients. The liquid film coefficient was given as a function of a characteristic liquid film thickness, δ_L.

$$k_L = \frac{D}{\delta_L} \qquad (2.20)$$

Although no predictive estimates of δ_L are available, it has been useful in predicting mass transfer rates with simultaneous chemical reaction based on data without reaction, as well as the impact of high mass transfer rates on heat transfer. Typical liquid films over which the concentration gradient occurs vary from 10 to 200 microns thick, depending on the level of turbulence in the bulk liquid (Hanratty, 1991).

The penetration theory by Higbie in 1935 assumes a small fluid element at concentration, C_0, is brought into contact with the interface for a short time, t, where diffusion into the element occurs as a transient process, decreasing with time. Equation (2.8) describes this process resulting in a value of the film coefficient as follows.

$$k_L = 2\sqrt{\frac{D}{\pi t}} \qquad (2.21)$$

The time of contact for bubble aeration is defined as the time for a single bubble to travel through liquid at a distance equal to its diameter, d_B, using the bubble velocity, v_B.

$$t = \frac{d_B}{v_B}$$

Mackay et al. (1991), summarizing results of Asher and Pankow from 1986 to 1990, illustrates the Higbie model gave a good description of CO_2 transfer through a clean air-water interface. The characteristic diffusional distance, given as

$$\delta_L = \sqrt{Dt}$$

was 42 μm at a contact time of 1 s. This thickness was much larger than the monomolecular interface thickness of 0.3 nm or 0.0003 μm.

Danckwertz (1951) expanded on the penetration theory by employing a wide spectrum of times instead of a single contact time, wherein an element of fluid would be exposed to the saturation concentration at the interface.

$$k_L = \sqrt{Dr} \qquad (2.22)$$

The parameter, r, is the fractional rate of surface renewal.

In the three above models for the liquid film coefficient, values are not generally available except in the case of bubble aeration for the penetration model. Therefore, experimental measurement of the film coefficient is required.

O'Connor and Dobbins (1958) defined the surface renewal rate as a function of fluid turbulence parameters, a characteristic mixing length, l, and vertical velocity fluctuation, $\bar{v}$, as:

$$r = \frac{\bar{v}}{l}$$

This definition led to two expressions for the reaeration coefficient of streams based on the stream characteristics. One was for shallow streams where there is a significant velocity gradient and shearing stress (nonisotropic turbulence), and the other was for deep streams where a significant velocity gradient and shearing stress do not exist (isotropic turbulence). In the case of deep streams, this expression led to the widely used equation for determining the stream reaeration coefficient based on stream velocity and depth.

$$\left. \begin{array}{l} k_L = \sqrt{\dfrac{DU}{H}} \\[4mm] k_L a = \dfrac{k_L}{H} = \dfrac{DU^{1/2}}{H^{3/2}} \end{array} \right\} \text{deep streams} \tag{2.23}$$

O'Connor (1983) went further to describe the overall resistance to oxygen transfer as two resistances in series, similar to the two-film theory but both in the liquid film. A viscous laminar sublayer is adjacent to the interface and the other a turbulent mixed zone between the laminar sublayer and the bulk fluid.

$$\frac{1}{k_L} = \frac{1}{k_\delta} + \frac{1}{k_\tau}$$

Brumley and Jirka (1988), pg 316, indicate that the above conceptual models are on the right track. They attempt "to describe a process where dissolved gas enters a boundary layer by molecular diffusion and is subsequently transported into the bulk by turbulent mixing in such a way that the boundary layer remains thin". Recent evaluations of the liquid film coefficient consider the hydrodynamics near the interface with the velocity fluctuations normal to the interface (Hanratty, 1991). Hydrodynamic models describing eddy motion are being developed for relatively smooth surfaces and are not capable of addressing the complex situations in aeration tanks where the interfacial area is not known.

Clearly, there is no simple theoretical expression for the liquid film coefficient that would be suitable for all types of aeration systems. It will be a function of the energy input to the system, the interfacial area developed, and the hydrodynamics and velocity profile at the interface. Thus, the interfacial area is generally combined with the overall liquid film coefficient and data from empirical correlations are used to design systems.

2.2 APPLICATION TO OXYGEN TRANSFER

2.2.1 BASIC EQUATION

The oxygen saturation concentration, C_∞^*, is defined as the value in equilibrium (at infinite time) with the concentration in the bulk gas phase, which is also the concentration at the interface since the gas side gradient is negligible.

$$C_\infty^* = \frac{C_G}{H} \tag{2.24}$$

Substituting Equations (2.18) and (2.24) into (2.17) yields the oxygen flux.

$$J = K_L\left(C_\infty^* - C_L\right) \tag{2.25}$$

Multiplying by the interfacial area per unit volume, the change in oxygen concentration with time, similar to Equation (2.10) results.

$$\frac{dC_L}{dt} = K_L a\left(C_\infty^* - C_L\right) \tag{2.26}$$

Equation (2.26) is the basic equation used to describe oxygen transfer in actual aeration systems. The maximum rate of transfer occurs when the dissolved oxygen concentration in solution is zero. No transfer occurs when the dissolved oxygen concentration has attained equilibrium with the gas phase.

The oxygen transfer coefficient, $K_L a$, is the product of the liquid film coefficient, K_L and the interfacial area exposed to transfer in a given liquid volume, a. In all but the simplest systems, the individual values, K_L and a, are impossible to individually measure. Incorporating them into one coefficient, $K_L a$, provides the ability to obtain a measurable value in complex field aeration systems.

The saturation value, C_∞^*, is also a measured value in aeration systems. Although oxygen saturation values in equilibrium with bulk atmospheric gas concentrations at the liquid surface have been tabulated, these conditions do not necessarily exist in aeration tanks. The actual values are impacted, especially for diffused aeration systems, by increased pressure from the release of gas below the water and by decreased bulk gas concentrations resulting from the transfer process of gas rising through the liquid.

2.2.2 FACTORS AFFECTING OXYGEN TRANSFER

From the basic equation defining oxygen transfer, Equation (2.26), the factors affecting each of the major parameters are discussed below.

2.2.2.1 Oxygen Saturation, C_∞^*

Using the Henry's law definition for the saturation value, Equation (2.24), the oxygen saturation value is a function of both the oxygen gas phase concentration and the Henry's constant. From the ideal gas law

$$C_G = \frac{nM}{V} = \frac{pM}{RT} \tag{2.27}$$

For dry air, oxygen is 20.95 percent by volume, thus the oxygen partial pressure, p, is related to the total pressure, p_t, by:

$$p = 0.2095\left(p_t - p_v\right) \tag{2.28}$$

For open systems, both surface and diffused, the vapor pressure, p_v, is assumed saturated at the liquid temperature, with gas phase temperature having no effect on the vapor pressure or C_G. Only in well mixed closed systems, where there are significant differences in gas and liquid phase temperatures, would vapor pressures at the gas phase temperature be utilized (Mueller, 1979).

The total pressure is related to both the barometric pressure, P_b, and increased pressure from aerator submergence.

$$p_t = P_b + p_{d_e} \tag{2.29}$$

An effective pressure, p_{d_e}, is determined from shop or field data for specific equipment. Previous theoretical relationships for this term have proven faulty due to the complexity of mixing patterns in aeration systems.

2.2.2.1.1 Temperature

The Henry's law constant, H, increases with increasing temperature and dissolved solid concentrations, which causes a reduction in the oxygen saturation value. The Henry's constants for oxygen in Table 2.1 are back calculated from the observed oxygen saturation values from Benson and Krause (1984) and *Standard Methods* (APHA et al., 1995) at one atmosphere total pressure and no dissolved solids (0 chlorinity), C_s^*.

In specifying aerator performance, 20°C is used as a standard condition with the saturation value at one atmosphere total pressure. The temperature correction factor for the saturation value, τ, is then given by the following equation and illustrated in Figure 2.10.

$$\left. \begin{aligned} \tau &= \frac{C_{st}^*}{C_{s20}^*} \\ C_{s20}^* &= 9.09 \frac{mg}{L} \end{aligned} \right\} \tag{2.30}$$

2.2.2.1.2 Wastewater

To account for the effect of wastewater constituents on oxygen saturation, a β factor is introduced as the ratio of saturation in wastewater to tap water.

$$\beta = \frac{C_{s\,\text{wastewater}}^*}{C_s^*} \tag{2.31}$$

The major impact on wastewater saturation value is the inorganic dissolved solids. The chlorinity data in *Standard Methods*, (APHA et al., 1995), was scaled up to total dissolved solids using NaCl (1.65 × chlorinity) from 0 to 20,000 mg/L TDS. As indicated in *Standard Methods*, this scale-up, shown in Figure 2.11, assumes that the wastewater inorganic composition is similar to that in seawater. It is the

TABLE 2.1
Henry's Constants for Oxygen as a Function of Temperature

Temperature, °C	C_s^*, mg/L	$H, \dfrac{(mg/L)_{air}}{(mg/L)_{water}}$ (*)
0	14.62	20.3
10	11.29	25.1
20	9.09	29.8
30	7.56	34.0
40	6.41	37.6

$$(*)H = \frac{5530\left(14.7 - p_v(psia)\right)}{C_s^* T(°K)}$$

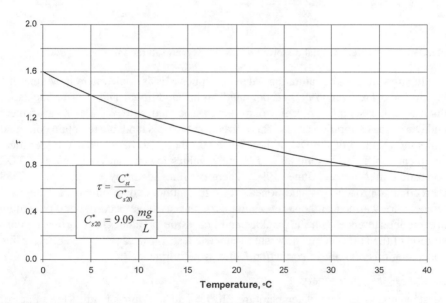

FIGURE 2.10 Effect of temperature on oxygen saturation.

consensus of the ASCE Committee on Oxygen Transfer Standards that this scale-up factor is sufficiently accurate for practical use (ASCE, 2001).

$$\beta = 1 - 5.7 \times 10^{-6} \times TDS \qquad (2.32)$$

For municipal wastewater at TDS<1500 mg/L, β is commonly taken as 0.99. For industrial wastewater such as pharmaceutical waste at a TDS of 10,000 mg/L, β will be as low as 0.94.

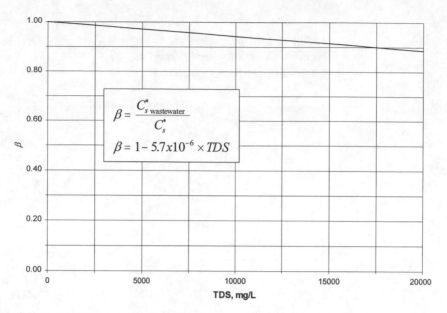

FIGURE 2.11 Effect of total dissolved solids on oxygen saturation.

In an evaluation of saturation values in upper Mississippi waters in Minnesota, Parkhill and Gulliver (1997) recommend taking additional DO measurements on distilled water samples to remove any bias associated with Winkler measurement and DO probe calibration errors. This correction factor is laudable to obtain continual calibration update of the DO probes. However, they also have taken Winkler measurements on the river samples and found β values lower than predicted by a TDS correction in May and June, 1994. Therefore the adequacy of the above TDS correction approach was questioned. It is the authors' opinion that as the river temperature warmed in the spring, algae growth may have occurred and caused an organic interference with the Winkler and not a true β value. Until further demonstration of the ability to run accurate titrametric tests on water with differing organic concentrations, the above correction factor is recommended.

2.2.2.1.3 Submergence

At standard conditions of temperature (20°C) and pressure (1 atm), the effect of diffuser submergence on oxygen saturation is given by δ.

$$\delta = \frac{C^*_{\infty 20}}{C^*_{s20}} = \frac{P_s + p_{d_e} - p_v}{P_s - p_v} \qquad (2.33)$$

Since δ is the measured value, the effective pressure can be defined.

$$p_{d_e} = (\delta - 1)(P_s - p_v) = \gamma_w d_e \qquad (2.34)$$

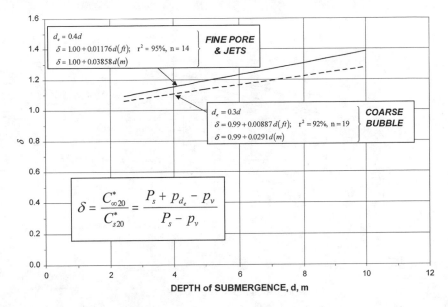

FIGURE 2.12 Effect of diffuser submergence on oxygen saturation.

The term d_e, representing the effective depth, is typically given as a fraction of the total depth of submergence, d.

To determine δ in the field, seven types of diffusers were used in clean water studies by Yunt et al. (1980), Mueller et al. (1982b), Mueller and Saurer (1986), and Mueller and Saurer (1987). Coarse bubble units provided significantly lower saturation values than fine pore and jet diffusers, as shown in Figure 2.12 and given below.

$$\left.\begin{aligned} & d_e = 0.4d \\ & \delta = 1.00 + 0.01176d(ft) \qquad r^2 = 95\%, \ n = 14 \\ & \delta = 1.00 + 0.03858d(m) \end{aligned}\right\} \text{ Fine Pore and Jets} \quad (2.35)$$

$$\left.\begin{aligned} & d_e = 0.3d \\ & \delta = 0.99 + 0.00887d(ft) \qquad r^2 = 92\%, \ n = 19 \\ & \delta = 0.99 + 0.0291d(m) \end{aligned}\right\} \text{ Coarse Bubble} \quad (2.36)$$

2.2.2.1.4 Barometric Pressure

The impact of barometric pressure on saturation is given by Ω, shown in Figure 2.13 and given as follows:

$$\Omega = \frac{C^*_{\infty P_b}}{C^*_{\infty P_s}} = \frac{P_b + p_{d_e} - p_v}{P_s + p_{d_e} - p_v} \approx \frac{P_b}{P_s} \qquad (2.37)$$

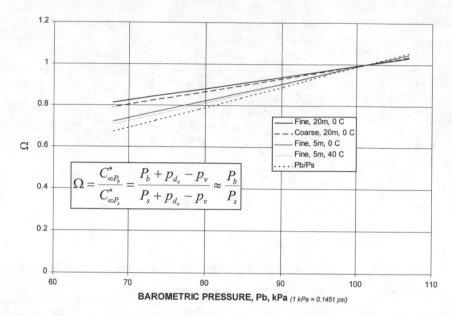

FIGURE 2.13 Effect of barometric pressure on oxygen saturation.

The approximation, which always gives lower values than actual, is satisfactory for tanks under six meters (20 feet) (ASCE, 1991) at barometric pressures and within five percent of standard atmospheric pressure. At greater aeration tank depths, the impact of reduced barometric pressure is less due to the large effect of hydrostatic head. Coarse bubble diffusers give slightly lower Ω values than fine pore, and vapor pressure has a minimal effect at normal temperatures. Since barometric pressure decreases with altitude, the following factor can be used for high altitude locations (Metcalf and Eddy, 1972).

$$P_b = P_{b0}\left(1 - \frac{alt(ft)}{30,000}\right) = P_{b0}\left(1 - \frac{alt(m)}{9100}\right)$$

For example, Denver, Colorado, at an altitude of 1500 meters (5000 feet), will have a surface saturation value of 7.6 mg/L at 20°C compared with 9.09 mg/L at sea level.

Using the above correction factors, the actual saturation value in an aeration tank under process conditions is given as $C_{\infty f}^*$.

$$\left. \begin{array}{l} C_{\infty f}^* = \tau\beta\Omega C_{\infty 20}^* \\ C_{\infty 20}^* = \delta C_{s20}^* \end{array} \right\} \tag{2.38}$$

2.2.2.2 Oxygen Transfer Coefficient, $K_l a$

Both the liquid film coefficient and the interfacial area through which transfer occurs are affected by the type of aeration equipment employed and the turbulence level in the system. Data available from manufacturers on their specific equipment operating in tap water are typically given as a function of gas flow or power input at a temperature of 20°C. Therefore, the data must be adjusted to account for the temperature in the aeration tank and for the wastewater constituents.

2.2.2.2.1 Temperature
Increasing temperature increases $K_L a$ similar to the effect on diffusivity and liquid film coefficient using the following relationship.

$$K_L a_t = K_L a_{20} \theta^{t-20} \tag{2.39}$$

In the above equation, θ is dimensionally not homogeneous requiring a temperature in °C. An alternative would be to express the temperature impact in the exponential form.

$$K_L a_t = K_L a_{20} e^{\kappa(t-20)}$$

$$\kappa = \ln \theta$$

At present, the above has not been used in the aeration field, but would be a logical direction for the future. The value of θ is commonly taken as 1.024 (ASCE, 1993; Jensen, 1991), equivalent to κ of 0.0237/°C.

As indicated previously, the liquid film coefficient, K_L, is a function of diffusivity raised to the power of 0.5 to 1.0. Using the Wilke-Chang correlation, this equation would result in θ values of 1.028 and 1.029 respectively. For diffused saran tube and sparger aeration units, Bewtra et al. (1970) have measured a value of 1.02 while Landberg et al. (1969) have found a lower θ of 1.012 for surface aeration units. Figure 2.14 and 2.15 show θ for static mixers and dome diffusers (Mueller et al., 1982a; Mueller et al., 1983a) to vary from 1.028 at low gas flows (low turbulence levels) to 1.017 at high gas flows. Metzger and Dobbins (1967) have determined the average θ values for the liquid film coefficient to be 1.032 for low intensity mixing and 1.006 for high intensity mixing. Jensen (1991) correlating K_L data over three orders of magnitude has shown θ to decrease from a value of 1.047 at low turbulence to 1.006 at higher turbulence levels.

Since temperature also affects viscosity and surface tension, changes in the interfacial area as well as K_L may also result. Lacking information on this relatively complex impact of temperature, data on specific aeration systems is required from manufacturers if accurate temperature corrections are to be obtained.

2.2.2.2.2 Wastewater
The presence of dissolved organics in wastewater can have a significant effect on $K_L a$, typically much greater than all the other factors combined. An experimentally measured parameter, α, is defined to account for the wastewater effects.

$$\alpha = \frac{K_L a_{wastewater}}{K_L a_{tap\ water}} = \frac{K_L a_f}{K_L a} \tag{2.40}$$

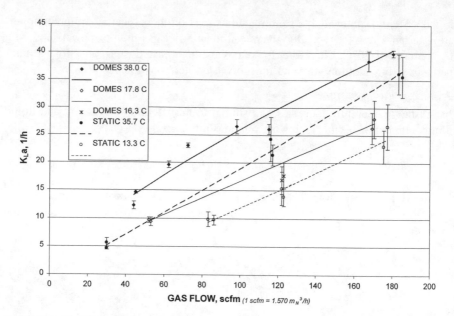

FIGURE 2.14 Effect of temperature on clean water K_La for dome and static aerators in a 9.1 m deep pilot plant.

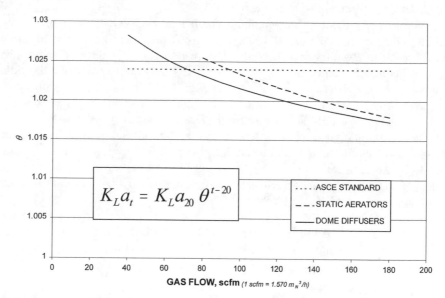

FIGURE 2.15 Effect of gas flow on θ for dome and static aerators in a 9.1 m deep pilot plant.

Surface active agents affect K_La due to a reduction in the liquid film coefficient but with an increase in surface area due to lowered surface tension (Wagner and Poepel, 1995). According to Mancy and Okun (1965), the resistance to oxygen transfer is caused mainly by a viscous hydration layer at the water surface and to a lesser extent by the interfacial film of adsorbed surfactant molecules.

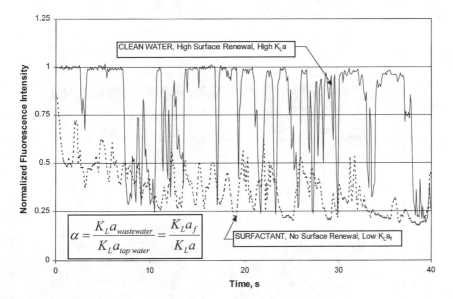

FIGURE 2.16 Impact of wastewater on $K_L a$ shown by velocity fluctuations at a water surface using fluorescence measurement (Asher and Pankow, 1991b; Asher, 1998).

Asher and Pankow (1991a,b) using surface fluorescence fluctuations with a vertically oscillating grid at different turbulence levels, showed a marked difference in surface renewal rates at the interface with and without a surfactant. Figure 2.16 shows the greater frequency and magnitude of the surface renewal rates in the clean water compared with that of the surfactant (Asher, 1998). As the peaks approach a value of 1.0, a high degree of surface renewal is occurring, typical of the clean water data. Figure 2.17 and 2.18 show the impact of turbulence and surfactants on the transfer process. For clean interfaces at low turbulence levels, the eddy caused by bulk mixing does not reach the surface. The concentration boundary layer, δ_c, is greater than the diffuse sublayer thickness, δ_d (~40 microns), so diffusion does not have enough time to saturate the eddy before returning to the bulk solution. For high turbulence, the eddy reaches the surface where it becomes saturated in the exposure time (~1 s) and then mixes into the bulk fluid. At these turbulence levels, the concentration boundary layer, δ_c, is less than the diffuse sublayer thickness, δ_d, where diffusion has enough time to saturate the eddy before leaving the surface.

When a surfactant is present at low turbulence levels, the concentration profile attains a greater depth, and δ_c increases. The additional resistance due to organics reduces transfer rate when compared with clean water. Due to the surfactant damping the turbulent motion through an increased shear stress at the interface, concentration fluctuations were never observed in the diffusive sublayer of Asher and Pankow, regardless of turbulence intensity of the grid system.

Eckenfelder (1970) indicates that for quiescent or laminar flow conditions, the bulk resistance to oxygen transfer is high and masks the surface resistance caused by the surfactant. In an intermediate range, low turbulence conditions, the bulk resistance to transfer is reduced and the surfactant interfacial resistance causes a significant

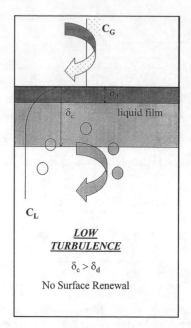
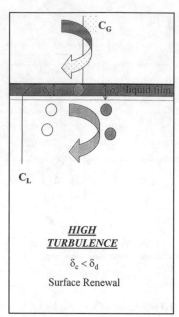

FIGURE 2.17 Turbulence impact on clean water O_2 transfer.

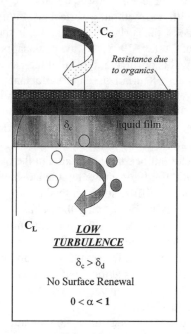
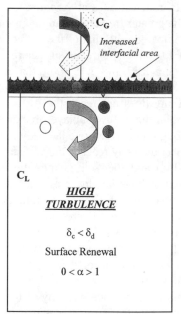

FIGURE 2.18 Turbulence impact on wastewater O_2 transfer.

reduction in transfer rate. At high turbulence levels, oxygen transfer depends on surface renewal and again, is not significantly affected by diffusion through interfacial resistances. Under these conditions, α may be >1.0 due to increased interfacial area (Figure 2.18). Both Lister and Boon (1973) and Otoski et al. (1979) contend that the increase in surface area does not offset the decrease in K_L with α always being less than one, which is most likely the case in full-scale systems.

For bubble systems, nonionic surfactants reduce oxygen transfer more strongly than anionic surfactants (Wagner and Poepel, 1995). They also show that surface tension measurements alone cannot be used to predict α values. Masutani and Stenstrom (1991) show that a measurement of dynamic surface tension was a potentially useful tool to determine the impact of surfactants on α. They also indicate that use of antifoam agents significantly decrease α.

During the course of biological oxidation of wastewater, the substances causing variations in $K_L a$ are being removed. Thus, in a plug flow aeration tank, α will normally increase as flow progresses down the tank. Completely mixed, step feed, and selector processes (Mueller et al., 1996, 2000) will tend to minimize this large variation in α and operate closer to the effluent value.

After an aeration system has been operational for a time, field-measured $K_L a_f$ values include not only the effect of the dissolved organics in the wastewater but also any deterioration in aerator characteristics. This effect is frequently found in fine pore diffusers when clogging or embrittlement occurs. An additional factor, F, is used to account for this diffuser aging process.

$$F = \frac{K_L a_{f\,\text{service}}}{K_L a_{f\,\text{new}}} \tag{2.41}$$

2.2.2.3 Dissolved Oxygen Concentration in Bulk Liquid, C_L

In setting a C_L value, two factors must be considered: the minimum dissolved oxygen concentration required by the activated sludge to maintain the maximum oxygen utilization rate, and the varying oxygen demands due to flow and organic load variations.

Activated sludge consists of microorganisms, the majority of which exist in biological floc particles. Data by Borkowski and Johnson (1967) indicate that a low oxygen concentration of 0.0004 mg/L is sufficient to maintain full activity of dispersed cells oxidizing carbonaceous organics. For oxygen to reach the active sites at the bacterial cell membranes, it must penetrate the liquid film surrounding the floc particle and diffuse through the floc matrix to the individual bacteria. Assuming a uniform oxygen uptake rate in the floc, the drop in dissolved oxygen concentration from the floc surface to the center of a spherical floc is given as follows (Wuhrmann, 1963).

$$C_L = C_m + \frac{A \gamma_f d_f^2}{24 D_f}$$

Larger size floc particles and higher oxygen uptake rates require higher dissolved oxygen values as shown in Figure 2.19 (Mueller 1979). The greater floc sizes had

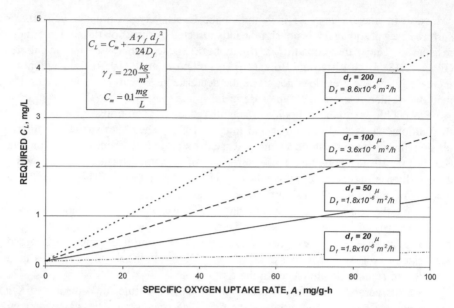

FIGURE 2.19 Impact of activated sludge mass transfer resistance on required O_2 concentration.

larger effective diffusivities. Argaman et al. (1995) shows that the effective diffusivity increases with increasing sludge volume index and specific surface area probably due to an increase in floc porosity. Activated sludge from an aeration tank at the Nancy (France) Metropolitan wastewater treatment plant had a mean diameter of 125 μ (Snidaro et al., 1997). Analysis after sonification revealed that the large floc were made up of more tightly bound 13 μ size microcolonies, which were in turn composed of 2.5 μ bacteria. A gel-like matrix of exopolymers provides the cohesion for these units. The loosely bound large floc should have greater porosity than the smaller more tightly bound floc, resulting in higher diffusivities.

For the typical size of activated sludge floc, 20 to 115 μ (Mueller et al., 1966), a dissolved oxygen concentration between 0.2 and 1.5 mg/L, typically 0.5–0.7 mg/L, is desirable. This parameter will insure the oxygen uptake rates of bacteria oxidizing carbonaceous organics are not oxygen limited. For nitrification to proceed at optimum rates, dissolved oxygen values > 2.0 mg/L are required (EPA, 1975). Stenstrom and Song (1991) show that the DO concentration for nitrification ranges from 0.5 to 2.5 mg/L depending on operational parameters and mass transport resistance. This level can go as high as 4.0 mg/L during an organic shock load.

To allow for variation in oxygen demand due to changing loads, a design C_L value of 2.0 mg/L is often used based on average load. Maximum load conditions should be evaluated to insure that C_L is above 0.5 mg/L to avoid septic conditions.

2.3 DESIGN EQUATIONS

In designing aeration systems, the basic equation used for the analysis is Equation (2.26), which is modified to account for the conditions at which manufacturers

TABLE 2.2
Standard Conditions for Specification of
Aeration Equipment Performance

Parameter	Condition U.S. Practice	Condition European Practice
Type water	Tap water	Tap water
Water temperature	20°C	20°C
C_L	0 mg/L	0 mg/L
Barometric pressure	1 atm	1 atm
Air flow	20°C and	0°C and
	36% relative humidity,	0% relative humidity,
	$\gamma = 0.075$ lb air/ft³	$\rho = 1.293$ kg air/m³
	$= 0.01736$ lb O_2/ft³	≈ 300 g O_2/m³

specify the capabilities of their equipment. Specifications for aeration equipment are given based on clean water data under the conditions in Table 2.2 (ASCE, 1991; ATV, 1996).

2.3.1 STANDARD OXYGEN TRANSFER RATE, SOTR

The SOTR is the mass of oxygen transferred per unit time into a given volume of water and reported at standard conditions. The European literature also refers to this term as the oxygenation capacity (OC). The nomenclature used in the ASCE Standard is utilized throughout this text and the alternate value indicated as done here. Equation (2.26) is multiplied by the aeration tank volume and standard conditions employed.

$$SOTR = V\left(\frac{dC_L}{dt}\right)_{STD} = K_L a_{20} C^*_{\infty 20} V \qquad (2.42)$$

Note that at standard conditions, the dissolved oxygen concentration is taken as zero thus providing the maximum driving force for transfer. As these equations are developed, an example calculation is performed in both the English and SI systems so that the units' conversion factors are clear (Table 2.3).

TABLE 2.3
SOTR Example Calculation

SI	U.S.
$SOTR = 10.5\dfrac{mg}{L} \times 1000 m^3 \times \dfrac{8}{h} \times 10^{-3}\dfrac{kg \cdot L}{mg \cdot m^3}$	$SOTR = 10.5\dfrac{mg}{L} \times 0.264 MG \times \dfrac{8}{h} \times 8.34 \dfrac{lb \cdot L}{mg \cdot MG}$
$= 84.0\dfrac{kg}{h}$	$= 185\dfrac{lb}{h}$

TABLE 2.4
Example Calculation for Specific Oxygenation Capacity, oc

SI and U.S.

$$oc = 10.5 \frac{mg}{L} \times \frac{8}{h} = 84 \frac{mg}{L \cdot h} = 84 \frac{g}{m^3 \cdot h}$$

The conditions for this computation will be an aeration tank of 1000 m³ (0.264 MG) at a water depth of 4.57 m (15 ft) with fine pore diffusers located at 4.27 m (14 ft) below the water surface. The saturation value calculated from Equation (2.35) is 10.59 mg/L, a measured value of 10.5 mg/L used in the computation. The clean water oxygen transfer coefficient of 8.0/h will be utilized within the range of actual values.

2.3.2 Specific Oxygenation Capacity, oc

This parameter is often used in the European literature to designate the rate of change in oxygen concentration in an aeration tank. Simply put, it is Equation (2.26) at standard conditions.

$$oc = \left(\frac{dC_L}{dt} \right)_{STD} = K_L a_{20} C^*_{\infty 20} = \frac{SOTR}{V} = SOTR_V \tag{2.43}$$

In both systems, the calculation is the same as shown in Table 2.4.

This parameter has the same units as the oxygen uptake rate (OUR) of the system and gives a feel for reaction rate in the system. Note that both $K_L a$ and C^*_∞ are a function of temperature, the former increasing and the latter decreasing. When defining the ratio of specific oxygenation capacity at any temperature to that at 20°C, Figure 2.20 shows that the impact of temperature on this product is much less than on the oxygen transfer rate or the oxygen saturation value.

$$\frac{oc_t}{oc} = \frac{K_L a_t}{K_L a_{20}} \frac{C^*_{\infty t}}{C^*_{\infty 20}} = \theta^{t-20} \tau = \frac{OTR_v}{SOTR_v} \tag{2.44}$$

2.3.3 Standard Aeration Efficiency, SAE

The SAE is the rate of oxygen transfer per unit power input, which may be based on either delivered (DP) or wire power (WP).

$$\left. \begin{array}{l} SAE = \dfrac{SOTR}{DP} \\ SAE = \dfrac{SOTR}{WP} \end{array} \right\} \tag{2.45}$$

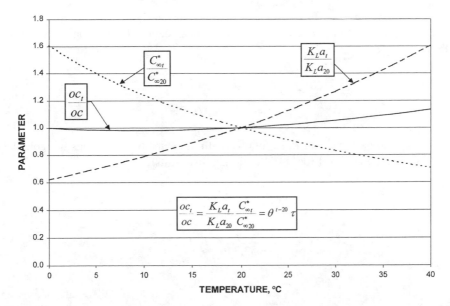

FIGURE 2.20 Impact of temperature on O_2 transfer at zero dissolved O_2 concentration.

The overall efficiency, e, of the aeration equipment is the product of the individual efficiencies of mechanical equipment. Typical efficiencies (EPA, 1983) of the individual components are: blowers (50 percent for older to 80 percent for newer units), motors (95 percent), coupling (95 percent) and gear box (95 percent). It is used to relate the consumed wire power to that which is delivered to the air for diffused aeration or to the liquid for mechanical aeration.

$$WP = \frac{DP}{e} \qquad (2.46)$$

For diffused aeration, the delivered power of blowers is typically based on the adiabatic compression equation, AP, (Yunt, 1979). The equations below for power are given under both SI and English units due to the difference in units and standard gas flow conditions.

$$DP = AP = \frac{wRT_a}{K}\left[\left(\frac{P_d}{P_a}\right)^K - 1\right] \qquad (2.47)$$

The value of K is 0.283 for air in the U.S. (36 percent relative humidity) and both pressures are in absolute units (gage + standard atmospheric) as is temperature. Modern German literature on turbo compressors applies adiabatic compression with a K of 0.2857 for dry air. A note of caution must be expressed with respect to using the adiabatic compression equation for all blowers. Although many blowers are nearly adiabatic, some may be closer to polytropic in operation (Yunt, 1979).

The mass flow rate of air, w, is related to the air density and the volumetric flow rate of the influent air, which will be specified at standard conditions as given in Table 2.2.

$$w = \rho_s G_s \quad \text{(SI)}$$
$$w = \gamma_s G_s \quad \text{(US)} \tag{2.48}$$

Using the gas constant, R, as follows with the standard conditions in Table 2.2 provides the power level for both SI and English units.

$$R = 287 \frac{J}{kg \cdot {}^\circ K}$$

$$= 53.346 \frac{ft \cdot lb}{lb_m \cdot {}^\circ R}$$

$$AP(kW) = 0.100 G_s \left(m_N^3/h \right) \left[\left(\frac{P_d}{P_a} \right)^K - 1 \right]$$
$$AP(hp) = 0.227 G_s \left(scfm \right) \left[\left(\frac{P_d}{P_a} \right)^K - 1 \right] \tag{2.49}$$

Note that the gas flows are given in terms of their standard conditions as (Normal) m_N^3/h and (standard) scfm. The pressures are expressed as follows. The discharge pressure includes the depth of water at the diffuser submergence as well as all the losses in the air piping and diffuser system. The inlet pressure at the blower is somewhat less than atmospheric due to losses in the air filtering system and inlet piping.

$$P_d = P_s + \gamma_w d + \Delta p_d$$
$$P_a = P_s - \Delta p_a$$

To illustrate use of these concepts, an example in the form of a tabular summary is given in Table 2.5.

Observing the 7.5 percent difference in power requirements using the U.S. and SI designations for standard gas flow conditions shows that the actual inlet air conditions are required to get an accurate estimate of power consumption.

For all aeration devices, wire power can be measured accurately using a recording polyphase wattmeter. An ammeter measuring current can also be used if both the voltage and power factor are known. For squirrel cage induction motors, a power factor of 0.9 is typical (Perry et al., 1984).

TABLE 2.5
SAE Example Calculation

Parameter	SI	U.S.
P_s	101.325 kPa	14.7 psi
Δp_d	6.89 kPa	1.0 psi
d	4.27 m	14.0 ft
γ_w	9.81 kN/m³	62.4 lb/cf
$\gamma_w d$	41.85 kPa	6.07 psi
P_d	150.1 kPa	21.8 psia
Δp_a	0.69 kPa	0.10 psi
P_a	100.6 kPa	14.6 psia
G_s	1000 m$_N$³/h	637 scfm*
AP = DP	12.0 kW	17.37 hp*
$K = 0.283$		
e	0.6	0.6
WP	20.0 kW	28.95 hp*
SAE (delivered)	7.0 kg/kWh	10.65 lb/hp-h*
SAE (wire)	4.2 kg/kWh	6.39 lb/hp-h*

* Not a direct scale-up (approximately 7.5 percent higher) from SI value due to the U.S. standard requiring compression at a temperature of 20°C and 36 percent relative humidity compared with 0°C for the SI with bone dry air. Gas flow based on similar SOTE values. Note that scfm × 1.570 = m$_N$³/h.

$$kW = \frac{\sqrt{3}EI\,pf}{1000}$$

2.3.4 STANDARD OXYGEN TRANSFER EFFICIENCY, SOTE

The SOTE is the fraction of oxygen supplied to the aeration tank, which is actually transferred or dissolved into the liquid at standard conditions. It is a major design parameter for diffused aeration systems.

$$SOTE = \frac{SOTR}{w_o} \qquad (2.50)$$

The mass fraction of oxygen in dry air is as follows.

$$\frac{w_o}{w} = 0.2095\,\frac{\text{mole } O_2}{\text{mole air}} \times 32\,\frac{\text{g } O_2}{\text{mole } O_2} \times \frac{\text{mole air}}{28.964 \text{ g air}} = 0.2315\,\frac{\text{g } O_2}{\text{g air}}$$

In the English system, taking into account the water vapor at 36 percent relative humidity provides a slightly lower value, 0.23 (ASCE, 1991).

TABLE 2.6
SOTE Example Calculation

SI	ENGLISH
$SOTE = \dfrac{84.0\,\dfrac{kg}{h}}{0.30 \times 1000\,\dfrac{m_N^3}{h}} = 0.280$	$SOTE = \dfrac{185\,\dfrac{lb}{h}}{1.04 \times 637 scfm} = 0.2793$
$\%SOTE = 28.0\%$	$\%SOTE = 27.9\%$

Using Equation (2.48) provides the oxygen supply rate.

$$w_o(kg/h) = 0.2315 \times 1.293 G_s = 0.30 G_s \left(m_N^3/h \right)$$

$$w_o(lb/h) = 0.23 \times 0.075 G_s \times 60 \frac{min}{h} = 1.04 G_s (scfm)$$

Inserting the above into Equation (2.50) provides the SOTE as a function of gas flow.

$$\left.\begin{array}{c} SOTE = \dfrac{SOTR(kg/h)}{0.30 G_s \left(m_N^3/h \right)} \\[2ex] SOTE = \dfrac{SOTR(lb/h)}{1.04 G_s (scfm)} \end{array}\right\} \tag{2.51}$$

Using the results of the prior example calculations, the SOTE is expressed in Table 2.6. The slight difference in SOTE values is due to the roundoff in Equation 2.51.

2.3.5 APPLICATION TO PROCESS CONDITIONS

Under process conditions, the oxygen transfer rate must meet the demand of the biomass in the aeration tank. The dissolved oxygen level in the tank will always move toward a concentration that balances the transfer rate with the demand. At a steady state condition, these two rates will be equal and will serve as the basis for design.

The actual oxygen transfer rate under process conditions is defined similar to Equation (2.42).

$$OTR_f = V \left(\frac{dC_L}{dt} \right)_{PROCESS} = K_L a_f \left(C_{\infty f}^* - C_L \right) V \tag{2.52}$$

Dividing Equation (2.52) by (2.42) provides the ratio of the actual to the standard oxygen transfer rate.

$$\frac{OTR_f}{SOTR} = \frac{K_L a_f \left(C_{\infty f}^* - C_L \right)}{K_L a_{20} C_{\infty 20}^*}$$

TABLE 2.7
OTR$_f$ and OTE$_f$ Example Calculations

$$\tau = \frac{7.56\frac{mg}{L}}{9.09\frac{mg}{L}} = 0.83$$

$$\beta = 1 - 5.7 \times 10^{-6} \times 12000\frac{mg}{L} = 0.93$$

$$P_b = 101.325kPa\left[1 - \frac{1000m}{9100m}\right] = 90.19kPa$$

$$P_{d_e} = 0.4 \times 41.85kPa = 16.76kPa; \qquad p_v = 4.24kPa$$

$$\Omega = \frac{90.19 + 16.76 - 4.24}{101.325 + 16.76 - 4.24} = 0.90$$

$$\frac{OTR_f}{SOTR} = 0.45 \times 1.024^{30-20}\left[\frac{0.83 \times 0.93 \times 0.90 \times 10.5\frac{mg}{L} - 1.5\frac{mg}{L}}{10.5\frac{mg}{L}}\right] = 0.31$$

$$OTR_f = 0.31 \times 84\frac{kg}{h} = 26.1\frac{kg}{h} = 57.5\frac{lb}{h}$$

$$\%OTE_f = 0.31 \times 28.0\% = 8.7\%$$

Employing the previously defined correction factors for the oxygen transfer coefficient and saturation value yields the following ratio for the commonly used design equations.

$$\frac{OTR_f}{SOTR} = \frac{oc_f}{oc} = \frac{AE_f}{SAE} = \frac{OTE_f}{SOTE} = \frac{\alpha\theta^{t-20}\left(\tau\beta\Omega C_{\infty 20}^* - C_L\right)}{C_{\infty 20}^*} \qquad (2.53)$$

Assuming an industrial wastewater with an α of 0.45, a TDS concentration of 12,000 mg/L being treated at 30°C, C_L of 1.5 mg/L and an altitude of 1000 m provides the results in Table 2.7.

The remaining process values use the same ratio as the OTR$_f$ and % SOTE calculations.

2.4 NOMENCLATURE

a	m^{-1}	interfacial area/unit liquid volume
A	m^2	interfacial area
A	mg/g-h	specific oxygen uptake rate
AE$_f$	kg/kWh, lb/hp-h	aeration efficiency under process conditions
AP	kW, hp	adiabatic delivered power
C	mg/L	oxygen concentration
C$_0$	mg/L	DO concentration at time zero

C_G	mg/L	bulk gas phase oxygen concentration
$C_{G,i}$	mg/L	gas phase oxygen concentration at interface
C_L	mg/L	bulk liquid phase oxygen concentration
$C_{L,i}$	mg/L	liquid phase oxygen concentration at interface
C_m	mg/L	oxygen concentration at center of floc
C_s	mg/L	DO saturation concentration
C_s^*	mg/L	surface saturation concentration
C_{s20}^*	mg/L	surface saturation concentration at 20 °C, 9.09 mg/L
C_∞^*	mg/l	oxygen saturation concentration
$C_{\infty 20}^*$	mg/l	clean water oxygen saturation concentration at diffuser depth and 20 °C
$C_{\infty f}^*$	mg/l	oxygen saturation concentration under process (field) conditions
D	m²/s	coefficient of molecular diffusion of oxygen in (waste)water
d	m	tank depth
D_{AB}	m²/s	coefficient of molecular diffusion of solute A into solvent B
d_B	m	bubble diameter
D_f	m²/h	diffusivity in floc
d_f	m	floc diameter
DP	kW, hp	delivered power
e	–, %	overall efficiency of blower or compressor
E	volts	measured voltage
F		diffuser aging factor on oxygen transfer coefficient
G_s	m_N^3/h, scfm	airflow rate at standard conditions
H	(mg/L)$_{gas}$/(mg/L)$_{liquid}$	Henry's constant
H	m	stream depth
I	amps	measured current
K		coefficient in adiabatic compression equation
J	g/m²-s	mass flux of oxygen
k_G	m/s	gas film coefficient
k_L	m/s	liquid film coefficient
K_L	m/s	overall liquid film coefficient
$K_L a$	h⁻¹	oxygen transfer coefficient
$K_L a_{20}$	h⁻¹	clean water oxygen transfer coefficient at 20°C
$K_L a_t$	h⁻¹	clean water oxygen transfer coefficient at temperature t
kW	kW	measured wire power
k_δ	m/s	liquid film coefficient in viscous laminar sublayer
k_τ	m/s	liquid film coefficient in turbulent sublayer

ℓ	m	characteristic mixing length
M	g	mass of oxygen transferred
M	g/mole	molecular weight
M_B	g/mole	molecular weight of solvent B
n	moles	number of moles in ideal gas law
oc	mg/L-h	specific oxygenation capacity in clean water = $SOTR_v$
oc_t	mg/L-h	specific oxygenation capacity in clean water at temperature, t, = OTR_v
OTE_f	–, %	oxygen transfer efficiency under process conditions
OTR_f	kg/h, lb/h	oxygen transfer rate under process conditions
p		partial pressure of oxygen
P_a	kPa, psia	absolute pressure upstream of blower
P_b	kPa, psia	barometric pressure
P_{b0}	kPa, psia	barometric pressure at zero altitude
P_d	kPa, psia	absolute pressure downstream of blower
P_{d_e}	kPa, psi	effective pressure
P_s	kPa, psia	standard barometric pressure, 101.325 kPa, 14.696 psia
p_t	kPa, psia	total pressure
p_v	kPa, psi	vapor pressure
r	s^{-1}	surface renewal rate
R	J/(kg·K)	universal gas constant (286.88 J/kg·K)
R_G	s/m	resistance to oxygen transfer in gas phase
R_L	s/m	resistance to oxygen transfer in liquid phase
R_T	s/m	total resistance to oxygen transfer
SAE	kg/kWh, lb/hp-h	standard aeration efficiency
SOTE	–, %	standard oxygen transfer efficiency
SOTR	kg/h, lb/h	standard oxygen transfer rate
T	°K	absolute temperature
t	°C	temperature
t	s	time
T_a	°K, °R	absolute temperature of influent gas to blower
TDS	mg/L	total dissolved solids concentration
U	m/s	stream velocity
V	m^3	tank volume
V_A	m^3	total volume of solute A
$\bar{v}$	m/s	vertical velocity fluctuation
w	kg/h, lb/h	mass flow rate of air
w_o	kg/h, lb/h	mass flow rate of oxygen
WP	kW, hp	wire power
y	m	depth of penetration
Δp_a	kPa, psi	pressure drop in inlet filters and piping to blower
Δp_d	kPa, psi	pressure drop in piping and diffuser downstream of blower

α		wastewater correction factor for oxygen transfer coefficient
β		wastewater correction factor for oxygen saturation
δ		depth correction factor for oxygen saturation
δ_c	m	concentration boundary layer thickness
δ_d	m	diffuse sublayer thickness
δ_L	m	liquid film thickness
ϕ		association parameter of solvent B, for water $\phi = 2.6$
γ_f	kg/m^3	specific weight of dry floc
γ_s	lb/ft^3	specific weight of standard gas, $0.075\ lb/ft^3$
γ_w	$N/m^3, lb/ft^3$	specific weight of water
κ		temperature correction factor for oxygen transfer coefficient expressed in exponential form
μ	g/m-s	absolute viscosity
μ_B	g/m-s	absolute viscosity of solvent B
θ		temperature correction factor for oxygen transfer coefficient
ρ_s	kg/m^3	density of standard gas
τ		temperature correction factor for oxygen saturation
Ω		pressure correction factor for oxygen saturation

2.5 BIBLIOGRAPHY

Aiba, S., Humphrey, A. E., and Millis, N. F. (1965). *Biochemical Engineering*, Academic Press, New York.

APHA, AWWA, and WPCF. (1995). *Standard Methods for the Examination of Water and Wastewater*, A. D. Eaton, L. S. Clesceri, and A. E. Greenberg, eds., American Public Health Assn. (APHA).

Argaman, Y., Eliosov, B., and Papkov, G. (1995). Mass Transfer and Effluent Quality in Activated Sludge Systems." *WEFTEC'95-68th Annual Conference of the Water Environment Federation*, Miami Beach, FL, 191–199.

ASCE. (1991). *Standard-Measurement of Oxygen Transfer in Clean Water-ANSI/ASCE 2–91*, American Society of Civil Engineers, New York.

Asher, W. E. (1998). Raw data on normalized fluorescence intensity for clean and surfactant influenced surfaces, personal communication.

Asher, W. E. and Pankow, J. F. (1991a). "The Effect of Surface Films on Concentration Fluctuations Close to a Gas/Liquid Interface." *Air-Water Mass Transfer: Selected Papers from the Second International Symposium on Gas Transfer at Water Surfaces*, Minneapolis, MN, 68–80.

Asher, W. E. and Pankow, J. F. (1991b). "Prediction of Gas/Water Mass Transport Coefficients by a Surface Renewal Model." *Environ. Sci. Technol.*, 25(7), 1294–1300.

ATV. (1996). *Messung der Sauerstoffzufuhr von Beluftungseinrichtungen in Belebungsanlagen in Reinwasser und in belebtem Schlamm, Merkblatt ATV-M209*, ATV-Regelwerk, Abwassertechnische Vereinigung.

Benson, B. B. and Krause, D. J. (1984). "The Concentration and Isotopic Fractionation of Oxygen Dissolved in Freshwater and Seawater in Equilibrium with the Atmosphere." *Limnology and Oceanography*, 29, 620.

Bewtra, J. K., Nicholas, W. R., and Polkowski, L. B. (1970). "Effect of Temperature on Oxygen Transfer in Water." *Water Research*, 4, 115–123.

Bird, B. R., Stewart, W. E., and Lightfoot, E. N. (1960). *Transport Phenomena*, John Wiley & Sons, Inc., New York.

Blank, L. (1982). *Statistical Procedures for Engineering, Management, and Science*, McGraw-Hill International Book Company, Auckland.

Borkowski, J. D. and Johnson, M. J. (1967). "Experimental Evaluation of Liquid Film Resistance in Oxygen Transport to Microbial Cells." *Applied Microbiology*, 15, 1483–1488.

Brumley, B. H. and Jirka, G. H. (1988). "Air-Water Transfer of Slightly Soluble Gases: Turbulence, Interfacial Processes and Conceptual Models." *PhysioChemical Hydrodynamics*, 10(3), 295–319.

Carslaw, H. S. and Jaeger, J. C. (1959). *Conduction of Heat in Solids*, Oxford at the Clarendon Press, Oxford.

Danckwertz, P. V. (1951). "Significance of liquid-film coefficient in gas absorption." *Ind. Eng. Chem.*, 43(6), 1460.

Eckenfelder, W. W., Jr. (1970). *Water Quality Engineering for Practicing Engineers*, Barnes & Noble, New York.

Eckenfelder, W. W. and O'Connor, D. J. (1961). *Biological Waste Treatment*, Pergamon Press, Elmsford, NY.

EPA. (1975). *Process Design Manual for Nitrogen Control*, USEPA.

EPA. (1983). "Development of Standard Procedures for Evaluating Oxygen Transfer Devices." *EPA-600/2-83–102*, USEPA, MERL.

Hanratty, T. J. (1991). "Effect of Gas Flow on Physical Absorption." *Air-Water Mass Transfer: Selected Papers from the Second International Symposium on Gas Transfer at Water Surfaces*, Minneapolis, MN, 10–33.

Jensen, N. A. (1991). "Effect of Temperature on Gas Transfer at Low Surface Renewal Rates." *Air-Water Mass Transfer: Selected Papers from the Second International Symposium on Gas Transfer at Water Surfaces*, Minneapolis, MN, 106–115.

Landberg, G., Graulich, B. P., and Kipple, W. H. (1969). "Experimental Problems Associated with the Testing of Surface Aeration Equipment." *Water Research*, 3, 445–455.

Lister, A. R. and Boon, A. O. (1973). "Aeration in Deep Tanks: An Evaluation of a Fine Bubble Diffused-Air System." *J. Institute Sewage Purification*, 72(5), 3–18.

Mackay, D., Shiu, W.-Y., Valsaraj, K. T., and Thibodeaux, L. J. (1991). "Air-Water Transfer: The Role of Partitioning." *Air-Water Mass Transfer: Selected Papers from the Second International Symposium on Gas Transfer at Water Surfaces*, Minneapolis, MN, 34–56.

Mancy, K. H. and Okun, D. A. (1965). "The Effects of Surface Active Agents on Aeration." *JWPCF*, 37, 212–227.

Masutani, G. K. and Stenstrom, M. K. (1991). "Dynamic Surface Tension Effects on Oxygen Transfer." *Journal of Environmental Engineering*, 117(1), 126–142.

Metcalf and Eddy. (1972). *Wastewater Engineering: Treatment and Disposal*, McGraw Hill, New York.

Metzger, J. and Dobbins, W. E. (1967). "Role of Fluid Properties in Gas Transfer." *Environ. Sci. & Technol.*, 1, 57–65.

Mueller, J. A. (1979). "Kinetics of Biological Flocs." *Prog. Water Tech., Suppl.*, 1, 143–155.

Mueller, J. A. and Saurer, P. D. (1986). "Field Evaluation of Wyss Aeration System at Cedar Creek Plant, Nassau County, NY." Parkson Corp., New York.

Mueller, J. A. and Saurer, P. D. (1987). "Case History of Fine Pore Diffuser Retrofit at Ridgewood, NJ." Manhattan College Environmental Engineering and Science, New York.

Mueller, J. A., Voelkel, K., and Boyle, W. (1966). "Nominal Diameter of Floc Related to Oxygen Transfer." *JASCE, SED*, 93, 920.

Mueller, J. A., Donahue, R., and Sullivan, R. (1982a). "Dual Nonsteady State Evaluation of Static Aerators Treating Pharmaceutical Waste." *37th Annual Purdue Industrial Waste Conference*, Purdue University, Lafayette, IN.

Mueller, J. A., Kim, C., and Court, N. (1982b). "Ridgewood Aeration System Analysis, Phase I. Coarse Bubble Sparger System." Frank Burde & Assoc., New York.

Mueller, J. A., Donahue, R., and Sullivan, R. (1983). "Comparison of Dome and Static Aerators Treating Pharmaceutical Waste." *38th Annual Purdue Industrial Waste Conference*, Purdue University, Lafayette, IN.

Mueller, J. A., Krupa, J. J., Shkreli, F., Nasr, S., and FitzPatrick, B. (1996). "Impact of a Selector on Oxygen Transfer-A Full Scale Demonstration." *WEFTEC'96–69th Annual Conference of the Water Environment Federation*, Dallas, TX, 427–436.

Mueller, J. A., Kim, Y.-K., Krupa, J. J., Shkreli, F., Nasr, S., and Fitzpatrick, B. (2000). "Full-Scale Demonstration of Improvement in Aeration Efficiency." *ASCE J. Environ. Engr.*, 126(6), 549–555.

O'Connor, D. J. (1983). "Wind Effects on Gas-Liquid Transfer Coefficients." *ASCE, J. Environ. Eng.*, 109(3), 731–752.

O'Connor, D. J. and Dobbins, W. E. (1958). "Mechanism of Reaeration in Natural Streams." *Trans. ASCE*, 123, 655.

Otoski, R. A., Brown, L. C., and Gilbert, R. G. (1979). "Bench and Full-Scale Tests for Alpha and Beta Coefficient Variability Determination." *Proc. Purdue Industrial Conf.*, Purdue University, Lafayette, IN, 835–852.

Parkhill, K. L. and Gulliver, J. S. (1997). "Indirect Measurement of Oxygen Solubility." *Water Research*, 31(10), 2564–2572.

Perry, R. H., Green, D. W., and Maloney, J. O. (1984). *Perry's Chemical Engineers' Handbook*. McGraw-Hill Book Company, New York.

Reid, R. C., Prausnitz, J. M., and Poling, B. E. (1987). *The Properties of Gases & Liquids*, McGraw-Hill, Inc., New York.

Sherwood, T. K., Pigford, R. L., and Wilke, C. R. (1975). *Mass Transfer*, McGraw-Hill, Inc., New York.

Snidaro, D., Zartarian, F., Bottero, J.-Y., and Manem, J. (1997). "New Statements in Activated Sludge Floc Structure." *WEFTEC'97-70th Annual Conference of the Water Environment Federation*, Chicago, IL, 429–437.

Stenstrom, M. K. and Song, S. S. (1991). "Effects of Oxygen transport Limitation on Nitrification in the Activated Sludge Process." *Research Journal Water Pollution Control Federation*, 63(208), 208–219.

Wagner, M. R. and Poepel, H. J. (1995). "Influence of Surfactants on Oxygen Transfer." *WEFTEC'95-68th Annual Conference of the Water Environment Federation*, Miami Beach, FL, 297–306.

Weast, R. C., Lide, D. R., Astle, M. J., and Beyer, W. H. (1989). "CRC Handbook of Chemistry and Physics.", CRC Press, Inc., Boca Raton, FL.

Wise, D. L. (1963). "The Determination of the Diffusion Coefficients of Ten Slightly Soluble Gases in Water and a Study of the Solution Rate of Small Stationary Bubbles,", PhD Thesis, U. of Pittsburg.

Wuhrmann, K. (1963). "Effect of Oxygen Tension on Biochemical Reactions in Sewage Purification Plants." *Advances in Biological Waste Treatment*, W. W. J. Eckenfelder and J. McCabe eds., Pergamon Press, Oxford, 27–40.

Yunt, F. (1979). "Gas Flow and Power Measurement." *Proceedings of the Workshop Toward an Oxygen Transfer Standard*, EPA-600/9-78-021, Asilomar Conference Grounds, Pacific Grove, CA, 105–127.

Yunt, F., Hancuff, T., Brenner, R., and Shell, G. (1980). "An Evaluation of Submerged Aeration Equipment Clear Water Test Results." Presentaton at the *WWEMA Industrial Pollution Conference*, Houston, TX.

3 Diffused Aeration

3.1 INTRODUCTION

Diffused aeration is defined as the injection of air or oxygen enriched air under pressure below a liquid surface. All of the equipment discussed in this chapter meets this definition. However, certain hybrid equipment that combines gas injection with mechanical pumping or mixing is also covered under this topic. These hybrid devices include jet aerators and U-tube devices. Other devices, such as sparged turbine aerators and aspirating impeller pumps, are covered under mechanical aeration systems.

Although the aeration of wastewater began in England as early as 1882 (Martin, 1927), major advances in aeration technology awaited the development of the activated sludge process by Arden and Lockett in 1914. A review of the history of aeration technology is most interesting and instructive. Early investigators were aware of the importance of bubble size, diffuser placement, tank circulation and gas flow rate on oxygen transfer efficiency. Perforated tubes and pipes provided the material framework for early aeration methods. One of the earliest patents for a diffuser was granted in 1904 in Great Britain for a perforated metal plate diffuser (Martin, 1927). In Great Britain, porous tubes, perforated pipes, double perforated tubes with fibrous material in the annular space and nozzles were used in early methods (Federation of Sewage and Industrial Wastes Associations, 1950). Investigators sought more efficient aeration through the development of finer bubbles. In England, experiments were conducted with sandstone, firebrick, mixtures of sand and glass and pumice. Most of these early materials were dense, creating high head losses. A secret process employing concrete was used to cast porous plates that were placed in cast iron boxes by Jones and Atwood, Ltd. around 1914. This system was used for many years by Great Britain and its colonies.

Meanwhile, in the U.S., porous plates produced by Filtros were widely used in newly constructed activated sludge plants. In Milwaukee, research was conducted using grids of perforated black iron pipes, basswood plates, Filtros plates and air jets. The Filtros plates were selected for the plant placed in operation in 1925 (Ernest, 1994). The Filtros plates, patented in 1914, were constructed from bonded silica sand and had permeabilities (see Section 3.4.1) in the range of 14.1 to 20.4 m^3_N/h (9 to 13 scfm) at 5 cm (2 in) water gage. Similar plates were installed in the Houston North-Side plant in 1917, as well as at Indianapolis; Chicago; Pasadena, CA; Lodi, CA; and Gastonia, NC (Babbitt, 1925). Ernest (1994) provides an excellent history of the development of the aeration system at Milwaukee where siliceous plates from Ferro Corporation (Filtros) are still used. Over time, aluminum oxide that was bonded with a variety of bonding agents, as well as silica became the major media of choice. Permeabilities continued to rise as well, up to as high as 188 m^3_N/h (120 scfm). In addition, new shapes were introduced, including domes and tubes and more recently, discs.

In Great Britain, the sand-cement plates were predominately used until approximately 1932. In 1932, Norton introduced porous plates bolted at either end. Norton introduced the first domes in 1946 with permeabilities in the range of 62.8 to 78.5 m^3_N/h (40 to 50 scfm). In Germany, early aeration designs (commencing about 1929) incorporated the Brandol plate diffusers produced by Schumacher Fabrik. Later they developed a tube design, and the material was modified as silica sand bonded by a phenol formaldehyde resin (Schmidt-Holthausen and Bievers, 1980).

Diffuser configuration was considered to be an important factor in activated sludge performance even as early as 1915. The Houston and Milwaukee plants were designed with a ridge and furrow configuration. In 1923, Hurd proposed the "circulatory flow" or spiral roll configuration for the Indianapolis plant. The Chicago North-Side plant also employed this diffuser configuration (Hurd, 1923). The design was promoted on the belief that the spiral roll would provide a longer contact time between wastewater and air than the full floor coverage. One set of basins at Milwaukee was converted to spiral roll in 1933, but even the 1935 database suggested that the spiral roll configuration required more air per unit volume of wastewater treated. The spiral roll configuration was abandoned at Milwaukee in 1961 after extensive oxygen transfer studies (Ernest, 1994). It is also interesting to note that the early plants employed a range of diffuser densities (percent of floor surface area covered by diffusers, $A_d/A_t \times 100$) ranging from about 25 percent at Milwaukee and Lodi, CA to 7 to 10 percent at the spiral roll plants (Babbitt, 1925).

Clogging of diffusers appears to have been a problem in some cases according to the earliest studies. Generally speaking, the porous diffusers produced the greatest concern but examples of clogging of perforated pipes can be found (Martin, 1927; Ernest, 1994). Early work by Bushee and Zack (1924) at the Sanitary District of Chicago prompted the use of coarser media to avoid fouling. Later, Roe (1934) outlined in detail numerous diffuser clogging causes. Ernest (1994) detailed cleaning methods adopted by Milwaukee in maintaining porous diffusers at their installations. Nonetheless, by the 1950s, many plants were using the large orifice type of diffuser. The newer designs improved upon their earlier counterparts and were designed for easy maintenance and accessibility. In general, these devices produced a coarser bubble, thereby sacrificing substantial transfer efficiency. The Air Diffusion in Sewage Works manual (Committee on Sewage and Industrial Wastes Practice, 1952) provides an excellent summary of air diffusion devices proposed and tested between 1893 and 1950. It should be emphasized that the trend toward coarser diffuser media was followed in the U.S. but not in Europe, where the porous diffusers continued to predominate in many designs.

An alternative to the diffused aeration systems was the mechanical aeration designs, which had been introduced in the early 1900s. These, too, began to replace some of the older diffused aeration systems where fouling was considered to be a problem. A more detailed discussion of the mechanical aeration systems is presented in Chapter 5.

With the emphasis on more energy-efficient aeration in the 1970s, porous diffuser technology received greater attention in the U.S. Since about 1970, the wastewater treatment industry has witnessed the introduction of a wide variety of new diffuser materials and designs. Many of the lessons learned with this technology in the early part of the century were revisited. Improvements in materials of construction, blower designs, and measurement technology have resulted in a new generation of highly efficient diffuser systems and the methodologies for maintenance of these systems.

This chapter addresses the current state of technology for diffused aeration. Although diffused aeration devices are often referred to as fine, medium and coarse bubble based on the perceived or measured bubble size, such classifications are often confusing and differentiation between devices is difficult. Therefore, in this chapter, diffused aeration devices are discussed based on the physical characteristics of the diffuser device. Two general categories are used, porous and nonporous devices. The reader is cautioned, however, to avoid drawing generalities about equipment performance based on these labels alone. These classifications are intended more as a guide for organization than as a categorical statement of performance.

3.2 DESCRIPTION OF DIFFUSED AERATION SYSTEMS

3.2.1 POROUS DIFFUSER DEVICES

Porous diffuser devices are defined in this text based on the current high efficiency devices now on the market as diffusers that will produce a head loss due to surface tension in clean water of greater than about 5 cm (2 in) water gauge. These devices are often referred to as fine pore diffusers and typically produce bubbles in the range of 2–5 mm (0.08–0.20 in) when new. An excellent reference on fine pore aeration technology is the USEPA's *Design Manual, Fine Pore Aeration Systems* (1989).

3.2.1.1 Types of Porous Media

Although several materials are capable of serving as effective porous media, few are being used in the wastewater treatment field because of cost, specific characteristics, market size, or other factors. Porous media used today may be divided into the following three general categories: ceramics, porous plastics and perforated membranes.

3.2.1.1.1 Ceramics

Ceramics are the oldest and currently the most common porous media on the wastewater market. Ceramic media consist of irregular or spherically shaped mineral particles that are sized, blended together with bonding materials, compressed into various shapes, and fired at elevated temperatures to form a ceramic bond between the particles. The result is a network of interconnecting passageways through which air flows. As air emerges from the surface pores, the pore size, surface tension, and airflow rate interact to produce a characteristic bubble size.

Ceramic materials most often used include alumina, aluminum silicate and silica. Alumina is refined from naturally occurring bauxite and subsequently crushed and screened to provide the appropriate size. Synthetic or naturally occurring aluminum silicates may also be used and are often referred as mullite when consisting of three parts alumina and two parts silica. The alumina and aluminum silicate particles are ceramically bonded to form the appropriate diffuser material. Silica is typically a mined material although crushed glass may be used. It is less angular and available in somewhat more limited particle sizes than the aluminum minerals. Silica minerals are normally vitreous-silicate bonded although resin bonding of pure silica is also practiced. It has been claimed that silica materials may be more resistant to fouling and more easily cleaned (Schmidt-Holthausen and Bievers, 1980), but no scientifically controlled experiments have been conducted to support this claim. No studies have been published that suggest there is a difference in process performance between diffusers made with different materials. Performance would be more a function of grain size, binding agent, shape of the unit, and other factors. Alumina may be the most abrasion resistant, but actual strength and abrasion resistance depends on the ceramic bond. Silica porous media are generally considered to have the lowest overall strength, thereby requiring greater thickness.

Sources of ceramic diffuser media include companies supplying industrial abrasives or refractories. They may provide diffusers to aeration equipment manufacturers who specify the characteristics of the media, or they may market finished diffuser assemblies. Ceramic diffusers have been used since the turn of the century, as described above, and their advantages and operational characteristics are well documented. As a result, they have become the standard for comparison. Each new generation of porous diffusers reportedly offers some advantages in cost or operation over ceramics. However, as in the past, the new diffusers have not always met expectations. As a result, ceramic diffusers continue to capture a significant share of the porous diffuser market.

3.2.1.1.2 Rigid Porous Plastics

Rigid porous plastics are made from several thermoplastic polymers, including polyethylene, polypropylene, polyvinylidene fluoride, ethylene-vinyl acetate, styrene-acrylonitrile (SAN), and polytetra-fluoroethylene (EPA, 1989). The two most common types of plastic media used in wastewater aeration are high-density polyethylene (HDPE) and SAN. Relatively inexpensive and easy to process, HDPE diffusers are typically made from a straight nonpolar homopolymer in a proprietary extrusion process. SAN diffusers have been made from small copolymer spheres fused together under pressure. The material is brittle, however. SAN diffusers have been used for more than 20 years in U.S. wastewater treatment plants. Although plastics have advantages of lighter weight and lower costs as compared with ceramic materials, their use has fallen out of favor in the U.S. due to lack of quality control and the emerging cost competitiveness of other fine pore diffuser devices.

3.2.1.1.3 Perforated Membranes

Membrane diffusers differ from the first two groups of diffuser materials in that the diffusion material does not contain interconnecting passageways for transmitting gas. Instead, mechanical means are used to create preselected small orifices in a membrane material that allows passage of air through the material. The earliest of this type diffuser was introduced in the 1960s and was referred to as a sock diffuser. Made from plastics, synthetic fabric cord, or woven cloth, a woven sheath of this material was supported by a metallic or plastic core. The diffuser design allowed easy removal from retrievable aeration piping for cleaning or replacement. These socks were capable of high transfer efficiencies but readily fouled and were often removed by operators and not replaced. There is virtually no market for these socks today.

In the late 1970s, a new generation of perforated membranes was introduced. They consisted of a thin flexible thermoplastic, polyvinyl chloride (PVC). The membrane was perforated with a pattern of small slits. The plastic PVC membrane was found to undergo dramatic changes while in service, which significantly affected oxygen transfer. Consequently, the material was found to have relatively short operating life in many wastewaters.

A new type of membrane material was introduced in the mid 1980's identified as an elastomer. The predominant elastomers used in perforated membrane diffusers today are ethylene-propylene dimers (EPDMs). These new copolymers promise to address many of the material deterioration problems of the earlier plasticized PVC membranes. Different rubber fabricators have developed EPDM elastomers independently, and the manufacturing process, ternomer, and catalyst systems employed can vary significantly. These factors can affect molecular weight distribution, chain branching and cure rate. Furthermore, EPDM master batch formulas can contain varying amounts of EPDM, carbon black, silica, clay, talc, oils, and various curing and processing agents. By varying these components and their method of manufacture, it is possible to obtain a product for a specific application. This engineering of EPDM (and other membrane materials) has resulted in significant improvement of product performance and resistance to environmental attack. As a result, membranes have been engineered for several industrial applications including pulp and paper, textile, food and dairy and petrochemical wastewater.

Today, several equipment manufacturers are actively engaged in engineering new and improved perforated membrane materials. Polyurethane that provides high modulus of elasticity and contains no oils has been used in wastewater applications (Messner in Europe and marketed in the U.S. by Parkson as panels). Although no chemical changes are observed with this material, the thinner membrane is sensitive to creep under stress of air pressure. The hydrophobic silicones, which also contain no oils, are claimed to be chemically resistant to a number of wastewater chemicals. Yet, once perforated, early designs exhibit little tear resistance. With more experience, these materials and others will be improved and may serve important niches in the wastewater treatment business.

An important feature of the new perforated membranes is the perforation number, size and pattern. Perforations are produced by slicing, punching, or drilling small holes or slits in the membrane. Each hole acts as a variable aperture opening. The slit or hole size will effect bubble size (and therefore, oxygen transfer efficiency) and back pressure; smaller slits will generate smaller bubbles at a sacrifice of some head loss. Typical slit or hole size is 1 mm, although manufacturers continue to experiment with opening size and pattern to optimize performance. The current panel system marketed in the U.S. employs a very fine perforation. Several manufacturers offer both a fine and coarse perforation in their membrane diffuser offerings. Most perforated membrane devices are designed so that when air is off, the membrane relaxes down against a support base, and a seal is formed between membrane and support plate. This closing action will reportedly eliminate or at least minimize the backflow of liquid into the aeration system.

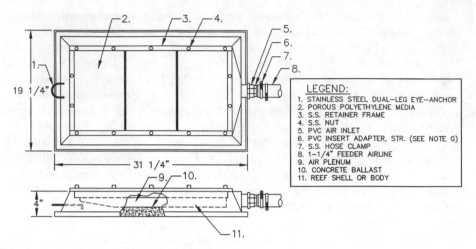

FIGURE 3.1 Typical plate diffuser (courtesy of EDI, Columbia, MO).

3.2.1.2 Types of Porous Media Diffusers

There are five general shapes of porous diffusers on the market: plates, panels, tubes, domes and discs. Each is briefly described below.

3.2.1.2.1 Plate Diffusers

One of the original designs for porous diffusers was the plate as described above. These plates were usually 30 cm (12 in) square and 25–38 mm (1–1.5 in) thick. Most were constructed of ceramic media. Installation was completed by grouting the plates into recesses in the basin floor or cementing them into prefabricated holders. Air was introduced below the plates through a plenum. Typically, no airflow control orifices were used in these designs. Although their use has declined since 1970, these ceramic plates are still used in Milwaukee and Chicago. A newer plate design was introduced in the late 1980s that employs either a ceramic or porous plastic media. They are marketed in sizes of 30 cm × 61 cm (12 × 24 in) and 30 cm × 122 cm (12 × 48 in). These units are typically mounted on ABS plastic plenums and subsequently placed on the basin floor. Air is introduced to each module by means of rubber tubing, and individual orifices control airflow. (See Figure 3.1.) Depending upon the layout, plate diffusers are typically operated at flux rates ranging from 0.09 to 0.18 m^3_N/h/m^2 of diffuser surface area (0.6 to 1.2 scfm/ft^2).

3.2.1.2.2 Panel Diffusers

Currently, the only panel marketed in the U.S. uses the perforated polyurethane membrane. The membrane is stretched over a 122 cm (48 in) wide base plate of variable length ranging from 183–366 cm (6–12 ft) in 61 cm (24 in) increments. The base plate may be constructed of reinforced cement compound, fiber-reinforced plastic, or Type 304 stainless steel. Air is introduced via tubing and an airflow control orifice attached at one end. The panels are placed on the flat bottom surface of the aeration basin and fastened with anchor bolts (Figure 3.2). These plates are designed to operate over a range of airflows from 0.007 to 0.111 m^3_N/h/m^2 (0.05 to

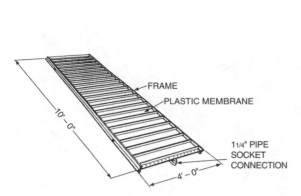

FIGURE 3.2 Typical panel diffuser (courtesy of Parkson Corp., Fort Lauderdale, FL).

0.76 scfm/ft) of membrane surface. Pressure loss across the panels ranges from 50 to 100 cm (20 to 40 in) water gauge (4.8 to 9.6 kPa [0.7 to 1.4 psi]).

3.2.1.2.3 Tube Diffusers

Like plates, tube diffusers have been used for many years in wastewater applications. The early tubes, Saran wound or aluminum oxide ceramic, have now been followed by SAN copolymer, porous HDPE and more recently, by perforated membranes. Most tubes on the market are of the same general shape, typically 51 to 61 cm (20–24 in) long with a diameter of 6.4 to 7.7 cm (2.5 to 3.0 in). The "magnum" tubes may range from 1 to 2 m (39 to 78 in) in length with diameters ranging from 6.4 to 9.4 cm (3.0 to 3.7 in). Diffusers may be placed on one (single band) or both (wide band) sides of the lateral header, which delivers the air to the units. An orifice inserted in the inlet nipple to aid in distribution typically controls airflow.

Whereas ceramic and porous plastic tubes are strong enough to be self-supported with aid of end caps and a connecting rod (Figure 3.3), perforated membranes require an internal support structure (Figure 3.4). The support is usually constructed from plastic (PVC or polypropylene) and has a tubular shape. The tube provides support either around the entire circumference or only the bottom half. Holes in the inlet connector, specially designed slots, or openings in the tube itself allow air distribution to the membrane surface. The membrane is usually not perforated at the air inlet points, so when airflow is off, the membrane collapses and seals against the support structure.

Most components of the tube assemblies are made of either stainless steel or a durable plastic. The gaskets are usually of a soft rubber material. Tubes are normally designed to operate at airflows ranging from 1.6 to 15.7 m^3_N/h (1–10 scfm) per diffuser, although most are operated at the lower end for optimum efficiency. It should be noted that because of the shape, it is difficult to design tubular diffusers to discharge around the entire circumference of the unit. The air distribution is a function of airflow rate and head loss across the media, usually improving with increased head loss. Fouling may occur in those regions where airflow is low or zero. New designs have developed internal air distribution networks that provide more uniform distribution of air around the entire circumference (Figure 3.5).

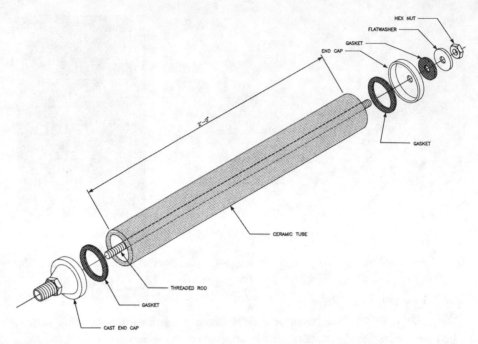

FIGURE 3.3 Ceramic tube diffuser (courtesy of Sanitaire, Brown Deer, WI).

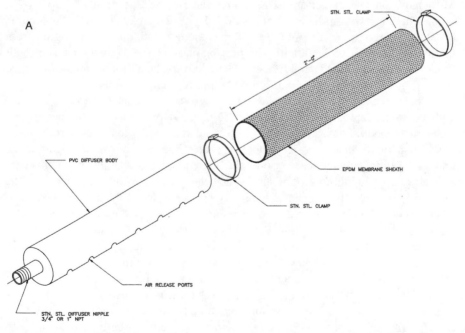

FIGURE 3.4 Membrane tubes [(A) courtesy of Sanitaire, Brown Deer, WI; (B) courtesy of EDI, Columbia, MO].

B

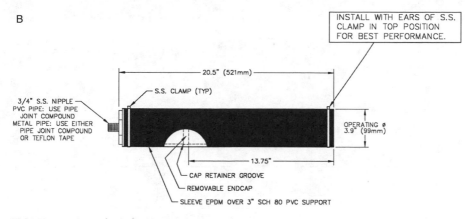

INSTALL WITH EARS OF S.S.
CLAMP IN TOP POSITION
FOR BEST PERFORMANCE.

20.5" (521mm)

S.S. CLAMP (TYP)

3/4" S.S. NIPPLE
PVC PIPE: USE PIPE
JOINT COMPOUND
METAL PIPE: USE EITHER
PIPE JOINT COMPOUND
OR TEFLON TAPE

OPERATING ø
3.9" (99mm)

13.75"

CAP RETAINER GROOVE

REMOVABLE ENDCAP

SLEEVE EPDM OVER 3" SCH 80 PVC SUPPORT

FIGURE 3.4 (continued)

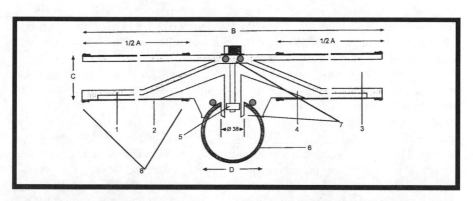

1. **SUPPORT TUBE** (non-buoyant)
2. **PERFORATED MEMBRANE**
3. **FLOODED COMPARTMENT**
4. **AIR DISTRIBUTION CHANNEL**
5. **SECURING CLIP** (19.7 ft• lb Torque required)
6. **AIR HEADER**
7. **EPDM "O" RING SEALS**
8. **316 ss OETIKER BAND CLAMP**

DIMENSIONS (subject to normal fabrication tolerance)

Active length, mm A	1000	1500	**2000**
Total length, mm B	1200	1700	**2200**
Diameter, mm C	Ø67	Ø67	**Ø67**
D	3" & 4"		

FIGURE 3.5 Membrane tube design (courtesy of OTT Systems, Inc., Duluth, GA).

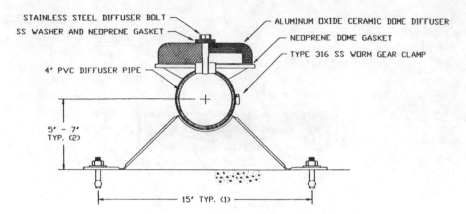

FIGURE 3.6 Ceramic dome (courtesy of Sanitaire, Brown Deer, WI).

3.2.1.2.4 Dome Diffusers

As described above, the porous dome diffuser was introduced in the U.K. in 1946 and was widely used in Europe prior to its introduction in the U.S. in the 1970s. The dome diffuser is a circular disc with a downturned edge. Today, these diffusers are 18 cm (7 in) in diameter and 38 mm (1.5 in) high. The media is ceramic, usually aluminum oxide.

The diffuser is normally mounted on a PVC or mild steel saddle-type baseplate and attached to the baseplate by a bolt through the center of the dome (Figure 3.6). The bolt is constructed from a number of materials including brass, plastics, or stainless steel. A soft rubber gasket is placed between the baseplate and the dome, and a washer and gasket are also used between the bolt head and the top of the diffuser. These gaskets are critical to the integrity of the diffuser as overtightening can lead to permanent compression set and eventual air leakage. Note that air pressure will force the dome upward off the baseplate. To distribute the air properly through the system, control orifices are located in the hollowed-out center bolt or drilled into the baseplate. Various means are used to fix the dome to the air distribution header. The baseplate may be solvent welded to the header in the shop or may be fastened to the header at the plant site by drilling a hole with an expansion plug.

Dome diffusers are normally designed to operate over a range of airflow rates from 0.8 to 3.9 m^3_N/h (0.5 to 2.5 scfm) per diffuser. Diffuser fouling and airflow distribution normally set the lower airflow rate and efficiency. Back pressure considerations normally dictate the higher rates.

3.2.1.2.5 Disc Diffusers

Disc diffusers, being relatively flat, are a newer innovation of the dome diffuser. Whereas dome diffusers are relatively standard in size and shape, available disc diffusers differ in size, shape, method of attachment, and type of diffuser material. Disc diffusers are available in diameters of 18 to 51 cm (7 to 20 in). The shape of porous plastic or ceramic media is normally two flat parallel surfaces with at least one exception whereby the manufacturer produces a raised ring sloping slightly downward toward both the periphery and the center of the disc. A step on the outer periphery is often built into the disc to improve uniformity of air flux and effectiveness of the seal at the diffuser edge (Figure 3.7).

As with the dome diffusers, porous plastic and ceramic disc diffusers are mounted on a plastic, saddle-type base plate. Two methods are used to secure disc media to the holder: a center bolt or a peripheral clamping ring. The center bolt and gasket arrangement is similar to that used for domes. Use of a screw-on retainer ring is more commonly the method of attachment. A number of different gasket arrangements may be employed, including a flat gasket below the disc, a U-shaped gasket that covers a small portion of the top and bottom and the entire edge of the disc, and an O-ring gasket placed between the top of the outer periphery of the disc and the retainer ring.

Two methods are used to attach the porous plastic or ceramic disc to the air header. The first method is to solvent cement the base plate to the header in the shop. The second type of attachment is completed through mechanical means using either a bayonet-type holder or a wedge section placed around the pipe. These mechanical attachments are performed in the field. Holes are drilled in the header and the disc assemblies are subsequently attached. Future expansion of the system is accommodated by predrilling and plugging holes or by drilling the required holes at the needed time. Individual control orifices in each diffuser unit are used to provide uniform air flux in the system. For bolted systems, the bolt may be hollowed and an orifice drilled in its side. Other designs incorporate either an orifice drilled in the base plate or a threaded inlet in the base where a small plug containing the desired orifice can be inserted.

Perforated membrane discs are designed to lie over a support plate containing apertures that allow air to enter between the membrane and the plate. The membrane is normally not perforated over the apertures and when the air is off, the membrane will seal against mixed liquor intrusion. The membrane may be secured to the base around the periphery by a clamping a ring, wire or a screw-on retaining ring. When the air is on, the membrane will flex upward approximately 6 to 64 mm (0.24 to 2.6 in). Flexing beyond the manufacturer's recommendations could lead to maldistribution of air. Therefore, some designs include additional means of support at the center to prevent overflexing. The base of the membrane support frame is usually threaded. A saddle that is also threaded is glued or clamped to the air header and receives the base plate. Several manufacturers utilize holders identical to that used for a ceramic or porous plastic disc. Such a design allows interchanging of membranes and porous diffuser discs. Several configurations of perforated membrane discs are shown in Figure 3.8 a and b and 3.9.

Ceramic and porous plastic diffusers typically have design airflow rates ranging from 0.8 to 4.7 m^3_N/h (0.5 to 3 scfm) per diffuser. The optimum airflow depends on disc surface area but continuous operation at airflows below about 0.8 m^3_N/h (0.5 scfm) per diffuser may lead to poor airflow distribution over the entire disc surface. In applications above 3.1 m^3_N/h (2 scfm) per diffuser, the control orifice must be properly sized so that the head loss produced does not adversely affect the economics of the system. For perforated discs, design airflows range from 1.6 to

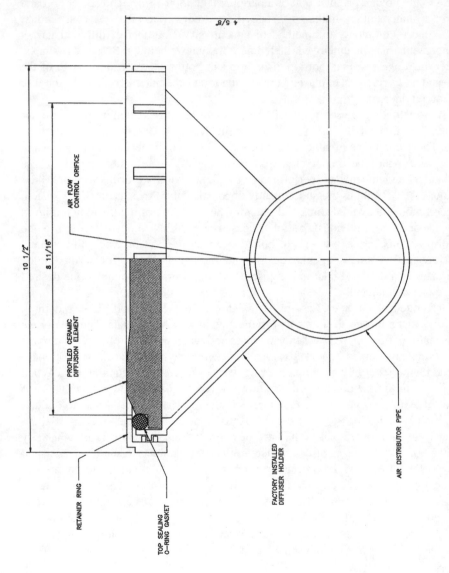

FIGURE 3.7 Ceramic disc (courtesy of Sanitaire, Brown Deer, WI).

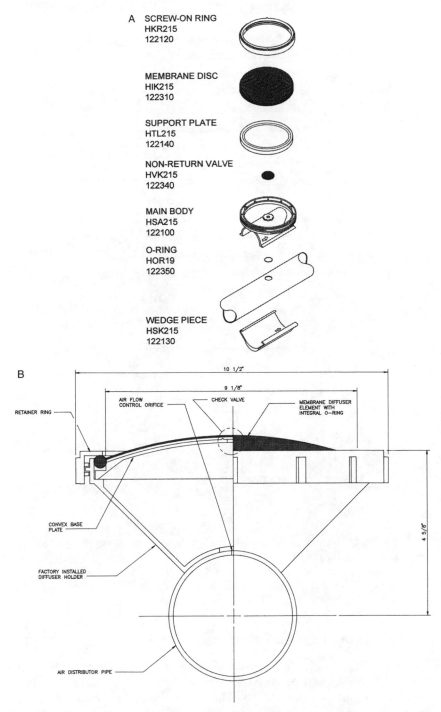

FIGURE 3.8 Several membrane disc configurations [(A) courtesy of Nopon Oy, Helsinki, Finland; (B) courtesy of Sanitaire, Brown Deer, WI].

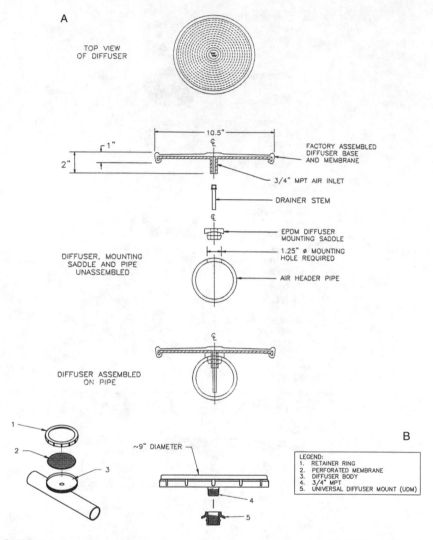

FIGURE 3.9 Several membrane disc configurations [(A) courtesy of Wilfey Weber, Inc., Denver, CO; (B) courtesy of EDI, Columbia, MO].

15.7 m^3_N/h (1 to 10 scfm) per diffuser for the discs up to 30 cm (12 in) in diameter and 4.7 to 31.4 m^3_N/h (3–20 scfm) per diffuser for the larger discs.

3.2.2 NONPOROUS DIFFUSER SYSTEMS

Nonporous diffusers differ from porous diffusers in that they use larger orifices or holes to discharge air. Introduced as early as 1893 these diffusers are available in a variety of shapes and materials. This section will describe these diffusers under the categories of fixed orifice, valved orifice, static tubes, perforated tubes, and other units.

3.2.2.1 Fixed Orifice Diffusers

Fixed orifice diffusers vary from simple openings in pipes to specially configured openings in a number of housing shapes. Historically, orifices much below 4 mm (0.16 in) were susceptible to rapid clogging in wastewater, although even the coarser openings clogged under some wastewater conditions. These devices typically employ hole sizes that range from 4.76 to 9.5 mm (0.1875 to 0.375 in) in diameter producing relatively coarse bubbles (6 to 10 mm). As a result, these diffusers are not efficient oxygen transfer devices but find use in grit separation processes, influent and effluent channel aeration, aerobic sludge digestion and aeration of certain wastewaters that have a propensity to precipitate or easily foul porous diffusers. Today, fixed orifice diffusers are usually molded plastic devices containing a number of holes or slotted stainless steel tubes containing rows of holes along the top or sides and an open slot on both sides of the tube below the holes (Figure 3.10 A and B). The slots in the tube are designed to carry air as airflow increases or as holes plug. One manufacturer produces a slotted tube constructed of plastic that may be converted to a porous membrane diffuser with the placement of a synthetic fiber sheath over the tube.

Many of the fixed orifice diffusers are saddle mounted on the air header. Most are equipped with airflow control orifices to balance airflow. Some contain blowoff legs to purge liquid or relieve back pressure in the event of fouling. Typical gasflow rates range from 9.4 to 47.1 m^3_N/h (6 to 30 scfm) depending on the unit. Perforated tubes normally are screwed into air headers in wideband configurations. Orifices are employed to control airflow distribution in the system.

3.2.2.2 Valved Orifice Diffusers

Valved orifice diffusers use a check valve to prevent backflow when the air is shut off. Some are designed to provide adjustment of the number or size of the air discharge openings. Orifice sizes are similar to those used in fixed orifice devices. Several designs incorporate a membrane (EPDM or other elastomer) as a diaphragm that opens and closes over orifices when air is on or off (Figure 3.11). Another uses a Delrin ball check valve that rides up and down a sleeve mounted inside a cylinder containing drilled holes. A third design employs a cast body with inner air chamber. A 7.6 cm (3 in) diameter plastic disc is retained in position by a steel spring wire that opens and closes over the air chamber depending upon airflow. All of these devices operate over a variety of airflows ranging from 9.4 to 18.8 m^3_N/h (6 to 12 scfm). The units are typically mounted on the crown of the air header thereby requiring header blowoff provisions to purge the system of water in the event of a check valve failure. As with fixed orifice diffusers, these devices exhibit lower oxygen transfer efficiencies than the finer bubble porous diffusers and typically find service in grit separation, inlet/outlet channel aeration, and aerobic digestion.

3.2.2.3 Static Tubes

Static tube diffusers consist of a stationary vertical tube placed over an air header that delivers bubbles of air through drilled holes. The static tube is similar to an airlift pump. As air rises through the vertical tube, interference devices within the

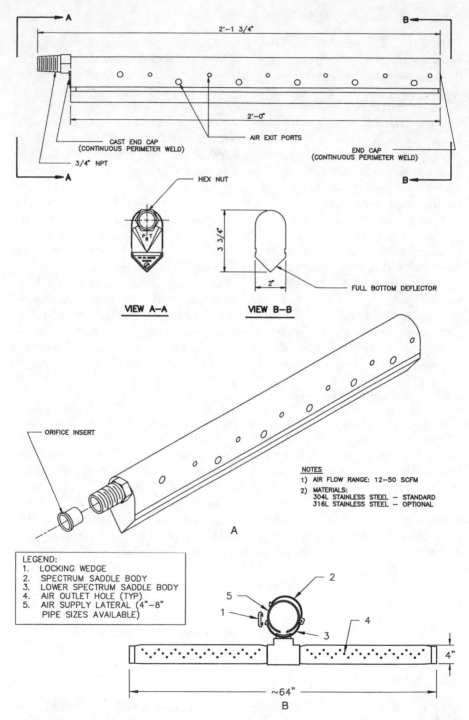

FIGURE 3.10 Coarse bubble diffuser [(A) courtesy of Sanitaire, Brown Deer, WI; (B) courtesy of EDI, Columbia, MO].

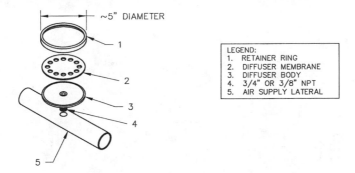

~5" DIAMETER

LEGEND:
1. RETAINER RING
2. DIFFUSER MEMBRANE
3. DIFFUSER BODY
4. 3/4" OR 3/8" NPT
5. AIR SUPPLY LATERAL

FIGURE 3.11 Selected coarse bubble diffusers (courtesy of EDI, Columbia, MO).

tube are designed to shear bubbles and mix the air and liquid, thereby promoting gas transfer. The vertical tubes are normally 0.3 to 0.45 m (12 to 18 in) in diameter and constructed of polypropylene or polyethylene. They are fixed to the tank bottom by stainless steel support stands. High-density polyethylene air piping is supported below the vertical tube. Holes drilled in the air header are normally of a size similar to fixed orifice diffusers. Airflow rates per tube vary with tube diameter but are typically in the range of 15.7 to 70.7 m^3_N/h (10 to 45 scfm). Static tubes are most often applied to aerated lagoon systems, although some may be used in activated sludge processes.

3.2.2.4 Other Devices

3.2.2.4.1 Jets

Jet aeration combines liquid pumping with gas pumping to result in a plume of liquid and entrained air bubbles. A pumping system recirculates the wastewater from the aeration basin and ejects it through a nozzle assembly. The nozzle configurations may include a venturi or mixing chamber whereby gas and liquid are mixed in the motive field. At least one manufacturer produces a jet aerator containing an inner and outer jet configuration with mixing chamber. Gas is pumped through a separate header and is introduced into the recycled wastewater at the venturi or within the mixing chamber (Figure 3.12 and 3.13). The resultant gas-liquid plume is then directed back into the aeration tank through the jet. Jet aerators may be configured as directional devices or as clustered or radial devices. The piping and jets are normally constructed of polypropylene, fiberglass, or stainless steel.

Typically the wastewater recirculation pump is a constant-rate device, and the power turndown for the aerator is accomplished by varying the airflow rate. Air is delivered under pressure by a low head blower. As such, power is consumed both in the recirculation of the liquid and the delivery of the air. The gas-liquid plume normally contains very fine bubbles of gas, thereby classifying jets as fine bubble devices. Depending upon basin geometry and jet exit velocity, the horizontal plume rises rapidly within the basin intermixing with the basin contents. It is significant to note that the air-head loss through the jet is very low or negative due to the ejecting action of the motive fluid. Although it has been used in rectangular basins,

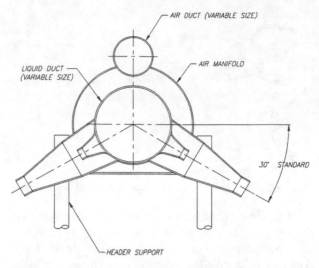

FIGURE 3.12 Unidirectional jet (courtesy of US Filter, Jet Tech Products, Edwardsville, KS).

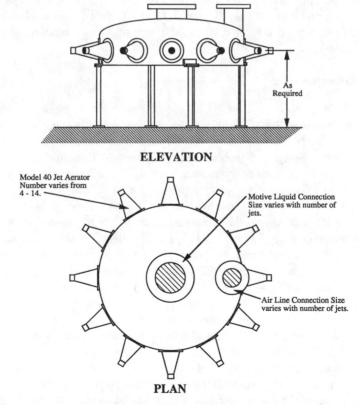

FIGURE 3.13 Radial jet (courtesy of US Filter, Jet Tech Products, Edwardsville, KS).

the directional feature of the device favors its application in oxidation ditches and circular basins.

3.2.2.4.2 Perforated Hose

Perforated hose typically consists of polyethylene tubing held on the floor of the basin by lead ballast. At least one manufacturer suspends the tubing from floats. The tubing contains slits or holes at the top of the tube to release air. Manifolds running along the basin length supply the air. Typically the tubing is mounted across the basin width. Applications of perforated tubing are limited to lagoon systems.

3.2.2.4.3 U-Tube Aeration

A U-tube system consists of a 9 to 150 m (30 to 500 ft) deep shaft that is divided into an inner and outer zone. As air is directed to the wastewater in the downcomer zone, the mixture travels to the bottom of the tube and then returns back to the surface for further treatment (Figure 3.14). The great depth to which the air-water mixture is subjected provides high dissolution due to the high oxygen partial pressures.

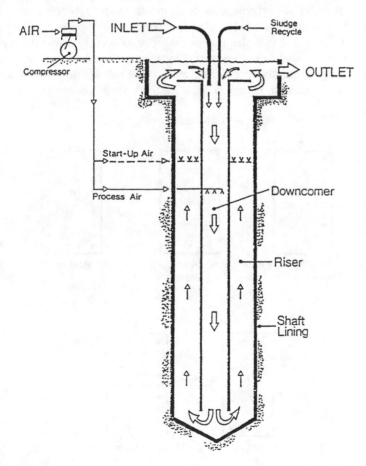

FIGURE 3.14 U-tube aerator.

The amount of air added depends on the wastewater strength and the depth of the shaft. For normal strength municipal wastewaters, the air requirement is dictated by the amount of air needed to circulate the fluid in the shaft since the air is the motive force for moving the wastewater around the shaft. At higher strengths (over 500 mg/L), the air required is governed by the oxygen demand of the wastewater. Under these conditions, all or most of the gas is dissolved. Thus, the economics of the deep shaft becomes more favorable as wastewater strength increases. Once this system is constructed, it is inflexible and not easily maintained or modified.

3.3 DIFFUSED AIR SYSTEM LAYOUTS

The layout of diffusers in a basin has an important influence on the performance of the system. Basin geometry, diffuser submergence, diffuser density and placement of the diffusers all must be considered in effective design of the system. Earliest layouts were in grid format, and basin depth was most often dictated by pressure requirements of air delivery systems. As described above, early experimentation with layout was tried, and depending upon the importance of maintenance and energy requirements, several configurations were adopted. Improvements in air delivery systems and the limitations on space also provided impetus to move to deeper basins where required. At the present time, several basin configurations are used in activated sludge designs. These include spiral roll, cross roll, mid-width, dual roll and full floor grid layouts (Figure 3.15). In addition, horizontal flow systems, ditch configurations, and deep

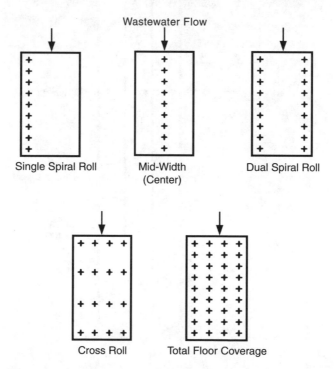

FIGURE 3.15 Typical diffuser layouts.

tanks are also considered during the design process. The sections that follow briefly describe these configurations and indicate which types of diffusers are most often used in them. Subsequent sections will discuss the effects of diffuser layout on performance.

3.3.1 FULL FLOOR GRID

Full floor grid arrangements are defined as any total floor coverage by diffusers whereby the diffuser positioning does not cause a roll pattern. In general, this pattern would result when the maximum spacing between diffusers in any direction does not exceed 50 percent of submergence. The pattern includes the once popular ridge and furrow layout, now all but abandoned, as well as closely spaced rows of diffusers running either the width (transverse) or length (longitudinal) of the basin. All porous diffusers and most nonporous diffusers may be placed in a full floor grid.

Ceramic and porous plastic plates are usually placed in full floor grids. Ceramic plates are often grouted into the basin floor. Downcomer pipes deliver air to channels below the plates. The newer plate designs are often not attached to the basin floor. These ceramic or porous plastic plates are furnished in rectangular sections each serviced by individual rubber air feed hoses. They may be placed as needed in a variety of patterns on the basin floor. This placement is limited only by the length of the tubing. Perforated membrane panels are most often placed in full floor grids. The panels are placed on the tank bottom and fastened with anchor bolts. Air is introduced at one end of the panel through flexible air tubes.

Although their shape and operating characteristics may differ, dome and disc diffusers are most often placed in full floor grids (Figure 3.16 and 3.17). The typical layout and air piping arrangements are identical. Air piping laterals are most often constructed of PVC in the U.S., while stainless steel piping is often specified in Europe. If PVC is used, it should be UV-stabilized with two percent minimum TiO2, or equivalent. In the U.S., the specifications, dimensions, and properties of the PVC pipe should conform to either ASTM D-2241 or D-3034, depending on pipe outside diameter. Where stainless steel is used, a light thin wall 304L or 316L stainless is preferred. The pipe is fixed to the basin bottom with PVC or stainless steel pipe supports. The diffusers are mounted as close to the basin floor as possible, usually within 23 cm (9 in) of the highest point of the floor. Air is delivered through downcomers mounted along the basin walls. Blow-offs are furnished at the ends of the laterals for purposes of purging water from the laterals in the event of power outages.

Tubular diffusers may also be placed in full floor grid configurations (Figure 3.18). Most tube diffuser assemblies include a threaded nipple (stainless steel or plastic) for attachment to the air piping system. Nonporous fixed and valved orifice diffusers often use a similar means of attachment and can also be placed in grid arrangements. The air headers are usually fabricated from PVC, CPVC, stainless steel, or fiberglass reinforced plastic. Extra strength is required for tubular diffusers as compared with discs/domes and some nonporous devices because of the cantilevered load. Threaded adapters or saddles are glued, welded, or mechanically attached to the headers at the points where the diffusers are to be attached. On the header itself, the diffusers may be installed along one side (single band) or both sides (wide band) of the pipe.

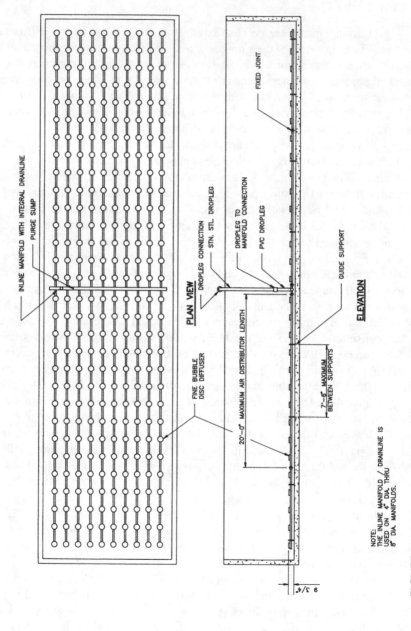

FIGURE 3.16 Fine pore grid layout (courtesy of Sanitaire, Brown Deer, WI).

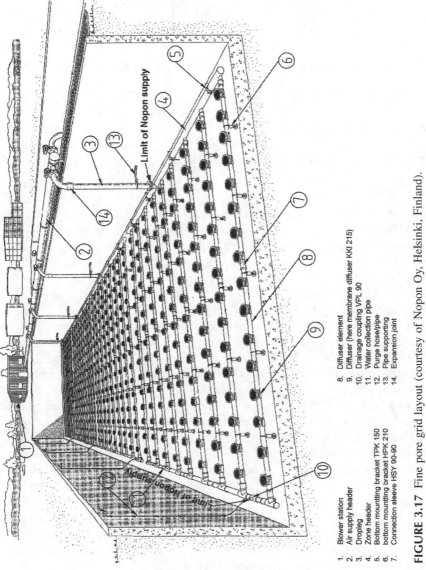

1. Blower station
2. Air supply header
3. Dropleg
4. Zone header
5. Bottom mountting bracket TPK 150
6. bottom mountting bracket HPK 210
7. Connection sleeve HSY 90-90

8. Diffuser element
9. Diffuser (here membrane diffuser KKI 215)
10. Drainage coupling VPL 90
11. Water collection pipe
12. Purge hose/pipe
13. Pipe supporting
14. Expansion joint

FIGURE 3.17 Fine pore grid layout (courtesy of Nopon Oy, Helsinki, Finland).

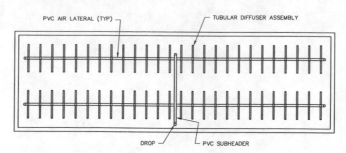

FIGURE 3.18 Tube grid layout (courtesy of EDI, Columbia, MO).

For full floor grid arrangements, fixed headers are almost always employed, and the distance between headers and the spacing between diffusers on the headers approach the same value. Drop pipes located along the sidewalls furnish the air. Laterals may run either a transverse or longitudinal direction. Diffusers are typically located approximately 30 cm (12 in) off the basin bottom.

3.3.2 SPIRAL ROLL

As discussed above, spiral roll was introduced in the U.S. at Indianapolis in 1923 (Hurd, 1923). It was believed that this configuration provided longer contact between the wastewater and the air due to the circulatory flow. Other advantages included lower construction costs and easy accessibility of the diffuser elements. Chicago North Side and Milwaukee Jones Island adapted the spiral roll for plates shortly thereafter. Later studies at Milwaukee and elsewhere indicated that spiral roll configurations were good bulk mixers but poor for oxygen transfer.

Plate and panel diffusers are very rarely placed in spiral roll configurations, although some plants use this arrangement. Rows of plates are placed along one side of the basin in a longitudinal direction. The plates may be grouted in special holders placed on the basin floor. The newer plates mounted on ABS or other plastic plenums may be placed within the tank and along one side.

Dome and disc diffusers are not normally placed in a spiral roll configuration, although some plants do use this arrangement where oxygen demand is low and mixing may control design. When used in this arrangement, tightly spaced rows of diffusers may be mounted on fixed longitudinal headers near the sidewall. A removable header or swing header arrangement typically used for tubes or nonporous diffusers may also be employed. In these applications, stainless steel is often used for the header system.

Tubular diffusers along with fixed and valved orifice diffusers are often placed in spiral roll patterns (Figure 3.19). They are typically mounted on removable or swing header arrangements for easy access. All other construction features are similar to those for these devices used in full floor grids.

3.3.3 DUAL SPIRAL ROLL

In an effort to improve oxygen transfer while retaining the advantages of good bulk mixing, lower construction cost, and ease of diffuser accessibility, a dual roll pattern

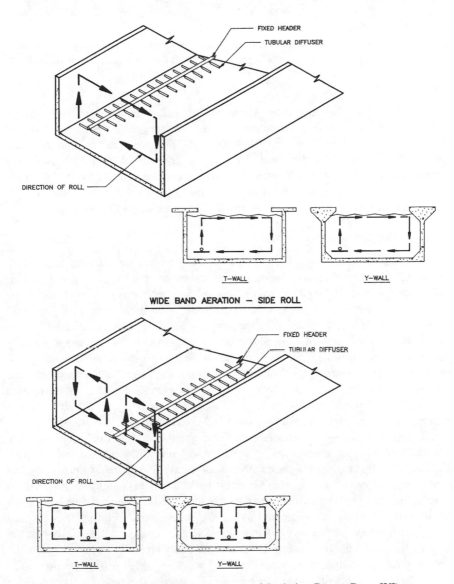

FIXED HEADER
TUBULAR DIFFUSER
DIRECTION OF ROLL
T-WALL
Y-WALL

WIDE BAND AERATION — SIDE ROLL

FIXED HEADER
TUBULAR DIFFUSER
DIRECTION OF ROLL
T-WALL
Y-WALL

FIGURE 3.19 Spiral roll configuration (courtesy of Sanitaire, Brown Deer, WI).

was devised. Plates, disc/domes, and tubes along with fixed and valved nonporous diffusers may be used in this arrangement. Most construction features are similar to spiral roll layouts with the exception that rows of diffusers are placed longitudinally on both sides of the aeration tank. Fixed, removable, and swing headers are used.

3.3.4 MID-WIDTH ARRANGEMENT

The mid-width diffuser arrangement provides an opposing dual roll pattern thought by some to offer a more efficient transfer system. This layout provides few advantages over those described above. Headers located along the centerline are most often fixed, and diffusers are not easily accessed. Less piping is employed (and fewer diffusers), however. This layout is most often found with tubular or nonporous diffusers.

3.3.5 CROSS ROLL

Cross roll patterns are produced by placing laterals perpendicular to the long axis of the basin. As with the spiral roll configuration, a circulatory pattern is established with return flow near the bottom of the basin back to the pumped water column. As such, bulk mixing is enhanced, although all designers do not agree that adequate mixing is developed by this arrangement. Tubular along with nonporous fixed and valved diffusers may be used in this configuration. The diffusers may be placed on fixed, removable, or mechanical lift-type headers. Other construction features are similar to other patterns.

3.3.6 HORIZONTAL FLOW SYSTEMS

In 1965, Pasveer and Sweeris (1965) introduced new insight into the aeration of wastewater by suggesting that imparting a horizontal velocity vector on diffused air bubbles would enhance oxygen transfer efficiency. They correctly deduced that diffuser pattern was an important variable in designing aeration systems. Spiral roll produced the poorest efficiency by virtue of the short bubble residence times resulting from the large velocity of ascent of the aerated mixture. They proposed that the ascent velocity was two to three times higher than the bubble rise velocity alone. Spreading the diffusers along the entire tank bottom would result in increased bubble residence time as a result of the lower vertical rise velocities of the air-water mixture. They proposed that a horizontal vector of flow might reduce or break up the fluid ascent velocities and thereby increase bubble residence time and concomitant oxygen transfer.

An experimental study was conducted using an oxidation ditch configuration. Selected horizontal velocities were imparted across a tube diffuser fixed to the bottom of the tank. Comparisons were made with typical spiral roll patterns of similar physical dimensions. In clean water tests, they were able to demonstrate that imposing a horizontal vector of flow past the diffuser significantly increased oxygen transfer for a given airflow rate per diffuser as compared with a spiral roll layout. Further, they showed that the magnitude of the oxygen transfer efficiency increased as the horizontal velocity increased up to a point. The demonstration typically revealed twice the efficiency rate as compared with spiral roll by providing this horizontal velocity.

Application of this finding was apparent in Europe by the early 1970s. Schreiber introduced the concept in the U.S. in the early 1980s. In the Schreiber design, bridge-mounted tubes were rotated through a circular aeration tank. Other European designs employ circular or ditch geometries. In these designs, the horizontal velocity is imposed by a mixing device, and the diffusers are fixed to the bottom of the basin (Figure 3.20 and 3.21). Results of testing of these configurations appear in the Performance section of this chapter.

3.3.7 DEEP TANKS

Deep tank aeration is being practiced on a limited scale in the U.S. and abroad. Limited land availability and the need for increased plant capacity have led to the

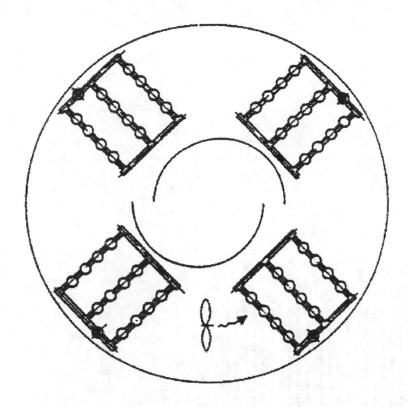

Placing of diffusers around basin equipped with flow generator

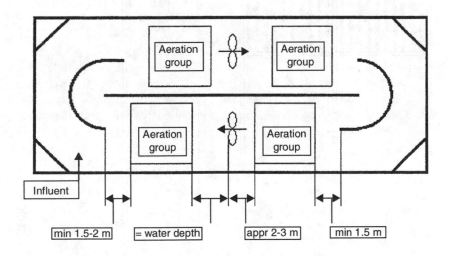

Placing of diffusers in an oxidation ditch (race track) equipped with flow generators

FIGURE 3.20 Mixer–diffuser horizontal configuration (courtesy of Nopon Oy, Helsinki, Finland).

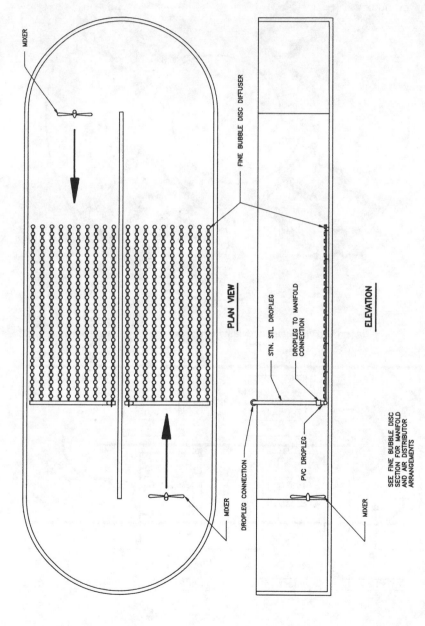

PLAN VIEW

ELEVATION

MIXER

FINE BUBBLE DISC DIFFUSER

STN. STL. DROPLEG

DROPLEG TO MANIFOLD CONNECTION

DROPLEG CONNECTION

MIXER

PVC DROPLEG

MIXER

SEE FINE BUBBLE DISC SECTION FOR MANIFOLD AND AIR DISTRIBUTOR ARRANGEMENTS

FIGURE 3.21 Mixer-diffuser horizontal configuration (courtesy of Sanitaire, Brown Deer, WI).

use of deep tanks in some locations. Other advantages to deep tanks include lower off-gas emissions of VOCs due to lower gas flux rates and, sometimes, greater aeration efficiencies. Deep tank aeration has generally found greatest application for industrial wastewaters. Very efficient aeration has been reported with jet injector aeration in industrial waste streams. However, salinities were high in these wastes, having a positive impact on oxygen mass transfer. Jackson (1982) and Jackson and Shen (1978) have reported successful application of deep tanks for industrial wastewater treatment. Nitrogen supersaturation was exploited as a means to achieve flotation separation of the mixed liquor. It is this phenomenon that can create a problem in treatment plants through the unwanted flotation of solids in the secondary clarifiers. A detailed discussion of deep tank aeration is found in Chapter 4.

3.4 PERFORMANCE OF DIFFUSED AIR SYSTEMS

3.4.1 Factors Affecting Performance

Equation (2.26) provides the basic equation describing the transfer of oxygen to water. As indicated in Chapter 2, the three fundamental parameters that describe oxygen transfer by a given aeration system are $K_L a$, C_∞^* and C_L. The variables that affect these parameters are also delineated in Chapter 2 and are included in the design equations. When evaluating a given aeration system, a number of factors intrinsic to the aeration device will affect oxygen transfer rates and efficiency including the process flowsheet, the mode of operation of the process, the control methodologies used, and the maintenance of the equipment. For diffused air systems these factors include

- diffuser type
- diffuser placement
- diffuser density
- gas flow rate per diffuser or unit area
- basin geometry and diffuser submergence
- wastewater and environmental characteristics
- process type and flow regime
- process loading
- DO control
- degree of diffuser fouling or deterioration
- mechanical integrity of aeration system

Most of these factors are under the control of the designer with the possible exceptions of wastewater and environmental characteristics along with diffuser fouling or deterioration. However, good design includes a careful evaluation of even these uncontrollable factors and provides for these uncertainties in the design.

The sections that follow will provide data on diffused air performance in both clean and process waters. The impact of the factors outlined above is illustrated as a part of this presentation. With many different types of diffused air systems, process geometries, and wastewater characteristics, it is not possible to realistically develop a general model incorporating all of these variables that will fit all situations. Rather, the trends that have been observed and the relative importance of these factors are discussed.

3.4.2 PERFORMANCE IN CLEAN WATER

Clean water performance provides the baseline for aeration system design in the U.S. and generally worldwide since clean water testing is relatively reproducible regardless of the geographical location. In 1984, the ASCE Oxygen Transfer Standards Committee developed a clean water test procedure that was shown to be reproducible (Baillod et al., 1986). That standard is now used throughout the world or has been adapted into other national standards such as the German ATV standards (ATV-Regelwerk, 1996). The clean water standard is discussed in more detail in Chapter 7.

The clean water performance data presented in this chapter in tabulations and graphical depictions were generated from 1975 to the present. Much was taken from the EPA *Design Manual, Fine Pore Aeration Systems* (1989) and the remainder from clean water test data. The data are presented to provide trends and ranges of performance of representative types of diffusers and are not intended for use in final design calculations.

The results of clean water oxygen transfer tests are reported in a standardized form as standard oxygen transfer rate (SOTR), standard oxygen transfer efficiency (SOTE), or standard aeration efficiency (SAE). These measures were described in detail in Chapter 2.

3.4.2.1 Steady-State DO Saturation Concentration

As described in Chapter 2, steady-state oxygen saturation concentration is one of the critical factors required in the calculation of oxygen transfer rate. For submerged aeration applications, this value is significantly greater than the surface saturation value published in standard tables. It is necessary to either measure this value in clean water tests or to calculate it based on comparable full-scale test data. The value is primarily dependent upon diffuser submergence and diffuser type and is often described by means of Equation (2.33). Alternatively, it may be described through the use of the term, effective depth, as given in Equation (2.34). Effective depth represents the depth of water under which the total pressure (hydrostatic plus atmospheric) would produce the steady-state saturation concentration observed for clean water with air at 100 percent relative humidity.

Figure 3.22 presents typical results for diffused air devices. An abbreviated survey of typical delta values for diffused aeration systems is given in Table 3.1.

The delta values presented in Table 3.1 which increase with increased depths are comparable to those described in Figure 2.12. They may be used for preliminary sizing, but final design calculations should be based on oxygen transfer tests of actual equipment and geometries. For diffusers submerged to approximately 90 percent or more of basin depth, effective depths are typically 21 to 44 percent of basin liquid depth for porous diffusers (Baillod et al., 1986).

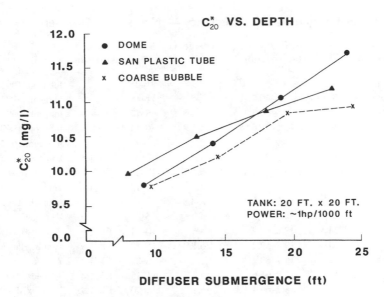

FIGURE 3.22 Diffuser submergence vs. DO saturation.

TABLE 3.1
Typical Delta Values for Diffused Aeration Devices

Diffuser Type	Range of Delta	Range of Depth (m)
Nonporous		
Static tube	1.08–1.16	4.2–5.2
Perf tube	1.05–1.15	2.7–7.3
Porous		
Plates	1.25–1.28	5.6
PM tubes	1.07–1.21	2.1–4.6
PM* disc	1.05–1.30	2.8–7.4
Cer disc	1.09–1.18	4.3–5.4
Cer dome	1.13–1.14	2.9

* PM- Perforated membrane.

3.4.2.2 Oxygen Transfer Data

Typical values of SOTE (and SAE for some nonporous diffusers) for various diffuser types are presented in Tables 3.2 through 3.5. With the continuous changes occurring in the development of diffuser materials and shapes, it is difficult to make many generalizations about the performance of any given diffuser. However, as discussed previously, there are some factors that influence performance of an aeration system.

TABLE 3.2
Clean Water Oxygen Transfer Efficiency — Nonporous Diffusers

Type and Placement		Airflow Rate (m_N^3/h/unit)	Submergence (m)	SOTE (%)	SAE (kg/kWh)	Reference
Fixed orifice	S	9.3–32.8	7.3	21–25	—	Johnson, 1992
perforated tube	S	8.6–40.0	5.2–5.6	11–18	—	Johnson, 1992
	S	9.3–64.3	4.1–4.8	5–17	—	Johnson, 1992
	S	16.0–39.6	3.0–3.8	6–14	—	Johnson, 1992
	S	8.9–31.5	2.7	6–7	—	Johnson, 1992
	G	7.5–23.2	3.0	7–8	1.3–1.5	Yunt & Hancuff, 1988
	G	8.3–24.4	6.1	17–20	2.0–2.2	Yunt & Hancuff, 1988
	MW	6.6–18.8	4.6	11–13	1.5–1.6	Yunt & Hancuff, 1988
Sparger	S	12.9–51.3	4.1–4.8	9–13	—	Johnson, 1992
	MW	18.7–57.0	3.0	6–7	1.3–1.5	Yunt & Hancuff, 1988
	MW	19.8–60.2	4.6	10–11	1.5–1.6	Yunt & Hancuff, 1988
	MW	19.8–59.2	6.1	15–17	1.8–1.9	Yunt & Hancuff, 1988
Static tube	G	15.7–60.2	3.0	6–8	1.1–1.5	Yunt & Hancuff, 1988
	G	15.7–65.6	4.2	11–15	1.5–1.8	Semblex, 1987
	G	15.7–66.4	6.1	13–20	1.7–1.9	Semblex, 1987
	G	24.4–51.0	4.2–4.6	8–12	—	Johnson, 1992
	?	37.0–68.3	5.2	12–15	—	Johnson, 1992

1 m = 3.28 ft; 1.0 m_N^3/h = 0.64 scfm; 1.0 kg/kWh = 1.644 lb/hp-h

G = Grid; S = Spiral roll; MW = Mid-width

TABLE 3.3
Clean Water Oxygen Transfer Efficiency — Aspirators and Jets

Type and Placement		Airflow Rate (m_N^3/h/unit)	Submergence (m)	SOTE (%)	SAE (kg/kWh)	Reference
Jets	Dir	21.1–119	4.6	15–24	1.7–2.0	Yunt & Hancuff, 1988
	Clu	7.1–86.3	3.0	8–14	1.1–1.6	Yunt & Hancuff, 1988
	Clu	7.7–50.5	6.1	21–33	1.6–2.2	Yunt & Hancuff, 1988
Aspirator tube	5.5 kw	—	2.0	—	0.5–0.9	Kayser, 1992
	15 kw	—	2.5	—	0.4–0.8	Kayser, 1992

1 m = 3.28 ft; 1.0 m_N^3/h = 0.64 scfm; 1.0 kg/kWh = 1.644 lb/hp-h

Dir = Directional; Clu = Cluster

Some of these factors are discussed in further detail in the following sections. Since the power consumed in transferring oxygen to the liquid is most important in assessing system performance, estimates of SAE are presented in this section for a variety of devices. For diffused air devices, this figure typically requires a calculation

TABLE 3.4
Clean Water Oxygen Transfer Efficiency — Porous Tubes

Type	Placement	Airflow Rate (m_N^3/h/unit)	SOTE (%) at Depth 2.1 m	3.0m	4.6m	6.1m	Reference
Porous plastic	G	3.8–6.3	—	—	28–32	—	EPA, 1989
	DS	4.7–11.0	—	10–16	16–24	22–32	EPA, 1989
	DS	14.1–17.3	—	10–14	15–17	21–26	EPA, 1989
	S	3.1–11.0	—	12–15	15–20	22–25	EPA, 1989
	S	12.6–18.8	—	10–15	10–17	22	EPA, 1989
Perforated	G	4.7	—	—	—	45	GSEE, Inc., 1998
membrane	G	3.0–10.0	—	27–28	—	—	Pöpel, 1991
	S	0.8–10.0	13–19	17–21	26–35	—	Johnson, 1993; Pöpel, 1991
	DS	0.8–18.8	10–20	15–21	21–36	27–36	EPA, 1989; Johnson, 1993

1 m = 3.28 ft; 1.0 m_N^3/h = 0.64 scfm

G = Grid; DS = Dual spiral roll; S = Spiral roll

of power required by a given blower under a given set of environmental conditions. In this case, the blower wire power consumption is related to the discharge pressure and the mass rate of air by the adiabatic compression of air. A discussion of this calculation is found in Chapter 4. The assumed values of system head loss, blower inlet and discharge temperatures, and combined blower/motor efficiency are presented as required for these calculations.

3.4.2.2.1 Diffuser Type

In diffused aeration, air bubbles, which are typically formed at an orifice (exceptions are jet and aspirator systems) near the bottom of the aeration basin, break off and rise through the liquid finally bursting at the surface. As the bubble begins to emerge from the orifice, the air-water interface is continuously being replenished causing a high surface renewal rate and thus, a high transfer rate. Once it breaks away from the orifice and theoretically reaches a terminal rise velocity, the effective liquid film thickness or surface renewal rate becomes constant. In an aeration tank, eddy currents normally will affect rise velocities, which are the sum of the terminal or "slip" velocity, v_s, of the bubble and the fluid velocity for the rising gas-liquid stream, v_w. As the bubble bursts at the surface, it sheds an oxygen-saturated film into the surface layers. Some surface aeration also occurs due to surface turbulence.

The size of the bubble released by a diffuser is related to the orifice diameter, surface tension, and liquid density when gas flows are low (typically less than 100 bubbles per minute). At the higher airflow rates used in wastewater aeration practice, bubble diameter is a function of gas flow rate, G_s, while frequency of formation remains constant yielding the following empirical expression.

TABLE 3.5
Clean Water Oxygen Transfer Efficiency — Porous Disc/Domes in Grid

Type	Diffuser Density (%)	Airflow Rate (m³ₙ/h/unit)	SOTE (%) at Depth			Reference
			3.0 m	4.6 m	6.1 m	
Plastic plates	10	35.6–84.7	—	—	30–40	Johnson, 1993
Ceramic disc, 24-cm	7.5	1.4–4.7	20–22	27–33	34–37	EPA, 1989
	11.7	1.3–4.6	21–24	30–34	35–41	EPA, 1989
	15.1	1.1–4.1	22–25	31–34	38–41	EPA, 1989
Ceramic disc, 22-cm	6.0–6.3	2.3–5.0	—	25–29	32–38	EPA, 1989
	6.9–7.7	0.9–3.9	—	25–30	33–40	EPA, 1989
	8.9–10.2	0.9–5.3	—	27–34	31–40	EPA, 1989; Johnson, 1993
	12.0–12.8	0.6–4.4	—	25–36	34–39	EPA, 1989
	16.4–21.6	1.1–4.9	—	27–38	31–38	EPA, 1989; Johnson, 1993
Ceramic disc, 23-cm	12.0	1.9	—	32–33	—	Johnson, 1993
Ceramic dome, 18-cm	4.8	0.8–3.9	—	23–31	28–40	EPA, 1989
	6.1–6.3	0.8–3.9	16–23	25–32	30–41	EPA, 1989
	8.1–8.4	0.8–3.9	20–24	27–37	31–44	EPA, 1989
	10.7–12.1	0.8–3.9	17–23	27–35	33–47	EPA, 1989
	17.3	0.8–3.9	18–26	27–34	—	EPA, 1989; Johnson, 1993
Plastic disc, 18-cm	3.9	0.9–5.5	15–18	22–27	—	EPA, 1989
	5.8	0.9–5.5	16–21	24–28	—	EPA, 1989
	6.8	0.8–3.6	—	25–31	—	EPA, 1989
	9.2	0.6–2.3	19–22	26–32	—	EPA, 1989

1 m = 3.28 ft; 1.0 m³ₙ/h = 0.64 scfm

$$d_b = C_1 G_s^n \qquad (3.1)$$

C_1 and n are empirical constants. For porous diffusers (fine pore) where pore size is typically 0.1 to 0.3 mm, n is usually less than 1.0, and bubble diameters range from 1.5 to 3.0 mm. For nonporous diffusers where orifice sizes typically range from 5 to 25 mm or larger, n may be greater than 1.0 and, bubble diameters range from 20 to 40 mm. For these coarse bubble diffusers, it is believed that as gas flow increases, the turbulence tends to redivide the larger bubbles into smaller ones (Eckenfelder, 1959). An intermediate group includes diffusers that have pore sizes that may range from 2 to 5 mm, and bubbles exhibit diameters typically intermediate between the fine pore diffuser and the nonporous diffuser.

Bubble size and shape affect oxygen mass transfer in several ways. Barnhart (1966, 1969) has shown that about 25 percent of the total oxygen transferred in a 3.65 m (12 ft) deep tank occurred at bubble formation for a fine pore diffuser system. Using coarse bubble diffusers, considerably less transfer occurred during bubble formation. Barnhart has shown that the liquid film coefficient, k_L, increases as bubble

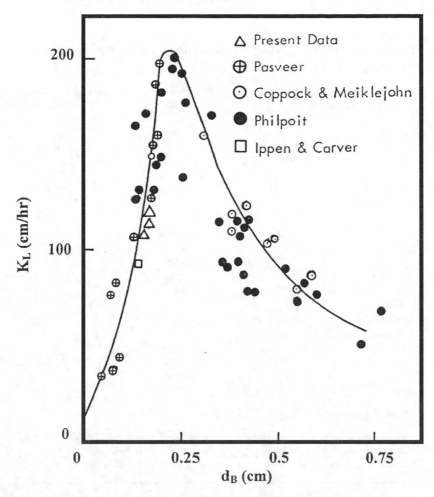

FIGURE 3.23 Relationship between bubble size and liquid film coefficient (adapted from Barnhart, 1966).

size increases up to a diameter of approximately 2 mm. At that point, the coefficient decreases with increases in bubble diameter (Figure 3.23). There is some controversy about the lower limit on bubble size where k_L decreases. Several investigators have found that k_L reaches a maximum value and remains relatively constant thereafter. The individual bubble surface area to volume ratio will decrease with increased bubble size, thereby directly affecting the overall mass transfer coefficient, K_La. Finally, the residence time of the bubble in the basin depends on bubble shape and size. The terminal bubble velocity, v_s, and its shape are related to Reynolds Number. At $R_e < 300$, the bubbles are spherical, and bubble rise is helical or rectilinear (Aiba et al., 1973). Between 300 and 4000, the bubbles are ellipsoidal and rise with a rectilinear, rocking motion. The bubbles form spherical caps at $R_e > 4000$. Since the basin total bubble surface area is the product of the discrete bubble area at time, t,

TABLE 3.6
Typical Clean Water Standard Aeration Efficiencies — Porous Diffusers
(Submergence 4.6 m)

Type/Configuration	Airflow Rate (m_N^3/h/diffuser)	SAE[a] (kg/kWh)
Plastic tube		
Grid	3.8–6.3	4.5–5.2
Spiral	3.1–11.0	2.4–3.2
Spiral	17.6–18.8	1.6–2.7
Dual	4.7–11.0	2.6–3.9
Perforated membrane		
Spiral	0.8–10	4.2–5.7
Dual	0.8–18.8	3.4–5.8
Ceramic disc		
18-cm grid	0.6–5.5	3.6–5.2
22-cm grid	0.6–5.0	4.1–6.1
24-cm grid	1.1–4.7	4.4–5.5
Ceramic dome		
18-cm grid	0.8–3.9	3.4–6.0
Perforated membrane disc[b]		
51-cm	24–172	2.7–4.6
30-cm	13–237	2.7–6.1
23-cm	13–280	2.4–7.1
Panel[b]	4–74	3.1–6.9

1.0 m_N^3/h = 0.64 scfm; 1.0 kg/kWh = 1.644 lb/hp-h

[a] Wire power calculated from adiabatic compression relationship for T = 20°C, P = 1 atm, blower/motor efficiency = 70%, discharge pressure varies with diffuser type

[b] Airflow rate — m_N^3/h-m^2

and the bubble residence time distribution, the total gas surface area in the basin decreases as the bulk bubble velocity increases.

Oxygen transfer efficiencies can therefore be related to diffuser type by means of the system parameters of bubble size and shape along with gas flow rate for a given basin geometry. Typically, for bubbles larger than about 1 to 2 mm, efficiency will decrease with increased bubble size down to some asymptotic value. Tables 3.2 through 3.5 illustrate that porous diffusers, which generally produce fine bubbles, will produce significantly higher efficiencies than nonporous large orifice diffusers. It should be noted that jet diffusers also generate fine bubbles due to cavitation and/or turbulence occurring in the region where gas is introduced into the recirculated liquid stream. Aspirating devices generally produce an intermediate bubble size that is less efficient than the porous diffuser or the jet.

An examination of Tables 3.4 through 3.8 indicate that among the porous diffuser systems, all appear to be similar in oxygen transfer efficiencies with the possible exception of certain membrane panel and high-density membrane disc configurations.

TABLE 3.7
SOTE vs. Airflow for Selected Fine-Pore Diffusers in Clean Water (EPA, 1989)

Diffuser Type	Layout	Diffuser Submergence (m)	Diffuser Density (No. units/m²)	SOTE (%)	Exponent "m"[a]
Ceramic dome	Grid	4.3	3.4	29.6	−0.150
Ceramic disc	Grid	3.7	2.8	31.7	−0.133
Ceramic disc	Grid	3.7	1.6	26.0	−0.126
Rigid porous plastic disc	Grid	4.0	3.7	27.9	−0.097
Rigid porous plastic tube	Double spiral roll	4.0	1.1	26.7	−0.240
Nonrigid porous plastic tube	Spiral roll	4.6	0.9	27.1	−0.276
Perforated membrane disc	Grid	4.3	0.9	29.2	−0.195
23-cm perforated membrane disc	Grid	3.0	2.1[b]	18.9	−0.110
EPDM perforated membrane tube	Grid	3.0	2.1[c]	21.0	−0.150

[a] Equation 3.2
[b] One 23-cm-diameter disc in a 76-cm-diameter column
[c] One 61-cm-long tube in a 76-cm-diameter column
1 m = 3.28 ft

Reasons for these higher levels of performance are elaborated further in this section. A comparison of diffuser performance based on SAE is provided in Tables 3.2, 3.3 and 3.6. It can be seen that most of the devices generating the finer bubbles will also require significantly less power for a given transfer rate than the coarser bubble devices. What is also clear from this tabulation is that those devices requiring power for both the delivery of air and liquid will suffer lower values of SAE even though SOTE values may be high.

3.4.2.2.2 Diffuser Airflow Rate

As seen from Equation (3.1), bubble size depends on airflow rate. The airflow rate also affects bubble shape, bubble rise velocity, and system turbulence. As described above, airflow influences overall bubble surface area and therefore, oxygen transfer rate. It also will influence surface renewal rates and bubble size distributions. For porous diffusers, an increase in G_s will produce larger bubbles and higher bubble velocities, thereby decreasing total bubble surface area and oxygen transfer rate. Over the normal range of operation for a given basin geometry, aeration system, and diffuser type, the relationship between SOTE and diffuser airflow rate can be described by the following empirical relationship.

$$SOTE_a/SOTE_b = \left[G_{sa}/G_{sb}\right]^m \tag{3.2}$$

In this equation $SOTE_a$ and $SOTE_b$ equals SOTE values at gas flow rates G_{sa} and G_{sb} respectively, and "m" is a constant for a given diffuser and system configuration.

TABLE 3.8
Clean Water Oxygen Transfer Efficiency — Perforated Membrane Panels/Discs in Grids

Type	Diffuser Density (%)	Airflow Rate[b] (m_N^3/h-m²)	Specific SOTE[c] (%/m)	Reference
Panels	5.0	37.2–74.4	4.6	Pöpel & Wagner, 1991
	8.0	45.6–92.9	5.9–6.2	Pöpel & Wagner, 1991
	31.0	4.7–16.9	7.5–10.1	Pöpel & Wagner, 1991
	44.6	4.1–12.5	7.9–9.5	Parkson, 1991
	98+	0.8–12.3	10.8–17.0[a]	GSEE, 1986
Disc — 51 cm	6	27.0–172	3.6–5.6	Huibregtse, 1987
	17.7	23.7–162	3.9–6.2	Huibregtse, 1987
Disc — 30 cm	1.5–3.0	13.5–27.0	5.3–8.0	Johnson, 1993
	4.1	49.0–312	3.6–6.2	Eimco, 1986
	6.9–7.6	27.0–346	4.5–6.0	Johnson, 1993
	6.8	59.1–237	4.5–7.8	Johnson, 1993
	13.6	59.1–237	4.1–8.2	Johnson, 1993
Disc — 25 cm (fine)	4.7	15.5–217	4.4–7.2	Wilfey, 1998
	12.6	15.5–217	5.6–8.2	Wilfey, 1998
Disc — 25 cm (coarse)	4.7–12.6	15.5–217	5.1–5.9	Wilfey, 1998
Disc — 23 cm	1.6	20.2–292	3.2–7.9	Wilfey, 1987
	3.2	20.2–255	4.6–6.9	Wilfey, 1987
	4.4	66.0–140	5.4–6.7	Johnson, 1993
	5.8–7.6	20.3–280	4.9–6.2	Wilfey, 1987
	12.4–12.8	13.5–140	6.1–8.5	Pöpel et al., 1993; Johnson, 1993
	24.9	13.5–69	8.8–9.5	Pöpel et al., 1993
Disc — 18 cm	22.0	14.5	8.1	Stenstrom, 1997

[a] for diffuser submergence of 1.75 m
[b] airflow rate per diffuser surface area
[c] SOTE/H_s where H_s is diffuser submergence
1 m = 3.28 ft; 1.0 m_N^3/h-m² = 0.059 scfm/ft²

Gas flow rates are often reported on a per diffuser element basis for discs, domes, tubes, and nonporous diffusers. For plate and panel diffusers, airflow per effective projected surface area is used. In some cases, tubes are rated on a per tube length basis. When comparisons are made between diffusers of different shape or size, it is most useful to express airflow on an effective area basis. This expression is not difficult to apply for ceramic and plastic discs and plates, but requires an understanding of the contributing surface area for perforated membrane diffusers. For tube diffusers, the contributing area is often difficult to assess since airflow distribution is not only dependent upon the perforated (or porous) area but also on the means for distributing air to the media and the airflow rate.

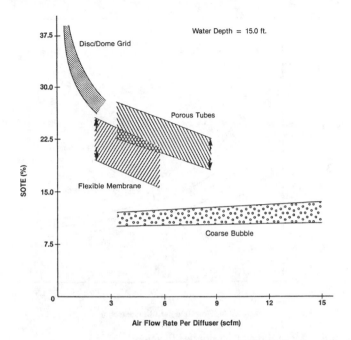

FIGURE 3.24 Efficiency vs. airflow for selected diffusers (US EPA *Summary Report on Fine Pore Aeration Systems*, EPA/62518-85/010,Water Environmental Research Lab, Cincinnati, OH, 1985).

Values of "*m*" for a number of porous diffuser systems appear in Table 3.7. It is useful to note that the values for "*m*" in the grid systems ranged from about –0.11 to –0.19 whereas the values for the spiral roll configurations produced significantly higher values of "*m*" (–0.24 to –0.27). These differences in slopes can have important design and operation implications that are addressed later in this chapter. Observation of the data in Tables 3.4 through 3.6 and 3.8 also confirm the effect of diffuser gas flow rates on oxygen transfer efficiency for porous diffusers.

For nonporous large orifice diffusers, gas flow rates have a significantly different impact. As gas flow increases, bubble size is not greatly influenced or may even decrease in size. Fluid turbulence will increase with gas flow rate that may increase both surface renewal rates and bubble surface area. The actual impact on efficiency will depend on placement and basin geometry. Studies by Bewtra and Nicholas (1964) indicated that gas flow had little effect on coarse bubble spargers. Figure 3.24, taken from an EPA summary report on fine-pore aeration systems (1985), summarizes the impact of gas flow rates on performance. It is immediately apparent that where high efficiencies are being sought with porous diffusers, low gas flow rates per diffuser should be considered.

3.4.2.2.3 Diffuser Densities
In this chapter, diffuser densities are defined as the percentage of the basin surface area covered by the total projected area of diffuser media, or $A_d/A_t \times 100$. The effects of diffuser density on SOTE for disc/dome diffusers, membrane panels, and discs

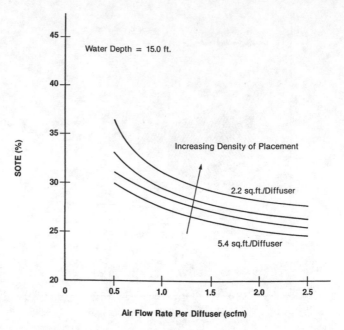

FIGURE 3.25 Impact of diffuser density on efficiency.

are illustrated in Tables 3.5 and 3.8. Generally, an increase in diffuser density results in an increase in SOTE for the same gas flow rate per diffuser. In 1976, Paulson tested dome diffusers in a 4.6 m (15 ft) deep tank and found a linear relationship between diffuser density and SOTE in the range of densities of 6.9 to 18.3 percent (Figure 3.25). Two airflow rates were evaluated in this work. Since that time, numerous other investigations have shown similar results (EPA, 1989). Huibregtse et al. (1983) evaluated the effects of density of disc and dome placements in a 6.1 × 6.1 m (20 ft × 20 ft) test tank. Grid placements of 23.8 cm (9.375 in) ceramic disc diffusers were studied at densities of 7.6, 11.6 and 15 percent. Header spacing was held constant at 0.76 m (2.5 ft). At all three test submergences they found that SOTE increased with diffuser density, but the increase was not linear in all cases. A comparison between dome diffusers (17.8 cm [7 in] in diameter) and the same disc diffusers indicated that, at the same density of diffuser number, the discs were more efficient. This result can be attributed to the higher projected surface area provided by the disc, which was about 70 percent greater than the dome. Yunt and Hancuff (1979) reported similar findings for dome and disc performance. There appears to be an upper limit to diffuser density where little improvement in SOTE will be found. This limit will depend on the diffuser size, airflow rate, and spacing. For example, a 23 cm (9 in) disc diffuser, at a submergence of 4.3 m (14.2 ft) and gas flow of 1.6 m^3_N/h (1 scfm) per diffuser, exhibited little increase in SOTE at densities > 14 percent (Sanitaire, 1976–1986). On the other hand, tests with a 51 cm (20 in) membrane disc indicated that SOTE increased to a density of 26 percent, but the increase was small. A 40 percent increase in the number of diffusers required to increase the density from 18 to 26 percent resulted in only a five percent increase

COLOR FIGURE 1.1 Original submerged turbine system for MCUA plant showing (**A**) aeration tank turbine drives, (**B**) gear reducer.

C

D

COLOR FIGURE 1.1 (continued) **(C)** high purity oxygen delivery piping, and **(D)** compressor room. (Courtesy of Middlesex County Utilities Authority, Sayreville, New Jersey.)

COLOR FIGURE 1.2 **(A)** New surface aeration system for MCUA plant showing **(B)** compact surface aeration drives.

C

D

COLOR FIGURE 1.2 (continued) (C) with elimination of most overhead piping and (**D**) elimination of most equipment from compressor room. (Courtesy of Middlesex County Utilities Authority, Sayreville, New Jersey.)

in SOTE. There appears to be no data available on similar comparisons between disc/domes and tube diffusers. The clean water tabulations presented above indicate that tubes typically produce similar OTEs to those for discs or domes. It is difficult to compare them, however, since the effective surface area of tubes is elusive for reasons stated above.

The increase in diffuser density, which is apparent with the application of perforated membrane panels, has produced high SOTE values, which is attributed to the lower airflow rates, higher densities, and fine slits. In an effort to be more competitive with the membrane diffusers, some disc manufactures are recommending higher placement densities and lower gas flow rates per diffuser than typically used in practice. Table 3.8 illustrates the impact of higher placement densities (12 to 25 percent) for 23 cm (9 in) discs as compared with the panel systems. This, of course, requires more diffusers and operation at lower than typical airflow rate per diffuser. The minimum airflow rate per diffuser typically cited by disc/dome manufacturers is based on concerns for uniform airflow distribution and fouling control. Use of a smaller diameter orifice will resolve that problem to some extent with little additional loss of head. It is important to note, however, that decreasing airflows may lead to mixing problems. This occurrence has apparently not been a problem in the range of airflows currently being used for these high-density disc diffuser systems. More is said about the proper selection of diffuser density, airflow rate, and mixing later on in this chapter.

3.4.2.2.4 Diffuser Placement

As described earlier in this chapter, there are a number of different diffuser placement configurations that may be used in aeration system design. The selection of the most appropriate placement may depend on maintenance considerations, mixing requirements, economies of construction, basin geometry, and efficiency of oxygen transfer. As early as the 1930s, it was found that configuration of diffusers dramatically affected performance. Studies at Milwaukee at that time (Ernest, 1994) demonstrated that a grid configuration was superior to spiral roll with respect to oxygen transfer. This finding was further confirmed in Milwaukee in the 1960s when process water off gas testing showed that the longitudinal and ridge and furrow placements of plates were more efficient than a spiral roll configuration (Leary et al., 1969). Downing et al. (1961) demonstrated that distributing dome diffusers along the basin floor produced transfer efficiencies 10 to 20 percent higher than placement along the centerline (mid-width) or along the wall (spiral roll). At a 3.4 m (11 ft) submergence in these tests, both the mid-width and spiral roll configurations produced similar efficiencies. A number of diffuser placements were evaluated in a 1.2 m (4 ft) long section of a full-scale aeration tank at the Philip Morgan Sanitary Engineering Laboratory at the University of Iowa (Bewtra and Nicholas, 1964). For tube diffusers, they demonstrated that multiple bands of diffusers were generally more efficient than either spiral roll or mid-width patterns. They concluded that configuration affects the bubble retention time when the velocity of the air bubble is the sum of the terminal rise velocity, v_s, and the velocity of the air-water mixture, v_w. In spiral roll placements, the value of v_w is much higher than v_s (three to five times higher), resulting in short bubble residence times and lower efficiencies. In full floor grid

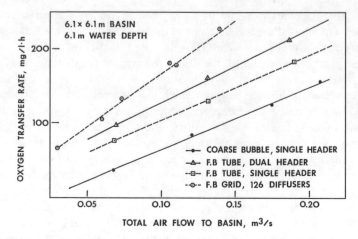

FIGURE 3.26 Efficiency vs. airflow for selected configurations. (From Huibregtse, G.L. et al. (1982). "Factors Affecting Fine Bubble Diffused Aeration," unpublished.)

TABLE 3.9
Clean Water Oxygen Transfer Efficiency Comparison for Selected Diffusers (EPA, 1989)

Diffuser Type and Placement	Airflow Rate $(m_N^3/h/diffuser)$[a]	SOTE (%) at 4.6-m Submergence
Ceramic plates — grid	35–85 m_N^3/h-m^2	26–33
Ceramic discs — grid	0.6–5.3	25–40
Ceramic domes — grid	0.8–3.9	27–39
Porous plastic discs — grid	0.9–5.5	24–35
Perforated membrane discs — grid	0.8–3.9	16–38
Rigid porous plastic tubes		
Grid	3.8–6.2	28–32
Dual-spiral roll	4.7–17.3	17–28
Single-spiral roll	3.1–18.8	13–25
Perforated membrane tubes		
Grid	1.6–6.2	22–29
Mid-width	3.1–9.4	16–19
Mid-width	3.1–18.8	21–31
Single-spiral roll	3.1–9.4	15–19
Coarse bubble diffusers		
Dual-spiral roll	5.2–15.5	12–13
Mid-width	6.6–7.1	10–13
Single-spiral roll	15.7–55.0	9–12

[a] Except for plates

1 m = 3.28 ft; 1 m_N^3/h = 0.64 scfm; 1 m_N^3/h-m^2 = 0.059 scfm/ft^2

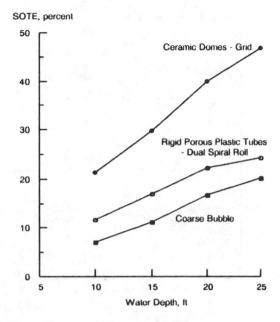

Tank: 20 ft x 20 ft
Power: ~1 hp delivered/1,000 cu ft for rigid porous plastic tubes
Power: ~5 hp delivered/1,000 cu ft for ceramic domes

FIGURE 3.27 SOTE vs. submergence for selected diffusers.

configurations the value of v_w is only one to two times greater than v_s producing longer bubble residence times and concomitant higher efficiencies. Schmit et al. (1978) showed that mid-width configurations were more efficient than spiral roll when the SOTR (and airflow rate) increased or when the submergence increased for the same basin width. Bewtra and Nicholas (1964) and, later, Rooney and Huibregtse (1980) observed the same phenomenon.

Clearly, there is no simple relationship that can be used to express the relationship between placement and performance. Diffuser type, gas flow rate, and basin geometry all play an important roll in the efficiency of the aeration system. Figure 3.26 taken from Huibregtse et al. (1983) summarizes the importance of diffuser pattern for several diffuser placements. Entering this curve for a given $SOTR_V$ will indicate the relative amount of gas flow required to achieve that value for given configurations in the same basin geometry. In this work, which confirms much of the earlier research, the grid configuration is most efficient, followed by dual and single roll configurations. Table 3.9 also provides typical results of clean water tests for a variety of diffuser system placements.

3.4.2.2.5   Diffuser Submergence

The influence of diffuser submergence on SOTE is primarily the result of the higher mean partial pressure of oxygen in the basin (and thus a greater driving force) and the longer residence time of the bubble in contact with the water. This influence is demonstrated in Tables 3.2 through 3.9 and illustrated in Figure 3.27 for three

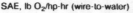

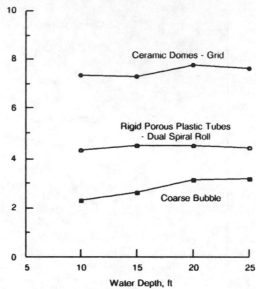

FIGURE 3.28 SAE vs. submergence for selected diffusers.

diffuser types. In the range of basin depths typically used in practice today, the effect of submergence appears to be approximately linear for many diffuser types. Often, investigators may report efficiencies of aeration systems as a percent per unit depth. Although useful for approximating performance, this practice is not recommended for final design calculations unless confirmed by actual measurements. Pöpel and Wagner (1994) have shown that this linear relationship might be valid for lower efficiency devices up to about 5.0 m (16.4 ft). The departure from linearity appears to occur at about 5 m for the more efficient dome/disc diffusers (Jackson, 1975).

The effects of submergence on SAE for a given diffuser appear to be relatively constant for the more efficient diffusers and may slightly increase for the more inefficient devices (Figure 3.28). This effect occurs because as depth increases, the energy required to drive the required air through the diffusers increases. This increase appears to approximately parallel the decrease in required energy needed at the lower airflow rates. This effect is apparently not the same for the coarser bubble devices. The impact of deep basins on diffuser performance is discussed in more detail in Chapter 4.

3.4.2.2.6 Horizontal Flow

Since its introduction in the U.S., a number of clean water oxygen transfer tests have been conducted for the Counter-Current Aeration (CCA) system. In this system, the diffusers are rotated from the bridge around a circular basin. Table 3.10 summarizes clean water field tests at five sites in the U.S. All used 76 cm (30 in) tube diffusers. Several points are worth noting from this presentation. The tests show that

TABLE 3.10
Clean Water Oxygen Transfer Efficiencies — Horizontal Flow Counter-Current Aeration

Diffuser Type	Airflow (m_N^3/h/unit)	Submergence (m)	SOTE (%)	Specific SOTE[c] (%/m)	SAE (kg/kWh)	Diffuser Density (No. units/m²)	Reference
Ceramic tube	1.3	4.2	26.0	6.2	3.17	0.7	Env. Leasing Corp., 1987
	1.8	4.2	25.6	6.2	3.06	0.7	
	2.5	4.2	23.5	5.6	2.98	0.7	
Ceramic tube	(1885)[a]	4.6 (SWD)	20.8	4.6	3.01	—	Donohue & Assoc., 1987
	(2510)[a]	4.6 (SWD)	20.2	4.3	2.79	—	
	(1260)[a]	4.6 (SWD)	24.7	5.2	3.42	—	
Perforated membrane	7.3	4.5 (SWD)	19.3	4.3	2.22	0.9	Donohue & Assoc., 1989
	4.5	4.5 (SWD)	22.6	4.9	2.83	0.9	
	3.0	4.5 (SWD)	26.9	5.9	3.31	0.9	
Ceramic tube	2.7	5.0 (SWD)	28.7	5.6	3.67	0.4	Marx & Redmon, 1991
	4.9	5.0 (SWD)	26.3	5.2	3.24	0.4	
	2.2	5.0 (SWD)	31.7	6.2	3.53	0.4	
Perforated membrane tube[b]	2.3	4.8 (SWD)	23.2	4.9	2.67	0.7	Marx & Redmon, 1991

[a] m_N^3/h total gas flow (no diffuser number available)
[b] Diffusers in service 2 years/cleaned
[c] SOTE/H_s where H_s is diffuser submergence or side water depth (SWD) where indicated

---

TABLE 3.11
Clean Water Oxygen Transfer Efficiency — Horizontal Flow Fixed Diffusers (Da Silva-Deronzier et al., 1996)

Horizontal flow rate (m/s)	0	0.17	0.33	0.45
SOTE /H_s (%/m)	5.9	7.1	8.3	8.7

Circular ditch 1364 m³
720 perforated membrane, 23-cm discs, uniformly distributed
Diffuser density 5.4%, gas flow rate 1.33 m_N^3/h/diffuser
Submergence, H^s = 2.75 m

---

all systems fall at the lower end of results for porous diffusers in grid patterns. The performance is similar to tube grid configurations and are somewhat higher than spiral roll patterns. However, it should be noted that SAE values are lower for the CCA system as compared with tube grid arrangements since energy is required to rotate the bridge. When conducting these field tests, Marx and Redmon (1991) noted that the horizontal velocity component of the fluid was close to that of the bridge. Thus, bubbles rose almost vertically rather than taking a diagonal flow pattern as suggested by the manufacturer. High airflow rates per unit area resulted in large fluid ascent velocities producing boils of air and water at the surface. Spreading the diffuser pattern over a larger area and providing more diffusers would likely improve performance of these systems.

In 1994, Da-Silva-Deronzier et al. (1994) described the influence of horizontal flow on performance of a porous diffuser system. Clean water performance was measured for a 1400 m³ (370,000 gal) annular ditch equipped with 23 cm (9 in) perforated membrane discs placed uniformly along the basin floor in 10 radial headers of 72 diffusers each. This measurement produced a diffuser density of 5.4 percent. Two-2 m (6.5 ft) banana blade mixers imparted horizontal flow. Results of this test appear in Table 3.11. It is apparent that horizontal flow across the fixed diffusers increased efficiencies by about 40 percent. The increased SOTE performance noted in this test approached that of a perforated disc grid system at high density and low airflow rate. It should be noted, however, that gains in efficiency would be offset by additional power requirements to drive the banana blade mixers. No calculated SAE values were presented in this work.

3.4.3 PROCESS WATER PERFORMANCE

3.4.3.1 Introduction

There is a substantial database for oxygen transfer devices in clean water. In designing aeration systems to operate under process conditions, clean water data are corrected to account for the influences of wastewater characteristics, process flow sheet, temperature and pressure. These corrections to process conditions are made using Equation (2.53) for estimating OTR_f, AE_f, or OTE_f. Although conceptually straightforward, this calculation is subject to considerable doubt because of the uncertainty of alpha and the influence of a number of process variables on alpha. Since oxygen transfer is a mass transfer operation involving both the dissolution of a slightly soluble gas into a liquid as well as the transport of the dissolved gas throughout the bulk of the liquid, it is necessary to examine the effect of contaminants on both of these components of the mass transfer coefficient, K_La. For a given aeration system, the differences noted between clean and process water are attributed both to contaminants in the process water and to changes in the properties of the diffuser due to fouling or material deterioration. Basically, it may be stated that contaminants do not usually alter the bulk transport of oxygen (eddy diffusion) to a great extent. Although some researchers have shown that suspended solids may alter pumping characteristics of an aeration tank, in the range of suspended solids concentrations found in most wastewater aeration systems, this effect is small compared with the impact of dissolved contaminants on the gas-liquid interface. These effects are described more fully in Chapter 2. Although surfactants appear to be the major class of compounds of concern, it should be noted that dissolved inorganics also play an important role on changes in the observed mass transfer coefficient.

Any process variable that influences the distribution or concentration of contaminants that affect K_La will have an effect on alpha. Such process variables include wastewater quality and quantity, intensity of mixing, process loading, and flow regime. Historically, alpha has been estimated by tests that ranged from laboratory scale to field scale. A survey of these test procedures by Stenstrom and Gilbert (1981) has indicated that the magnitude of the value of alpha that was estimated was dependent on the characteristics of the test method and often bore little resemblance to observed full-scale observations. The differences observed between the test and the full-scale measurements were due to differences in the levels of turbulence and surface renewal. Therefore, the problem is one of scale-up, and attempts to achieve both dynamic and geometric similarity from test to full-scale have not been entirely successful. Currently, some pilot scale alpha determinations for diffused aeration systems are being used with some success. The estimation of alpha for grid systems, using deep columns with the appropriate diffuser elements, airflow rates, and submergence, has been reported by several investigators (see Chapter 7). However, even with this rather simple system, scale-up may be troublesome (Hwang and Stenstrom, 1985; Doyle and Boyle, 1985). For systems that do not exhibit columnar airflow distribution (grids), tall, narrow columns will not be suitable for estimating alpha since bulk mixing, an important component in mass transfer, will be eliminated by restrictions of flow in narrow columns. As a result, the most reliable estimates of alpha today arise from field testing. The development of reliable field techniques for oxygen transfer testing (ASCE, 1996) has significantly advanced our understanding of alpha in process wastewaters.

The other major factors affecting the observed mass transfer of oxygen in process wastewaters are changes that occur to the diffuser element or the aeration system itself. These changes include fouling, material deterioration, or mechanical failures. They will influence measured oxygen transfer under field conditions and are typically lumped together with wastewater characteristics in reporting values of alpha. In 1989 (EPA, 1989), an effort was made to discriminate between wastewater effects and media/fouling effects on K_La through the use of the fouling factor, F.

TABLE 3.12
Sources of Information for Equation (2.53)

Parameter	Source of Information
$C^*_{\infty 20}$	Clean water test
delta	Clean water test
$K_L a_{20}$	Clean water test
SOTR, SOTE	Equation (2.53)
C_L	Process conditions
t	Process conditions
alpha	Field testing, experience
beta	Total dissolved solids
theta	Normally 1.024, clean water test
omega	Pressure correction for $C^*_{\infty 20}$
tau	Temperature correction for $C^*_{\infty 20}$

Although this factor provides a logical advancement in describing the independent effects of media deterioration and fouling on diffuser elements, an insufficient database is currently available to accurately delineate it. A discussion on diffuser fouling and deterioration is found in Section 3.4.5 and 3.4.6. In this text, alpha is used to describe the observed effects of both process wastewater as well as media deterioration and fouling, insofar as it has not been possible to readily separate these effects in reported field observations.

Table 3.12 provides a guide for applying Equation (2.53), indicating the source of information for the parameters needed to estimate process water performance.

As can be observed from this table, the engineer must rely on field tests and observations to estimate the value of alpha. The other parameters are either obtained through the engineer's clean water test specifications or straightforward calculations. To supply this information, field-testing has become an important element in the design of aeration systems. Process water testing has greatly accelerated over the past 10 years, due primarily to the development of several process water test procedures and their standardization (see Chapter 7 — Testing and Measurement).

3.4.3.2 Process Water Database

Whereas a substantial database exists for the clean water performance of many diffused aeration systems, the process water oxygen transfer data is limited. The in-process database presented here is from field-scale measurements using currently acceptable measurements (see Chapter 7). The majority of this information is for porous diffusers, primarily because most of the new and retrofit systems installed on municipal systems where information is a matter of public record have employed these high efficiency devices.

Summaries of process water performance data are presented for nonporous and porous diffuser systems in Tables 3.13 through 3.17. Many of the process variables described under clean water tests are provided in these tables. It should be noted that the values of alpha are the mean weighted values and the ranges that are reported

TABLE 3.13
Process Water Performance — Municipal Nonporous Diffusers

Diffuser Type and Placement		Flow Regime	Density (No./m²)	Submergence H_s (m)	G_{sd} (m³_N/h-diff)	$\alpha SOTE/H_s$ (%/m)	α Mean Weighted	α Min–Max	Nitrification	Reference
Fixed orifice tube	Dual[a]	Step	0.50	4.6	18.0	2.2	1.07	0.83–1.19	No	Redmon et al., 1983
Coarse bubble	Grid	—	0.35	4.1	15.7	1.9	0.94	—	Yes	Groves et al., 1992
Coarse bubble	Grid	—	0.35	4.1	15.5	1.6	0.80	—	Yes	Groves et al., 1992
Coarse bubble	Spiral	—	0.39	4.0	26.8	1.2	0.60	—	Yes	Groves et al., 1992
Coarse bubble	Spiral	—	1.25	3.8	18.7	2.3	0.88	—	Yes	Groves et al., 1992
Coarse bubble	Mid-width	—	0.31	4.3	15.4	1.2	0.57	—	Yes	Groves et al., 1992
Coarse bubble	Grid	—	0.53	5.8	23.6	1.6	0.55	—	Yes	Groves et al., 1992
Coarse bubble	Grid	—	0.36	5.2	22.5	1.9	0.64	—	Yes	Groves et al., 1992
Fixed orifice tube	Spiral	CSTR	—	4.0	15.5 m³_N/h-m²	1.9	0.75	0.67–0.83	—	EPA, 1985
Jet aerator	Directional	Plug	0.08	4.4	22–74	2.0	0.69	0.52–0.91	No	Yunt, 1990
Jet aerator	Directional	CSTR	0.19	3.8	11.0	2.9	0.45	0.40–0.50	No	Brochtrup, 1983
Jet aerator	Directional	CSTR	0.19	3.8	34.5	2.0	0.47	0.46–0.48	No	Brochtrup, 1983

a Third pass of aeration tank

(m³_N/h) × 0.64 = 1.57 scfm

m = 3.28 ft

From EPA (1985) Summary Report — Fine Pore Aeration Systems, USEPA, EPA/625/8-85/010, Oct. 1985, Water Engineering Research Laboratory, Cincinnati, OH.

TABLE 3.14
Process Water Performance — Municipal Porous Tube Diffusers

Diffuser Type and Placement		Flow Regime	Density (No./m²)	Submergence H_s (m)	$G_s{}^a$ (m^3_N/h-m²)	Nitrifying	αSOTE/H_s (%/m)	Alpha Mean Weighted	Alpha Low–High	Reference
PVC membrane	Grid	Plug	2.3	5.8	12.3	No	2.3	0.43	0.35–0.54	EPA, 1989
PVC membrane	Grid	Plug	1.2	—	11.5	?	1.6	—	—	EPA, 1989
Porous plastic	Grid	Plug	3.3	4.0	7.3	?	1.8	0.28	0.26–0.29	EPA, 1989
Porous plastic	Spiral	Plug	5.2	3.7	3.2	Some	1.8	0.56	0.42–0.67	EPA, 1989
Porous plastic	—	—	—	6.1	—	Yes	2.7	—	0.45–0.50	Stenstrom, 1997
EPDM membrane	—	—	1.2	3.4	8.9	Low	2.0	0.4	—	Stenstrom, 1997
EPDM membrane	—	—	1.4	6.6	2.5	Yes	3.7	0.73	—	Stenstrom, 1997
Ceramic	Spiral	Plug	0.5	4.0	4.1	No	1.5	—	—	Leary, 1969
Ceramic	Cross	Plug	0.6	4.0	4.4	No	1.5	—	—	Leary, 1969
EPDM membrane	Grid	—	1.2	5.3	4.5–5.9	Yes	1.7–2.4	—	0.32–0.55	Groves et al., 1992
EPDM membrane	Grid	—	0.8	4.1	1.5	Yes	2.7	0.46	—	Groves et al., 1992
EPDM membrane	Grid	—	0.8	4.1	3.9	Yes	2.1	0.73	—	Groves et al., 1992
EPDM membrane	Grid	—	1.9	4.0	9.6	No	1.6	0.28	—	Groves et al., 1992
EPDM membrane	Spiral	—	1.9	4.0	8.9–11.6	No	1.2	0.34	—	Groves et al., 1992
EPDM membrane	Grid	—	2.3	3.9	4.9–7.1	Yes	2.1–2.7	—	0.33–0.48	Groves et al., 1992
EPDM membrane	Grid	—	2.4	5.8	4.7–6.4	Yes	2.4–2.5	—	0.43–0.45	Groves et al., 1992

[a] Gas flow per unit tank surface area
$(m^3_N/h/m^2) = 0.059$ scfm/ft²
m = 3.28 ft

TABLE 3.15
Process Water Performance — Municipal Ceramic/Plastic Domes and Discs — Grids

Diffuser Type	Flow Regime	Diffuser Density (%)	Submergence H_s (m)	$G_s{}^a$ (m_N^3/h–m^2)	Nitrifying	αSOTE/H_s (%/m)	Alpha Mean Weighted	Alpha Low–High	Reference
Ceramic plate	Plug	18	4.3	3.3	No	3.6	—	—	EPA, 1989
Plate	Plug	23	4.6	2.9	No	2.6	—	—	EPA, 1989
Plate	Step	22	4.6	2.4	Some	4.1	—	—	EPA, 1989
Dome	Plug	26	4.6	4.7	No	2.4	0.43	0.31–0.57	EPA, 1989
Dome	Plug	26	4.6	3.9	Yes	3.7	0.66	0.56–0.79	EPA, 1989
Dome	Plug	7	4.3	6.6	No	2.2	0.41	0.23–0.58	EPA, 1989
Dome	Plug	9	3.8	5.4	No	1.8	0.24	0.11–0.39	EPA, 1989
Dome	Step	8	4.6	9.0	No	1.5	0.27	0.24–0.31	EPA, 1989
Dome	Plug	10	4.2	6.3	No	1.9	0.29	—	EPA, 1989
Dome	Step	5	4.1	7.3	No	2.3	0.43	—	EPA, 1989
Dome	Step	5	4.1	7.3	Yes	2.5	0.43	—	EPA, 1989
Dome	Step	14	4.1	6.6	Yes	3.3	0.52	0.45–0.59	EPA, 1989
Dome	Plug	6	3.0	3.6	No	2.2	—	—	EPA, 1989
Dome	Plug/Anoxic	7	3.0	3.6	Yes	3.3	—	—	EPA, 1989
Dome	?	7	3.75	5.2	No	1.9	—	0.10–0.35	Stenstrom, 1997
Dome	?	7	4.0	5.8	No	1.8	0.30	—	Stenstrom, 1997
Dome	?	7	3.5	6.1	No	2.8	0.42	—	Groves et al., 1992
Dome	?	7	4.3	4.2–9.1	No	1.7–2.2	—	0.28–0.41	Groves et al., 1992
Dome	?	7	4.3	4.9–6.1	Yes	1.4–4.3	—	0.24–0.57	Groves et al., 1992
Dome	?	8	4.6	6.9–8.0	Yes	2.5–2.8	—	0.39–0.46	Groves et al., 1992
Dome	?	8	4.6	5.2–11.2	No	1.4–1.7	—	0.24–0.31	Groves et al., 1992
Dome	?	12	4.3	2.2	Yes	3.0–3.9	—	0.49–0.64	Groves et al., 1992

TABLE 3.15 (continued)
Process Water Performance — Municipal Ceramic/Plastic Domes and Discs — Grids

Diffuser Type	Flow Regime	Diffuser Density (%)	Submergence H_s (m)	G_s^a (m_N^3/h–m²)	Nitrifying	αSOTE/H_s (%/m)	Alpha Mean Weighted	Alpha Low–High	Reference
Ceramic disc	Plug	8	4.9	3.4	Yes	3.0	0.2	0.19–0.22	EPA, 1989
Ceramic disc	Plug	9	3.8	4.2	No	2.4	0.31	0.21–0.40	EPA, 1989
Ceramic disc	Step	7	4.4	6.1	Yes	2.1	0.35	0.28–0.54	EPA, 1989
Ceramic disc	Plug	11	4.2	5.6	No	1.9	0.28	—	EPA, 1989
Ceramic disc	?	9	3.7	4.4	Yes	2.4	—	0.3–0.4	Stenstrom, 1997
Porous plastic disc	?	10	4.0	6.9	No	1.8	0.3	—	Stenstrom, 1997
Ceramic disc	?	11	5.1	9.7	No	2.1	0.35	—	Stenstrom, 1997
Ceramic disc	?	8	4.8	2.7	No	2.4–2.8	—	0.35–0.41	Groves et al., 1992
Ceramic disc	?	11	5.7	7.6	Yes	3.8	0.60	—	Groves et al., 1992
Porous plastic disc	?	6	4.0	12.4	No	2.1	0.33	—	Groves et al., 1992
Ceramic disc	?	10	4.4	8.3–11.3	No	2.9	0.50	—	Groves et al., 1992
Ceramic disc	?	10	4.5	3.1–3.9	Yes	3.3–4.2	—	0.5–0.61	Groves et al., 1992
Ceramic disc	?	10	5.2	3.9	Yes	3.6	0.52	—	Groves et al., 1992

[a] Airflows per unit tank surface area
1 m = 3.28 ft; 1 m_N^3/h-m² = 0.059 scfm/ft²

TABLE 3.16
Oxygen Transfer in Process Water — Municipal Perforated Membrane Discs/Panels — Grids

Diffuser Type	Diffuser Density (%)	Submergence H_s (m)	G_s[a] $(m_N^3/h\text{-}m^2)$	Nitrifying	$\alpha SOTE/H_s$ (%/m)	Alpha	Reference
Disc	5	4.2	105	Yes	3.9	0.62	Groves et al., 1992
Disc	8	4.8	35.6	No	2.4	0.40	Groves et al., 1992
Disc	6	4.6	40.7–57.6	No	3.0–3.1	0.47–0.50	Egan-Benck et al., 1992
Disc	7	4.6	45.8–54.2	Yes	2.6–2.8	0.42–0.45	Guard et al., 1990
Disc	33	4.0	7.5–11.1	Yes	6.0–6.4	0.68–0.76	Sanitaire, 1993
Disc	12	5.1	55.9–72.9	Yes	2.3–3.0	0.44–0.48	Currie & Stenstrom, 1994
Disc	—	4.0	—	Yes	3.0	—	Stenstrom, 1997
Disc	10	5.6	45.7	Some	3.5	0.53	Stenstrom, 1997
Disc	28	4.3	9.5	No	4.1	0.51	Stenstrom, 1997
Panel[b]	51	4.7	11.3	Yes	5.0	0.66	Dezham et al., 1992
Panel[b]	51	4.7	9.7	Yes	4.4	0.57	Dezham et al., 1992
Panel[b]	51	4.7	9.0	Yes	3.6	0.49	Dezham et al., 1992
Panel[c]	51	4.7	5.6	Yes	4.5	0.52	Dezham et al., 1992
Panel	38	5.1	12.5–16.4	Yes	2.9–3.6	0.42–0.52	Currie & Stenstrom, 1994
Panel	66	4.0	3.3–4.9	Yes	6.8–7.1	0.7–0.72	Sanitaire, 1993
Panel	40	4.6	8.1–8.6	?	3.5–4.0	—	BBS Corp., 1990
Panel	42	4.6	12.9	Yes	3.6	0.59	Stenstrom, 1997

[a] Gas flow per diffuser surface area
[b] Consecutively new, 6 months, and 11 months of service
[c] Following cleaning
1 m = 3.28 ft; 1 $m_N^3/h\text{-}m^2$ = 0.059 scfm/ft²

represent temporal variations of these mean weighted values and not spatial variations within the aeration system. Spatial variations in alpha (and αSOTE) are addressed later. The values of alpha were determined from clean water test data for similar tank geometries, airflow rates, diffuser densities, and placements. As described above, many of the data were collected after the diffusers were in service for significant periods of time. Therefore, the value of alpha reflects both the impacts of wastewater constituents and changes in diffuser characteristics. Values of αSOTE were calculated from field data by correcting to standard conditions of temperature, pressure, and basin DO of 0 mg/L.

It must be emphasized that this in-process oxygen transfer data represent the results of many oxygen transfer tests, each conducted over a period of several hours duration. The data should not be used for design purposes. It is provided to give some insight into the range of values observed in primarily municipal wastewater and to illustrate the effects of selected process variables on performance.

TABLE 3.17
Process Water Oxygen Transfer — Horizontal Flow

Layout[a]	Airflow (m_N^3/h/diffuser)	Side-Water Depth, H (m)	αSOTE/H (%/m)	Alpha	Nitrification	Reference
CC, ceramic tube	2.06–2.11	5.0	2.1–2.3	0.36–0.38	No	Marx & Redmon, 1991
CC, perforated membrane tube						
Stage 1	1.80	4.8	2.0		Yes	Marx & Redmon, 1991
Stage 2	5.06		3.7			
Total	2.30		3.1	0.65		
Stage 1	2.92	4.8	1.8		Yes	Marx & Redmon, 1991
Stage 2	7.98		3.4			
Total	3.70		2.8	0.65		
CC, ceramic tube	10.9–12.1	4.6	2.3–2.5	0.54–0.56	Yes	Groves et al., 1992
CC, ceramic tube	21.5–34.8	4.6	1.6–2.1	0.43–0.58	Yes	Groves et al., 1992
FD, perforated membrane disc	1.33	2.75	5.5[b]	0.62	Yes	Gillot et al., 1997

[a] CC — counter-current aeration; FD — fixed diffusers
[b] Horizontal velocity = 0.46 m/sec
1 m = 3.28 ft.; 1 m_N^3/h = 0.64 scfm

As discussed above, several design and operational variables affect the performance of aeration systems. The lack of controlled studies makes it difficult to draw strong conclusions regarding the impact of many of these variables. The following sections discuss the observations made to date from in-process test data.

3.4.3.3 Wastewater Characteristics

The presence of surfactants and dissolved solids in wastewater cause changes in bubble shape and size once the bubble begins to rise through the liquid. They also may change the rate of surface renewal at the air-water interface. The mechanisms causing both the changes in bubble geometry and the film surrounding the bubble have been addressed in Chapter 2. The effect on surface renewal rate of the air-water interface is most significant when bubble motion is either spiral or zigzag, characteristics most commonly found in fine bubble aeration systems. As a result, the impact of these contaminants is more pronounced in porous diffuser systems than in those producing coarser bubbles. In fact, systems that continuously form fresh air-liquid interfaces through violent mixing are usually not adversely affected by surfactants and may even exhibit alpha values above 1.0 by virtue of the production

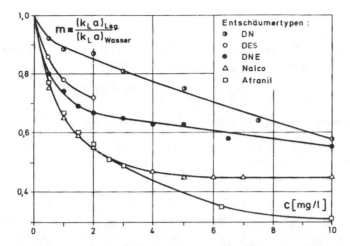

FIGURE 3.29 Effect of surfactant type and concentration on efficiency. (From M. Zlokarnik, *Korrespondenz Abwasser*, 11, p. 731, 1980. With permission.)

of smaller bubbles (and therefore higher surface area to volume). However, one cannot necessarily assume that coarse bubble diffusers will always produce higher values of alpha than those diffusers producing fine bubbles. Downing and Bayley (1961) demonstrated that both fine and coarse bubbles produced similar values of alpha when rising in a narrow column. Thus, the degree of bulk mixing and the eddy diffusivity of oxygen are important determining factors of the effect of surfactants on alpha. Tables 3.13 through 3.17 illustrate that porous diffusers generally produce lower mean weighted values of alpha than nonporous devices with the exception of jet diffusers that generate a fine bubble. Although the values of alpha presented in these tables depend on several process and design variables for the specific plants tested, it is apparent that the average mean weighted values of alpha are less than 0.5 for porous diffuser systems and perhaps closer to 0.7 for the nonporous systems. The impacts of process loading and flow regime are described in more detail in later sections.

Alpha in diffused air systems generally decreases with increased concentration of surface-active materials up to a point where further increases in concentration show little additional impact on alpha. The type of surfactant also plays an important role in the degree to which it affects the oxygen transfer coefficient (Figure 3.29). The removal of these agents by sorption or biodegradation will decrease the impact of the contaminant on oxygen transfer. The wide variation in alpha noted in the tables is likely due to variations in wastewater strength and composition, both in time and space. Examples of this variation are presented in Table 3.18 for several porous diffuser aeration facilities. It should be emphasized that these values are for typical municipal wastewater with only small contributions of industrial wastes. The impact of industrial wastewater on alpha is highly wastewater specific and may or may not have a greater impact on the porous diffuser systems. Attempts to correlate wastewater effects on $K_L a$ with organic matter content have not resulted in any generalizations that can be successfully applied from site to site. Masutani and

TABLE 3.18
24-Hr Alpha and Alpha (SOTE) Variations at Selected Porous Diffuser Municipal Treatment Plants (EPA, 1985)

Process Type	Flow Regime	Alpha			Alpha (SOTE)			Sampling Position in Basin
		Ave.	Min.	Max.	Ave.	Min.	Max.	
CS	Step	0.30	0.23	0.44	8.3	6.4	11.2	Influent Pass
C	Plug	0.24	0.22	0.29	8.7	7.7	10.4	Inlet End
C	Plug	0.46	0.44	0.59	10.7	9.5	13.1	Entire Basin Weighted
C	Plug	0.25	0.21	0.27	7.8	6.4	8.7	Influent Grid
C[a]	Plug	0.26	0.20	0.30	8.7	6.6	9.9	Middle Grid
C	Plug	0.45	0.41	0.50	12.2	11.1	13.5	Effluent Grid
C	Step	0.23	0.19	0.28	—	—	—	Influent Pass
C	Step	0.39	0.33	0.45	—	—	—	Effluent Pass

[a] Data for 6-hour period
CS — Contact Stabilization; C — Conventional

Stenstrom (1991) have demonstrated that dynamic surface tension was a potentially useful tool to determine the impact of wastewater on alpha. Observations of alpha values from different wastewater effluents have shown wide variations in the upper limit on alpha in porous diffuser systems even when quality is very high. It is apparent that very low concentrations of some surfactants may have a significant impact on oxygen transfer in these systems.

Although most effects of wastewater on alpha have been ascribed to surface-active materials, there is good evidence that salts also impact K_La. Hantz (1980) has shown that alpha significantly increases with increased specific conductivity. These laboratory studies were conducted with distilled water and mixtures of distilled water and tap water with a total dissolved solids concentration of about 600 mg/L. Stenstrom (1996) showed similar trends with the addition of sodium chloride to water. He demonstrated that the high salt concentrations cancelled the effects of surfactants added to the mixture. For many years those that have performed clean water oxygen transfer tests with porous, nonporous, and mechanical aeration systems have noted that additions of sodium sulfite will elevate measured values of K_La (ASCE, 1992). Attempts to model this effect have been successful for a given device but a rigorous model for all types of aeration systems has not yet been developed. The enhanced mass transfer coefficient occurs because higher salt concentration increases surface tension with concomitant finer bubbles (O'Connor, 1963; Marrucci and Nicodemo, 1967). The salt does not apparently affect surface renewal nor does it block transport at the air-liquid interface. Thus, K_La will increase as the surface area to volume ratio increases. Other salts, including the transition elements such as iron and manganese, may also affect the value of alpha.

The effects of wastewater on oxygen transfer also occur as a result of changes in the steady-state saturation concentration of oxygen as estimated by the factor beta. Dissolved salts and organics tend to lower the saturation concentration of oxygen in wastewater as compared with distilled water. Although the activity of an oxygen-saturated solution of water is by definition independent of the dissolved contaminants, the concentration of oxygen changes as the activity coefficient is altered by the salting-out effect. This fact has important implications in the measurement of DO saturation under field conditions. Direct measurement of DO by the Winkler Method (APHA, 1995) is often complicated by oxidizing or reducing compounds in the wastewater. Membrane probes theoretically respond to oxygen activity that depends on the degree of saturation, not the absolute concentration. Thus, a probe standardized in clean water will not necessarily yield a true reading of DO in contaminated water. As a result, the value of DO saturation in wastewater is usually estimated by means of a total dissolved solids concentration correction (Equation (2.32)). Typically this correction is small in most wastewater, and the error in this estimate will not be significant in estimating αSOTE (αSOTR). It can, however, be an important factor in some industrial wastewaters.

3.4.3.4 Diffuser Airflow Rate

The effect of diffuser airflow rate on the value of αSOTE is similar to that found in clean water testing. Equation (3.2) may be used to estimate process water efficiencies for porous diffusers at different airflow rates. The constant, m, will change, however, to reflect the impact of the process wastewater and changes in diffuser characteristics. The results of numerous process water tests have shown that the effect of process wastewater conditions is to shift the curves downward from the corresponding clean water curve. The slope of the process water curves appears to be site specific, however. In most cases, the process water curves were parallel indicating that alpha remained constant over the range of airflows tested. On the other hand, a few sites demonstrated a process water curve that had a steeper slope than that for clean water. In those cases, it may be presumed that alpha decreased with increased airflow. Hwang and Stenstrom (1985) found that alpha decreased with increased airflow in tall column studies on process wastewater in California (Figure 3.30). Redmon (1998) also has found that alpha appears to decrease with increased airflow rate in column tests. The reason for this apparent anomaly between column tests and full-scale measurements is not clear. Finally, some sites with porous diffusers have shown lower but constant values of αSOTE with increased airflow. Clearly, at this time, one cannot generalize on the impact of airflow on alpha in process wastewater. However, it is not unreasonable to presume that at least some of this variation from site to site may be due to changes in diffuser characteristics over time. It has also been speculated that when operating in the field at low airflow rates, poor airflow distribution might lead to circulation pattern changes (rolls) that would lead to lower efficiencies as compared with those observed in clean water test grids.

3.4.3.5 Diffuser Layout

At this time, there are insufficient data to demonstrate any impacts of diffuser layout or other characteristics of the diffuser system on alpha. As mentioned above, bubble

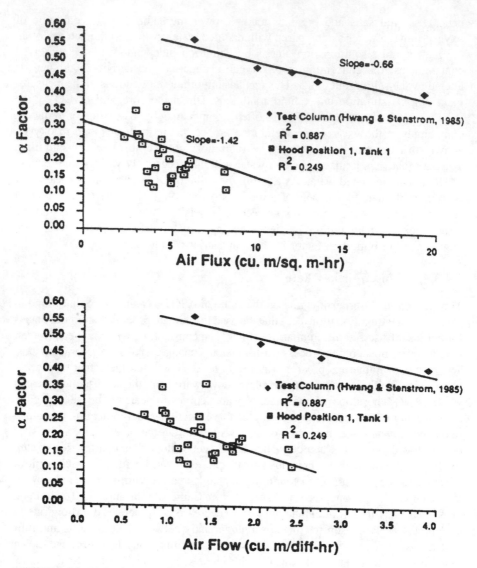

FIGURE 3.30 Alpha factor vs. airflow for fine pore diffusers (Stenstrom and Masutani, 1994).

size alone may not completely explain the effects of wastewater characteristics on alpha. Since oxygen transfer consists of gas to liquid transfer followed by transport throughout the bulk liquid, an aeration tank with nonuniform diffuser arrangement (such as spiral roll) has a significant transport component. Since surfactants have only a minor effect on oxygen transport in the pumping zone, these nonuniform arrangements may not exhibit the sensitivity to surfactants that grid systems would. Since most nonuniform arrangements are associated with nonporous diffusers, it is not unreasonable to presume that the diffuser arrangement rather than bubble size may play an important roll in the observed values of alpha. Thus, uniform grids of nonporous diffusers may, in fact, produce lower alpha values than nonuniform layouts. This presumption is speculative at this time but worthy of some consideration when translating clean water test data to field conditions.

The effect of diffuser submergence on observed alpha values is clearer. Doyle and Boyle (1985) have shown in column tests with porous diffusers that observed alpha values would decrease as submergence increases up to some asymptotic value. This decrease is attributed to the residence time distribution of the bubbles and the time required for the surfactant to adsorb and orient itself at the bubble air-liquid interface. Based on laboratory and field observations, this effect does not appear to be critical at depths above about 4 to 5 m (13 to 16 ft).

3.4.3.6 Flow Regime

Aeration basin flow regime affects the mixing pattern of the basin, and therefore, the residence time distribution of the influent wastewater. Since the composition and concentration of contaminants have an impact on alpha, it is reasonable to assume that flow regime will affect alpha. The impact of flow regime is illustrated by a study conducted at Madison, WI (Boyle, 1994) as shown in Figure 3.31. Single day αSOTE profiles, as a function of grid position, are shown for this ceramic dome diffuser system in both a step feed and plug flow regime. Both off-gas tests were conducted on the basins when operated at an SRT of approximately 2.2 days. As may be observed, the values of αSOTE increased downstream in the plug flow regime, whereas values of αSOTE decreased at each feed point where primary effluent was added. The effect was a greater mean-weighted αSOTE for the plug flow basin. Similar examples can be found in the EPA Design Manual, Fine Pore Aeration Systems (1989).

The alpha profiles along the length of an aeration basin will depend upon the degree of mixing that occurs within the basin. Typical results from a number of plants with differing basin geometries are shown in Table 3.19. As can be seen, the

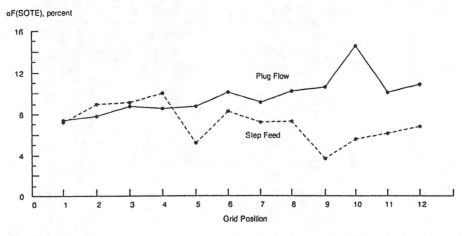

FIGURE 3.31 Effect of flow regime on diffuser performance. (From Boyle, W.C. et al. (1994). *Oxygen Transfer Studies at the Madison Metropolitan Sewerage District Facilities,* EPA 600/R-94/096, NTIS No. PB94-200847, EPA, Cincinnati, OH.)

TABLE 3.19
Alpha Profiles for Various Municipal Aeration Systems (Modified from EPA, 1989)

Flow Regime	F/M[a]	Process	Diffuser	L/W (pass)	AlphaF Zone 1			Zone 2			Zone 3			Total Basin		
					Mean	Min.	Max.	Mean	Min.	Max.	Mean	Min.	Max.	Mean	Min.	Max.
Plug	0.59	CS	MT	6.6	0.45	0.35	0.55	0.43	0.35	0.59	0.40	0.31	0.54	0.43	0.36	0.53
Plug	0.59	CS	CDi	6.6	0.49	0.41	0.68	0.50	0.34	0.67	0.46	0.30	0.64	0.49	0.36	0.64
Step	0.16	CS	CDo	9.7	0.37	0.18	0.49	0.37	0.28	0.49	0.35	0.24	0.45	0.36	0.24	0.48
Plug	0.77	C	CP	17.2	0.45	0.32	0.60	0.58	0.44	0.79	0.60	0.44	0.77	0.54	0.44	0.68
Plug	0.82	C	CP	10.0	0.34	0.18	0.46	0.40	0.27	0.49	0.43	0.25	0.60	0.39	0.23	0.52
Plug	0.63	C	CDo	4.6	0.32	0.24	0.44	0.44	0.29	0.62	0.52	0.36	0.76	0.43	0.31	0.57
Plug	0.12	C	CDo	4.6	0.40	0.33	0.47	0.64	0.54	0.78	0.92	0.72	1.00	0.66	0.56	0.79
Plug	0.15	C	CDi	10.0	0.33	0.26	0.40	0.54	0.52	0.56	0.55	0.52	0.58	0.48	0.44	0.51
Step	0.37	C	CP	12.3	0.64	0.50	0.92	0.62	0.47	0.83	0.64	0.51	0.83	0.63	0.51	0.75
Plug	0.61	C	CDi	10.0	0.25	0.15	0.42	0.30	0.15	0.40	0.38	0.22	0.51	0.31	0.21	0.40
Plug	0.61	C	CDo	10.0	0.16	0.09	0.27	0.23	0.08	0.40	0.31	0.17	0.49	0.24	0.11	0.39
Step	0.76	C	CDo	12.4	0.29	0.25	0.34	0.27	0.23	0.31	0.25	0.21	0.30	0.27	0.24	0.31
Step	—	C	CDo	12.5	0.36	0.32	0.40	0.36	0.23	0.42	0.37	0.24	0.45	0.37	0.29	0.45
Step	—	C	CDo	12.5	—	—	—	—	—	—	—	—	—	0.40	0.34	0.46
Step	—	C	CDo	12.5	0.50	—	—	—	—	—	—	—	—	0.52	0.45	0.59
Plug	0.19	CS	PPT	4.1	0.59	0.43	0.69	0.54	0.44	0.77	0.56	0.37	0.65	0.56	0.42	0.67

Each zone represents 1/3 of aeration volume

Reaeration volume not included in contact stabilization systems

SOTE for plate diffusers was assumed to be 6.6%/m submergence

MT — membrane tube, CDi — ceramic disc, CDo — ceramic dome, CP — ceramic plates, PPT — porous plastic tubes

CS = Contact Stabilization Process, C = Conventional Process

[a] 16 BOD_5/day – 16 MLSS

basins with large length to width ratios, operating as plug flow basins, generate significant alpha gradients. Conversely, plants with low length to width ratios exhibit much less change along the basin profile.

The use of selectors has significantly increased in the 1990s as a result of attempts to improve process stability and/or to achieve some level of biological nutrient removal. Insofar as selectors will achieve some biochemical transformation of wastewater contaminants, it is not surprising to find that they may have an impact on alpha. Rieth et al. (1995) showed that aerobic and anoxic upstream selectors improved the αSOTE of a complete mixed ceramic diffused air pilot plant operated at a 10 day MCRT. The pilot plants were nitrifying during this study. Mueller (1996, 2000) also demonstrated that the incorporation of an anaerobic selector at a porous diffuser contact stabilization facility significantly increased the mean weighted alpha value for the plant that was operating at an MCRT of six days. This facility was not nitrifying during the study. Field studies by Fisher and Boyle (1998) observed the effects of anaerobic and aerobic selectors (in series) by comparison with a parallel plug flow system without selectors. Both systems were operated with MCRTs between 7 and 10 days and were completely nitrifying. Their observations indicated that there was no effect of the selectors at this plant. In all three examples, the inclusion of selectors appeared to attenuate the variability of alpha (and αSOTE). The differences found in these studies relative to the impact of selectors on transfer efficiencies most likely are due to differences in wastewater characteristics and the level of treatment achieved prior to the addition of the selectors. The study by Fisher and Boyle (1998) was conducted at a facility that was producing a very high quality effluent, even without selectors. Furthermore, the plant load was low and most soluble COD was removed within the first 15 to 20 m (50 to 65 ft) of the aeration tank. Therefore, the addition of selectors likely had little impact on wastewater contaminants that would affect transfer.

3.4.3.7 Process Loading Effects

The presence of certain contaminants in a reactor has been shown to depress the value of K_La for systems using porous diffusers. Any chemical, physical, or biological reaction occurring within the aeration tank that results in the removal of these contaminants will directly affect K_La and alpha. This result is clearly seen in the spatial changes that occur in alpha with the level of treatment obtained. Studies conducted at the Madison, WI treatment plant equipped with dome diffusers revealed significant increases in αSOTE with increasing MCRT (Boyle, 1994). From 1984 to 1985, when the plant was not nitrifying, the MCRT averaged 2.4 days and the average αSOTE was 11.5 percent. In 1987, when the plant was nitrifying, the average MCRT was 14 days and the average αSOTE was measured at 17.1 percent. Rieth et al. (1995) showed that at a volumetric loading of 0.48 kg BOD_5/m^3d (30 lb/1000 ft^3d), a system that operated at an MCRT of eight days produced a significantly higher value of alpha than one that operated at two days. The wastewater treatment plant at Phoenix increased its MCRT from one day to 14 days to achieve nitrification. The diffusers were dome diffusers in the two parallel tanks. The αSOTE increased from a range of 6.9 to 7.2 percent for the one day MCRT to a range of 11.5 to 12.7 percent for the 14-day MCRT. The corresponding value of alpha increased from 0.24 to 0.39. In

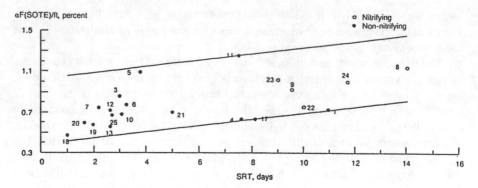

FIGURE 3.32 Effect of SRT on diffuser performance.

still another study, the Los Angeles–Glendale facility tested the same basin under almost identical operating conditions but with two different MCRTs, 1.6 days and 8.8 days. The lower MCRT operation produced an αSOTE of 7.5 percent versus 11.6 percent for the higher MCRT mode of operation. The corresponding alpha values for these two operating conditions were 0.33 and 0.46 (Groves et al., 1992).

Data from 21 operating ceramic diffuser plants were plotted to illustrate the effect of MCRT (SRT) on alpha SOTE (EPA, 1989) and are shown in Figure 3.32. Although wide variations in system design and operation, as well as wastewater characteristics, are evident at these sites, it appears that a trend does exist between process loading and αSOTE. Nitrification plants have been highlighted to indicate their relative importance to the relationship. Tables 3.13 through 3.17 also illustrate the apparent importance of process loading on alpha using nitrification as the measure of loading.

A review of the dynamics of αSOTE in a number of aeration systems suggests that several process variables affecting oxygen transfer are not clearly identifiable based on our current knowledge. For example, αSOTE data collected at Madison over an 800-day period (Figure 3.33) in the first pass of a three pass conventional plug flow system, demonstrate significant variability in SOTE with time. Some of this variability is attributed to wastewater characteristics but does not account for all of the variation. Multiple linear regression of the data including independent variables of MCRT, F/M, volumetric loading, MLVSS, and airflow rates could only account for up to about 60 to 70 percent of the variability. Similar findings were described by Stenstrom (1994) for the Whittier Narrows treatment plant where 30 to 74 percent of the variability in αSOTE could be accounted for by F/M, airflow rate and time-in-service.

3.4.4 MIXING CHARACTERISTICS

In aeration tanks sufficient mixing is required both to disperse DO throughout the basin and to provide reasonably uniform solids concentrations throughout the liquid. The former requirement is easier to meet than the latter. Deposition of suspended solids is undesirable in most aeration tanks (aerated facultative lagoons are one exception), and therefore, this requirement most often dictates mixing requirements. With the exception of the horizontal flow systems, where mixing and aeration are separate functions, the aeration device is expected to deliver adequate oxygen to satisfy

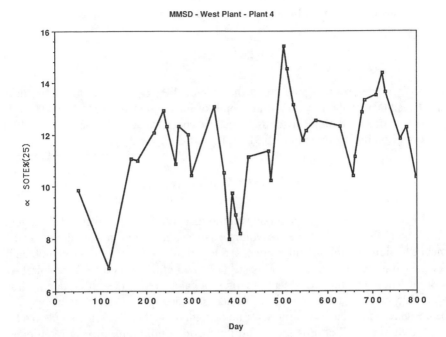

FIGURE 3.33 Variation in αSOTE in municipal plant. ((From Boyle, W.C. et al. (1994). *Oxygen Transfer Studies at the Madison Metropolitan Sewerage District Facilities,* EPA 600/R-94/096, NTIS No. PB94-200847, EPA, Cincinnati, OH.)

the oxygen demand and to provide sufficient energy to prevent solids deposition. In activated sludge systems that are completely mixed, oxygen demand typically dictates the aeration energy requirement. However, in plug flow activated sludge systems, mixing energy may dictate aerator design and operation at the effluent end of the process where oxygen demand is low and required airflow (or power input) is also low. This is more likely to be a problem with high efficiency aeration devices and/or with weaker wastewater.

In evaluating mixing requirements, different diffuser configurations exhibit very different mixing characteristics. Unfortunately, only very limited information has been published on minimum mixing requirements. The *Aeration-Manual of Practice FD-13* (WPCF, 1988) specifies that for degritted wastewater, a velocity of about 0.15 m/s (0.50 fps) across the tank bottom is required. This is a difficult parameter to measure for many aeration systems. Another mixing parameter often used is the root mean square velocity gradient, G, described by Equation (3.3).

$$G = (W/\mu)^{1/2} \tag{3.3}$$

Here, G is the velocity gradient, sec^{-1}, μ is the absolute viscosity, N-sec/m², and W is the power dissipation, W/m³, calculated by the following

$$W = E/V \tag{3.4}$$

where E is the power, W, transferred to the fluid, and V is the liquid volume, m³.

The power transferred by a gas to a liquid may be calculated as

$$E = 0.277\ P_1 G_s ln\left(P_2/P_1\right)$$ (3.5)

where P_1 is the absolute pressure at the surface, kP_a, G_s is the airflow rate, m^3/h, and P_2 is the absolute pressure at the depth of injection (Fair et al., 1968).

For mixing of biological solids, a recommended value of G ranges from 40 to 80 sec^{-1}. Combining Equations (3.3), (3.4), and (3.5) yields the following.

$$G_s/V = 3.61\ G^2 \mu/P_1\ ln\left(P_2/P_1\right)$$ (3.6)

Most often, rule-of-thumb mixing requirements are used for diffused air systems based on airflow per unit area or volume. For example, one manufacturer recommends a minimum mixing intensity of 0.6 to 0.9 m^3/h-m^3 (10 to 15 cfm/1000 cu ft) for grid systems and 0.9 to 1.5 m^3/h-m^3 (15 to 25 cfm/1000 cu ft) for a spiral roll system. These recommended values represent calculated values of G ranging from 80 to 125 sec^{-1} for a 4.6 m deep (15 ft) aeration tank. Spiral roll systems may also be designed on the basis of airflow per unit length of the header; for example, 16.6 to 38.9 m^3/h-m (3 to 7 cfm/ft). For a full floor grid, a minimum mixing requirement of 2.2 m^3/h-m^2 (0.12 cfm/sq. ft) is specified (calculated G value of approximately 70 sec^{-1} for a 4.6 m deep (15 ft) tank). The only data for aeration tank mixing reported in the recent literature was for an activated sludge dome grid configuration at Glendale, CA (Yunt, 1980). Measurements revealed no solids settling problems after two weeks of testing at airflow rates as low as 0.9 m^3/h-m^2 (0.05 cfm/sq. ft) (calculated G value of 45 sec^{-1}). An examination of Tables 3.13 through 3.17 indicates that average airflow rates per unit area are normally higher than the minimum mixing requirements for grid configurations. Presently, there have been no recorded problems with solids separation in aeration basins at these levels of mixing intensity. (It should be noted that upon basin dewatering, operators often notice the accumulation of some solids, usually high-density grit, below the diffuser headers. This is normal and of little real concern unless primary clarifiers or degritting facilities are overloaded. In that case, upstream retrofitting of degritting operations is far more cost effective than efforts to suspend this heavier material in the aeration tanks through the use of greater mixing intensity.) At the present time there is no standard method prescribed for specifying mixing requirements for aeration devices. Over time, operational experience will reveal whether the current rule-of-thumb values are acceptable.

3.4.5 DIFFUSER FOULING

All porous diffusers are susceptible to buildup of biofilms and/or deposition of inorganic precipitates that can alter the operating characteristics of the diffuser element. Porous diffusers are also susceptible to air-side clogging of pores due to particles in the supply air. There is a history of diffuser fouling problems in the U.S.

since the introduction of ceramic plate diffusers in the 1910s (Boyle and Redmon, 1983). Numerous mechanisms have been cited and foulants identified. The list includes the following:

Air Side

- dust and dirt from unfiltered air
- oil from compressors or viscous air filters
- rust and scale from air-pipe corrosion
- construction debris
- wastewater solids intrusion due to power outages or breaks

Liquor Side

- fibrous materials attached to sharp edge
- organic solids entering media at low pressures
- oils and greases in wastewater
- precipitated deposits, including iron and carbonates, on and within media
- biological growths on and within media
- inorganic and organic solids entrapped by biomass on or within media

The rate of fouling has historically been gauged by the rise in back pressure while in service. Since significant levels of fouling can take place with little or no increase in back pressure but with substantial reductions in OTE, this method provided only a crude and qualitative estimate at best. In fact, by the time back pressures were significantly high enough to observe, fouling may have reached serious proportions within the system. What is often observed is that as one diffuser becomes fouled and less air is distributed to that diffuser, others receive more air and little change is noted in line pressure. Maldistribution of air along the air header exacerbates the problem; the diffuser with low airflow fouls more rapidly, and grid airflow regimes deteriorate to major turbulence. All of this results in poor OTEs and increased power consumption. Better methods of measuring the degree of fouling and the effectiveness of cleaning have been developed (EPA, 1989). These methods include DWP, EFR, off-gas methods to evaluate OTE, and the use of portable diffuser headers that can be removed from the basin and examined for fouling potential. This latter method is recommended where wastewaters may be potentially problematic with respect to liquor-side fouling. DWPs may now be monitored *in situ* on selected diffusers. Off-gas measurements may be conducted routinely to evaluate changes over time in OTEs.

3.4.5.1   Air-Side Fouling

Although many early installations of ceramic diffusers exhibited fouling problems often attributed to air-side fouling, this type of fouling no longer appears to be the problem that it once was. Improvements in materials of construction, air delivery systems, construction practices and, perhaps, improved air filtration systems, have resolved most air-side fouling problems.

The effects of air-side fouling were determined during an EPA interplant fouling study of porous diffuser systems conducted in 1989 (Baillod and Hopkins, 1989). Results of this study at six treatment facilities indicated that over a 12- to 15-month period, the incidence of air-side fouling was negligible. The plants studied included those with a range of air filtration devices from electrostatic precipitators to coarse roll filters. Facilities operated with little or no air filtration have not experienced air-side fouling of porous diffusers (EPA, 1989). Today it is recommended that the air filtration that is required to protect the blower is adequate insurance against air-side fouling of porous diffusers due to particulates in the air. The major air-side fouling problem today is the intrusion of mixed liquor solids through the diffuser element during power outages or into the air header or plenum due to breakage. These solids may collect on air-side diffuser surfaces or may accumulate within the diffuser itself causing increased back pressures and, perhaps, some changes in airflow distribution along the diffuser. Clogging caused by mixed liquor intrusion can be minimized by carefully installing systems with good mechanical integrity and by providing careful preventative maintenance, i.e., inspecting the system on a regular basis and fixing leaks, operating the system at or above the manufacturer's recommended minimum airflow rate, and avoiding power outages that will interrupt airflow to the system.

3.4.5.2 Liquor-Side Fouling

Based on recent studies of diffuser fouling, several hypothetical fouling scenarios have been developed. The *Design Manual, Fine Pore Aeration Systems* (EPA, 1989) cites two types of fouling, Types I and II. Kim and Boyle (1993) have extended this scenario to an intermediate type which is likely more prevalent in most municipal wastewater treatment plants. Still another fouling type was identified by Hartley (1990) and expanded by Waddington (1995) and Hung and Boyle (1998). These scenarios have been developed based entirely on observations of ceramic diffuser elements, but observations of porous plastic and perforated membrane elements indicate similar mechanisms are also applicable to these diffusers.

Type I fouling is characterized by clogging of the diffuser element pore, either on the air-side by air-borne particulates or on the liquor side by precipitates such as metal hydroxides and carbonates. Figure 3.34 illustrates this type of fouling on the liquor side. During the fouling process, it is hypothesized that the areas of the diffuser with the highest local air flux will foul more rapidly. This occurrence serves to reduce flux in high-flow areas and increase flow in low-flow areas. The combined effect may improve uniformity of air distribution (EFR approaching 1.0). As fouling progresses, the DWP will rise as pore size decreases. The reduced effective pore size may produce smaller bubbles such that OTE may remain relatively constant or slightly increase. Figure 3.35 presents an idealized plot of how OTE and DWP may change with time for this type of fouling. Kim and Boyle (1993) experimentally demonstrated this scenario by precipitating carbonate salts on a ceramic diffuser placed in wastewater (See Figure 3.36). Photomicrographs indicated that this precipitate was surficial and only penetrated a few mm below the surface.

In the second type of fouling (Type II), the development of a biofilm layer on the liquor-side surface is the dominating feature, based on microscopic analyses by

FIGURE 3.34 Photomicrograph of type I fouling. (From Kim, Y.K., Mechanisms and Effects of Fouling in Fine Pore Ceramic Diffuser Aeration, PhD Thesis, University of Wisconsin, Madison, 1990. With permission.)

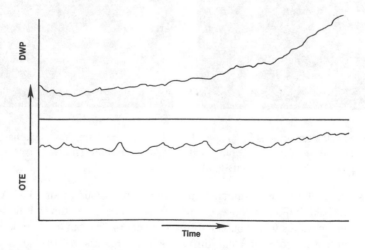

FIGURE 3.35 Impact of type I fouling.

Costerton (1994) and Kim and Boyle (1993). It was noted that the biofilms were not connected at all points to the diffuser surface so that large spaces existed within the film at the element surface. The biofilms were traversed by large structured air passages that originated at the diffuser surface and branched toward the top of the

Test 1 (Type A Fouling)

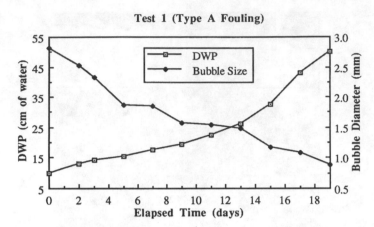

FIGURE 3.36 Bubble size vs. DWP during fouling type A. (From Kim, Y.K., Mechanisms and Effects of Fouling in Fine Pore Ceramic Diffuser Aeration, PhD Thesis, University of Wisconsin, Madison, 1990. With permission.)

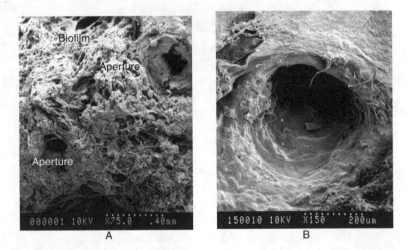

FIGURE 3.37 Photomicrographs of type II fouling. (From Kim, Y.K., Mechanisms and Effects of Fouling in Fine Pore Ceramic Diffuser Aeration, PhD Thesis, University of Wisconsin, Madison, 1990. With permission.)

biofilm surface where they terminated in large apertures (see Figure 3.37 A and B). It is hypothesized that air is conveyed from the diffuser pores through these spaces to the surface apertures where bubble formation occurs. The bubbles would be larger than those released from the clean diffuser surface because of the larger aperture size of the biofilm. As a result, OTE would generally decrease and the EFR would increase significantly above 1.0 as a result of the nonuniformity of the biofilm producing areas of high localized flux. The DWP may increase due to frictional losses through the biofilm, but since the effects of surface tension (which is the major force producing pressure differential in porous diffusers) may be minimized in those areas where the bubbles are released to an air pocket, the effects on DWP

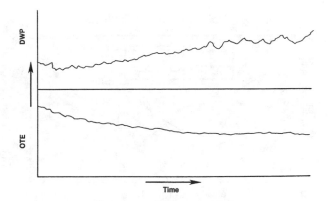

FIGURE 3.38 Impact of type II fouling.

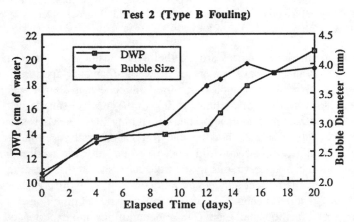

FIGURE 3.39 Bubble size vs. DWP during type B fouling. (From Kim, Y.K., Mechanisms and Effects of Fouling in Fine Pore Ceramic Diffuser Aeration, PhD Thesis, University of Wisconsin, Madison, 1990. With permission.)

may be small. Figure 3.38 depicts an idealized plot of the progression of DWP and OTE with Type II fouling. Kim and Boyle (1993) experimentally demonstrated the impact of biofilm development as well as progression of DWP and bubble size distribution as shown in Figure 3.39. Their data support the hypothesis of biofilm effects on performance. It should be emphasized that this type of fouling has been observed for both ceramic and perforated membrane diffusers.

A third type of fouling, postulated by Kim and Boyle (1993), involves both biofilm formation and entrapment/deposition of inorganic particulates. During the examination of foulants on different diffuser surfaces, it was often noted that a significant proportion of the foulant was inorganic, often high in silica. This matrix of biofilm and inorganic residue may modify biofilm properties and its concomitant effects on DWP and OTE. It is hypothesized that the inorganic particles may block smaller pores and partially clog larger pores within the biofilm, producing higher back pressures and smaller bubbles than found for typical Type II fouling. As foulants accumulate, it is speculated that the inorganic particles may serve as seed causing

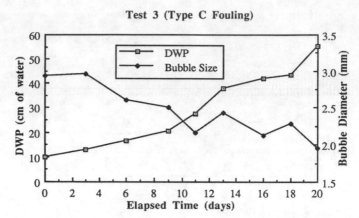

FIGURE 3.40 Bubble size vs. DWP during type C fouling. (From Kim, Y.K., Mechanisms and Effects of Fouling in Fine Pore Ceramic Diffuser Aeration, PhD Thesis, University of Wisconsin, Madison, 1990. With permission.)

cohesion of polymeric substances around them. The result would be a more rapid increase in DWP and the development of more rigid, smaller apertures producing smaller bubbles. Experimental studies supported this concept as shown in Figure 3.40.

The early observations of fouling leading to the mechanisms described above were based on surface foulant development. Scanning Electron Microscopy (SEM) supported these observations, showing that foulant generally accumulated on the surface of ceramic diffusers and did not penetrate very far within the profile. However, more recent observations in the field have shown that in some cases for ceramic diffusers, the foulant may penetrate deep within the diffuser cross section. Hartley (1990) reported penetration of foulant in some ceramic domes to a depth of 10 mm. These facilities had high TDS concentrations and experienced frequent power outages. X-ray diffraction identified the crystalline structure of these deposits to be calcium sulfate in one plant and calcium phosphate in another. Waddington (1995) examined a number of ceramic discs at the Madison, WI facility and found deposits 4 to 10 mm below the diffuser surface. Energy Dispersive X-Ray Spectroscopy (EDXS) identified the white crystalline structure as calcium phosphate. What is important about these investigations is that these internal foulants may significantly affect diffuser back pressures (DWP) over time. At Madison, back pressures in the influent grids of several aeration tanks were so high that it was not possible to supply sufficient air to the grid to meet oxygen demands. Diffuser cleaning, including hydrochloric acid spraying, was not effective in removing this deeper foulant Even kiln firing of individual diffusers did not completely restore the diffuser DWP. Hartley also observed this fouling problem.

The mechanism of this in-depth fouling is not entirely clear at this time. Although observed in a number of plants using ceramic diffusers, no common thread has been identified, but several plant conditions may be responsible for the phenomenon. In several plants, power outages were prevalent. Even more notable was that several of the wastewaters were high in total dissolved solids. A hypothesis that might explain this follows. During normal operation, ceramic diffusers (and other porous diffuser

FIGURE 3.41 In-depth fouling of fine pore diffuser.

elements) contain little moisture within the diffuser profile. Moisture will penetrate the diffuser cross section if airflow is reduced or discontinued. Although most of the suspended solids may be filtered out by the diffuser element, dissolved solids will penetrate the cross section. When airflow is increased or resumed, dissolved solids may be concentrated due to evaporation. These solids may then accumulate due to precipitation or sorption at nucleation sites within the diffuser. Over time, these accumulated solids will block air passages resulting in increased DWPs. Typical surficial treatment of the diffuser will not effectively penetrate deep enough into the diffuser to remove these solids, which continue to increase. Observations of fouled diffuser cross sections (Figure 3.41) indicate that even with acid cleaning at the surface, these solids will tend to remain about 5 mm or greater below the surface.

Although porous diffusers appear to be most susceptible to fouling as described above, it must be emphasized that even nonporous diffusers will foul to some extent depending upon the wastewater characteristics and application. Closure of large orifices with organic and inorganic foulants normally will have little impact on OTE but may eventually result in significant increases in back pressure and changes in mixing pattern within the aeration basin.

3.4.6 MEDIA DETERIORATION

The deterioration of diffuser media, which affects both OTE and DWP, is of concern to designers when seeking proper diffuser applications for a given wastewater. Ceramic and porous plastic diffusers are generally inert to chemical, biochemical or physical deterioration but may suffer breakage or mechanical failure of gaskets, piping, and support saddles. Examples of gasket failures and failures of plastic center bolts on dome diffusers are described in Houck and Boon (1981), Stenstrom (1989) and Gilbert (1989). Plastic hold-downs and center bolts on dome diffusers appeared to fail due to creep. Center bolts are typically constructed from metals today to avoid this problem.

Perforated membrane elements may show changes in character after use. Conditions that can substantially affect membrane performance and life include loss of plasticizer, loss of oils, hardening or softening of the material, loss of dimensional stability through creep, absorptive and/or extractive exchange of materials with wastewater, and chemical changes resulting from environmental exposure.

Plasticizer migration can cause hardening and reduction in membrane volume, resulting in dimensional changes. Studies at several municipal wastewater treatment facilities showed that the plasticized PVC perforated membrane tubes experienced changes in dimension, weight, and elasticity due to loss of plasticizers (EPA, 1989). These changes resulted in a widening of the slit perforations and sometimes produced

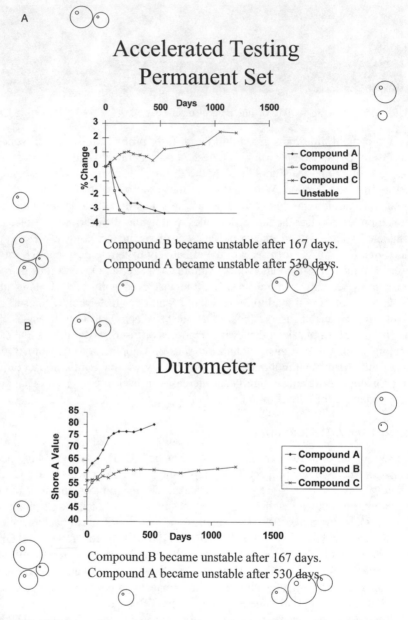

FIGURE 3.42 Impact of wastewater on membrane characteristics for selected membrane types (courtesy of Sanitaire, Brown Deer, WI).

tears due to the increased rigidity of the element. In some cases, significant changes in αSOTE were observed, while in others, no significant change in performance was noted. As slits open, however, DWP values will decrease. The effects of this hardening and creep are not reversible by known maintenance procedures.

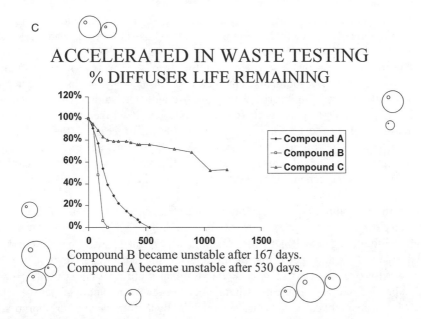

C

ACCELERATED IN WASTE TESTING
% DIFFUSER LIFE REMAINING

Compound B became unstable after 167 days.
Compound A became unstable after 530 days.

FIGURE 3.42 (continued)

There are a number of media properties that may be used to evaluate a particular membrane material to assess its performance and useful life. Changes in media properties that are useful indicators include:

- change in hardness (Shore A or B durometer)
- loss of dimensional stability by creep or chemical change
- change in specific gravity
- change in tensile modulus
- change in volume, either an increase or decrease.

The causes of these changes are not well understood although there are some things that are known. Loss of oils through chemical reaction, dissolution, or solvation will result in loss of dimensional stability (shrinkage) and increased hardness. This result will affect the performance properties of the membrane (as measured by dynamic wet pressure (DWP), effective flux ratio (EFR) and oxygen transfer efficiency (OTE)) as well as decrease the life of the material. DWP and EFR are described in more detail in Chapter 7.

An example of how engineering of the EPDM affects performance and life of the material is illustrated in Figure 3.42 (Sanitaire, 1998). Three different EPDM perforated membranes were installed in an activated sludge facility treating dairy wastewater, known to be aggressive to EPDM materials. Hardness (Shore A) and permanent set (changes in physical dimension) were monitored. As illustrated in these figures, hardness increased with service time but exhibited a much greater rate for two of the three materials. Permanent set rapidly decreased for two of the EPDM materials (shrinkage) whereas little change in set was observed for the newly engineered material. The changes observed in these three perforated membranes greatly influenced their useful life, ranging from 167 days to greater than 1,200 days for the new formulation (compound C).

Absorption of various constituents, including oils, can result in the softening of the membrane with volumetric changes and subsequent dimensional changes. For example, Ewing Engineering Co. (1989) conducted studies with plasticized PVC and two EPDM elements in vegetable oil. The PVC membrane lost weight and hardened due to lost plasticizer in the oil. On the other hand, the two EPDM elements gradually softened and one lost weight likely because of exchanges between plasticizer and oil. This study serves to illustrate the variety of mechanisms that may take place depending upon the characteristics of both the wastewater and the membrane material.

There are continuous changes taking place in the development of membrane materials for aeration system applications. To improve chemical resistance and prolong life, changes in EPDM formulations will result in many new choices for the designer. It is anticipated that membrane life for these materials will increase dramatically over the next few years. Polyurethanes are now being used in panel and tube diffuser arrangements. Chemically resistant and more expensive than EPDMs (on a weight basis), this material is typically thinner than EPDM membranes and is sensitive to creep under stress. Currently silicones are also being used in some perforated membrane systems, but there is not yet sufficient experience with this material to know how well it will hold up in wastewater applications.

An integral part of the perforated membrane is the hole size and pattern that affect both the OTE and the DWP of a given element. These perforations also affect membrane strength and tear resistance and must be carefully developed to balance performance against durability. Designers are advised to carefully review manufacturers' claims of performance and durability, especially with the newer products on the market. The use of test headers containing selected diffuser elements is very helpful for assessing effects of a given wastewater on performance and durability. Some engineers now specify specific tests on membranes to evaluate their integrity. Chapter 7 will provide some examples of these tests.

3.4.7 FOULING AND DETERIORATION CHARACTERIZATION

The fouling and deterioration of diffusers can be evaluated in several ways, the simplest of which is by visual observation. Visual observations, however, can be very misleading and result in inappropriate action. The best methods of characterization include measurements of foulant accumulations, physical changes in diffuser element, DWP, OTE and EFR. Measurement of DWP may be performed *in situ* or in the laboratory once the diffuser is removed from service. EFR can best be performed in the laboratory, although it is adaptable to field applications once the aeration tank is dewatered. OTE assessments are performed in the field or may be used to evaluate selected diffusers once removed from service.

In 1989, an effort was made to quantify these observations on a number of wastewater treatment plants by calculating a new term, F, called the fouling factor (EPA, 1989). The value of F describes the impairment of diffuser performance caused by foulants or media deterioration and is calculated as the ratio of the mass transfer coefficient, K_La, of a fouled diffuser to that of a new diffuser, both measured in the same process wastewater. The value of F was theorized to decrease from 1.0 with time in service, but the actual model of the dynamics of this decrease could not be identified. A linear model was assumed for simplicity, and the fouling rate, f_F, was estimated for a number of sites. This controlled study using portable headers equipped with ceramic disc diffusers demonstrated that values of F appeared to correlate with foulant accumulation and the changes in uniformity of operating pores. These values ranged from 0.99 to 0.56 over the 12-month study. The lower values of F were from plants that received a significant industrial waste contribution.

It is noted that there was significant temporal variation in foulant accumulations at these plants. Further, the effect of foulant (or deterioration) may depend on position within the aeration tank. Foulant accumulations have been found to be highest at the influent end of plug flow tanks in some instances and randomly distributed in others (EPA, 1989). No definitive studies have been performed, however, to quantify the independent effects of fouling/deterioration temporally or spatially on OTE. Clearly the dynamics of fouling are not understood well enough to effectively apply the fouling factor correction to the oxygen transfer relationship for aeration system design.

3.5 DIFFUSED AIR SYSTEM DESIGN

A typical diffused air system is illustrated in Figure 3.43. The air supply system consists of blowers, air filters, air piping, and airflow control equipment, including flow meters and flow control valves. The diffusion system consists of a series of headers and lateral piping in the aeration tank and the associated diffusers. The system may be arranged in a series of grids (as depicted in the figure) so as to allow for proper airflow distribution or in laterals running longitudinally along one or both sides of the basin with diffusers placed on one or both sides of the lateral in a tapered or regular spacing format. Other diffuser arrangements are also used on occasion as described earlier in Section 3.3. The basin may be rectangular, square, circular, or oval with a number of different l/w/d (or radius/depth) ratios. Aeration tanks may be laid out in series using common wall construction, folded arrangements, or individual, independent basins. This section presents the procedures and considerations required for the design and installation of a diffused air system. A number of steps are involved in the process. A brief outline of the process is first presented followed by a more detailed description of the design elements.

3.5.1 STEPS IN DESIGN

One suggested format used in the selection and design of a diffused air system is given below. It should be emphasized that there are any number of approaches that may be followed. The procedure given below has proven to be an effective approach for the design of most systems.

- determine flows and loads
- select a process flowsheet that meets the objectives of the system design

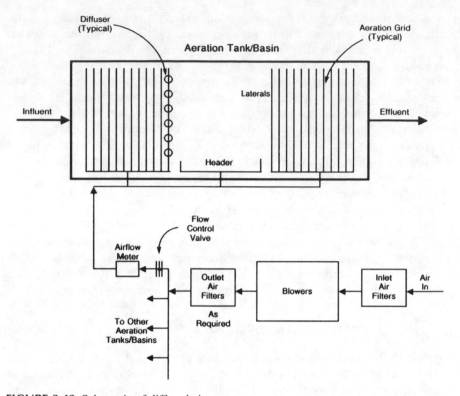

FIGURE 3.43 Schematic of diffused air system.

- establish design criteria for the process selected
- size the basins
- configure the basins
- determine the temporal and spatial oxygen demand for the process
- select the diffusers
- determine the appropriate airflow rates and their distribution
- check for mixing
- configure the diffuser system
- design the blower system
- review system flexibility
- design air piping
- select and design appropriate control system
- retrofit considerations

3.5.1.1 Determine Flows and Loads

Design wastewater flow and loads should be established for the entire range of operating conditions anticipated. From these, system oxygen requirements can be calculated. Load parameters of interest include carbonaceous oxygen demand, nitrogenous oxygen demand and any inorganic oxygen demand that might occur. Waste streams should include all return side-streams including sludge handling and internal recycle flows. Important load and flow conditions that must be determined are:

- minimum month to establish blower and diffuser turndown requirements
- average conditions (nitrifying and nonnitrifying), to establish normal operating conditions for blowers and other system components
- maximum month, to determine the maximum condition under which process oxygen requirements must be met to meet permit requirements
- peak day/ 4 hour peak (considering diurnal fluctuations), to establish the maximum operating point for all system components, including diffusers, air supply piping and blowers

3.5.1.2 Select Process Flowsheet

The selection of a process flowsheet depends on a number of factors. Among the more important considerations are:

- achievement of target pollutant removal (carbonaceous oxygen demand, nitrification, nitrogen removal, phosphorous removal, etc.)
- achievement of process stability (solid/liquid separation, qualitative or quantitative shock loads, etc.)
- site-related issues (footprint, near residential, etc.),
- low yield of biosolids
- low oxygen requirements
- the efficient removal of pollutants (e.g., plug flow vs. completely mixed)

The selection of the appropriate flowsheet will impact directly on the selection and design of an aeration system. Examples are cited below.

Conventional activated sludge processes designed for BOD and solids removal often use plug flow configurations or basins-in-series to achieve efficient removal of contaminants. The oxygen demand in these systems is highest near the influent end thereby requiring the highest transfer rate. If aeration is tapered by means of diffuser placement, the highest diffuser density, which is normally the most efficient, is used at the influent end. Counteracting this, however, is that the value of alpha is normally the lowest in this zone, and the requirement for airflow rates is the highest. Furthermore, there is a greater likelihood that diffuser fouling will take place where the load is highest. As a result, there may be a limit on the sizing and configuration of the basin due to the characteristics of the diffused air system that is selected and the wastewater that is being treated.

The requirements for ammonia oxidation will dictate longer MCRTs and greater oxygen requirements than the conventional carbonaceous systems. In fact, for many years, operators of conventional BOD removal facilities tried to avoid nitrification in the warmer seasons by maintaining low MCRTs to hold their oxygen requirements (and power consumption) down. It is now very evident that although nitrification will increase oxygen demand, the value of alpha in porous diffuser systems will significantly increase, resulting in oxygen transfer rates in nitrifying systems that are not much different (or even higher) than those for carbonaceous systems. Thus, the operation of nitrifying systems may not have any significant effect on blower sizing and power consumption. It is likely, however, that the distribution of oxygen demand through the system (plug flow) will differ significantly from the carbonaceous system.

Nitrogen removal may be accomplished in single-sludge systems by the incorporation of anoxic zones within the reactor system. This zone may be located at the influent or effluent end of the process and serves as the zone where nitrates are converted to nitrogen oxides and nitrogen. The flowsheet may have significant impacts on the aeration system where oxygen demand and alpha are concerned. If nitrate is reduced by organic matter in the influent stream, then some oxygen demand is satisfied reducing the requirements for oxygen (the nitrate serves as the electron acceptor in place of DO). Furthermore, the value of alpha for porous diffusers following the anoxic zone may be elevated by virtue of the removal of some organic matter. Whether to take advantage of these "credits" is a matter of engineering judgment. Often, they are ignored and presumed to add a degree of conservatism in the design. One important factor to consider in the aeration system design for this flowsheet is the type of diffuser. In some designs, a variable anoxic zone is used to provide greater flexibility in seasonal operation. Since these zones may be aerated or anoxic, diffusers may be idle for significant periods of time. Perforated membranes are often used for this application.

Phosphorus removal by biological methods will normally call for anaerobic zones located within the reactor system. Anoxic zones may also be incorporated into biological phosphorus removal plants where nitrification is required. The impact of these zones on alpha has been shown to be positive resulting in higher alpha values than observed for the carbonaceous removing facilities without these zones. It appears that alpha will approach the values found in high MCRT facilities that nitrify.

Process stability is often an important consideration in process design. The use of selectors has become popular in many new and retrofit designs to insure improved settleability of sludge. Inherent in biological nutrient removal schemes, aerobic, anoxic, or anaerobic selectors may be included in carbonaceous systems as well. These selectors will typically result in higher observed alpha values for porous diffuser systems as compared with systems without selectors. The magnitude of this improvement is not well documented, but it will be wastewater and selector design specific, likely approaching values found in high MCRT processes. The step aeration process may also be used to achieve process stabilization by attenuating the effects of load and flow on the system, approaching a completely mixed flow regime. As seen earlier, the step aeration process will even out spatial oxygen demand and alpha values but may result in somewhat lower mean-weighted alpha values. Often provided in plug flow systems to add operational flexibility, the engineer must evaluate the impact of this flow regime on oxygen transfer distribution. The ultimate in attenuating qualitative and quantitative shock loads to the aeration system is the completely mixed flow regime. This scheme is the easiest for designing and controlling the aeration system since there is little or no spatial variation in oxygen demand. The completely mixed flow regime generally results in lower mean-weighted alpha values for porous diffusers as compared with plug flow processes. However, it also requires lower volumetric oxygen transfer rates. One other flowsheet often selected to provide system flexibility is the contact-stabilization or sludge reaeration process. Like the step aeration process, this flowsheet is often designed as an option in conventional systems. Although there is insufficient data to support this contention, the value of alpha for porous diffusers in the reaeration section of these systems is often in the range of that found in effluent portions of the conventional plug flow system (Aeration Technologies, Inc., 1994; Donohue & Associates, Inc., 1994). It appears that the mean weighted values of alpha for porous diffusers are similar to those for conventional systems loaded at the same MCRT.

Site constraints may dictate flowsheet selection. Small footprints available for the facility may dictate the use of deep aeration tanks, the use of high purity oxygen systems, or deep shaft reactors. Each has unique characteristics that will affect aeration system design. All three systems will result in higher partial pressures of oxygen and therefore, higher transfer rates. The details of these systems are found elsewhere in this book.

Smaller communities may elect to use processes that are highly stable and require minimum operational requirements. Extended aeration systems, designed for high MCRT operation will have high total oxygen demands (mass of oxygen required per unit oxygen demand satisfied) where a significant portion of the oxygen is required for endogenous respiration. These systems may be designed in a number of configurations including oxidation ditches, aerobic or facultative lagoons, completely mixed processes, or conventional plug flow systems. Aeration system design for these processes will generally follow the same guidelines as that used for the flow regimes described above with the exception of the use of higher overall oxygen requirements. At the other extreme are the highly loaded, high-rate activated sludge systems sometimes used as a pretreatment step in industrial waste flowsheets. High-rate processes are characterized as systems with lower overall total oxygen requirements at the cost of higher biomass yields, as compared with conventional designs. Generally, they will exhibit lower porous diffuser alpha values than carbonaceous removal systems and will potentially produce a greater opportunity for diffuser fouling. Nonporous diffusers are excellent candidates for this process.

3.5.1.3 Establish Process Design Criteria — Oxygen Transfer Considerations

Several design criteria are important to the estimation of system oxygen requirements both temporally and spatially. They include:

- maximum wastewater temperature and the corresponding MCRT which are used to estimate maximum carbonaceous (and nitrogenous) oxygen requirements
- minimum wastewater temperature and the corresponding MCRT which are used to estimate minimum carbonaceous (and nitrogenous) oxygen requirements

- expected extent of denitrification, if system is designed to denitrify, to estimate oxygen "credits" in oxygen requirements calculations
- basin configuration, which will be used to estimate spatial distribution of oxygen demand
- wastewater flow distribution (step and recycle points and flows which will be used to estimate oxygen demand distribution
- design life and process growth patterns

3.5.1.4 Size the Basins

The required sizing of the aeration basins, the anoxic, aerobic and anaerobic zones, and selectors is determined by the biological process design methodology selected by the engineer and is outside the scope of this discussion.

3.5.1.5 Configure the Basins

Once total reactor volumes are calculated, the number, size and shape of the basins must be determined. Basin dimensions are important considerations in aeration system design. Depth of submergence influences both the OTE, the value of the steady-state DO saturation concentration, and the static pressure that the blowers must overcome. The basin length to width ratio will affect spatial oxygen demand and the physical layout of the diffused air system. Points of wastewater inflow, recycle flows, and return sludge will affect the magnitude and distribution of oxygen demand. The selection of a single basin severely constrains the selection of diffusers and diffuser layout in that porous diffusers require routine servicing and must be readily accessible. To avoid basin shutdown, diffusers need to be placed on retrievable lifts and should be capable of long-term operation without maintenance.

3.5.1.6 Determine Temporal and Spatial Oxygen Demand

Oxygen demand is dictated by the quality and quantity of wastewater treated and will vary over the life of the facility, normally being lower in initial years of operation and increasing to the design life of the facility. Hourly, daily, and seasonal variations will also occur and must be estimated to ensure that process oxygen requirements are properly met in accordance with the process design objectives. An evaluation of the potential impacts of periodic low mixed liquor DO on process performance and operating characteristics should be performed to determine the range of conditions that should be considered in estimating oxygen requirements. The loading conditions normally considered are outlined in Section 3.7.1.1 above. *Design of Municipal Wastewater Treatment Plants, Vol. 1, Manual of Practice 8* (WEF, 1991) provides an excellent discussion of wastewater flow and loading considerations for design and should be consulted.

Typically, oxygen demand calculations will be made for a variety of process loading conditions as appropriate for the particular system. For example, ammonia oxidation may be required from spring through fall but not the remainder of the year. The calculation of nitrogenous oxygen demand would only be necessary during this period and may or may not control aeration system design depending upon loads and temperatures during the fall to spring season. The seasonal discharge of a particular industrial waste that may impact oxygen demand in the plant must be considered in evaluating the flexibility of the aeration system.

There are several approaches to calculating process oxygen requirements for biological systems. Several factors are important in determining the procedure for a specific design situation. The most important factor is the confidence the designer has in the accuracy of the design database. Little is gained in using highly sophisticated modeling if the process loading and operating conditions are only approximately known. If, on the other hand, the database is quite accurate, a more elegant method for estimating oxygen demands may be justified. Empirical models exist that have been used for many years to estimate oxygen requirements for biological systems and are found in the *Aeration-Manual of Practice FD-13* (1988), the *Design of Municipal Wastewater Treatment Plants, Vol. 1, Manual of Practice* (WEF, 1991) and the *Design Manual, Fine Pore Aeration Systems* (EPA, 1989). Currently, there are a number of excellent biological treatment models that are available for estimating both steady state and dynamic process carbonaceous and nitrogenous oxygen requirements. The advantage of these models is that both temporal and spatial oxygen demand distributions can be estimated. The disadvantage is that the models must be calibrated to the system being designed. Most models involve a large number of variables and require substantial data collection to verify calibration. All too often, engineers do not calibrate these models and rely on default values provided in the model for their estimates. The accuracy of the models is critically dependent upon appropriate calibration. The details for estimating temporal and spatial variations in oxygen demand are beyond the scope of this text. The reader is referred to the manuals cited above for further details on these calculations.

3.5.1.7 Selection of Diffusers

Several factors should be considered in the selection of the diffusers to be used in a specific application. Cost considerations include the initial cost of the system, operation and maintenance costs, and life-cycle cost. Although the initial cost of the system is often considered paramount, it usually only represents 15 to 25 percent of the life-cycle cost of the system (EPA, 1989). The major cost element is operation and maintenance costs that include system OTE, operational flexibility, reliability and propensity to foul or deteriorate under process conditions.

The field OTE of a particular diffuser system depends on a number of factors described in detail above. Porous diffusers are generally more susceptible to wastewater constituents that will impede transfer (alpha) and may cause diffuser element fouling or deterioration. On the other hand, these diffusers are significantly more efficient in clean water and, typically, more efficient in many process wastewaters than most nonporous diffusers. The aeration efficiency of the diffuser system is also an important consideration when it is a measure of power that will be consumed. When OTE increases significantly with submergence, the SAE varies less in the range of 4 to 8 m (13 to 26 ft) (see Chapter 4). The performance of porous diffusers appears to be more sensitive to airflow rate per diffuser (OTE decreasing with increased airflow) than nonporous devices. This dependence on airflow is an important consideration when examining system flexibility under a variety of operating conditions. As described earlier, the influent end of plug flow basins produces high oxygen demands, low alpha values and greater opportunities for diffuser element fouling and deterioration. Since many of the nonporous diffuser systems are less susceptible to fouling and exhibit higher alpha values in wastewater, the use of hybrid aeration systems, which incorporate nonporous diffusers at the influent end and porous diffusers through the remainder of the aeration basin, is sometimes practiced. It should be noted, however, that most nonporous diffusers produce lower back pressures than porous diffusers and therefore require careful selection of airflow orifice controls to ensure appropriate airflow distribution throughout the system.

Designers attempt to provide sufficient process operational flexibility in their facility. This provision is often accomplished by providing several alternative flow regimes to handle a number of different process objectives and to improve system stability. Step feed or sludge reaeration may be used to supplement a conventional plug flow system to accommodate fluctuations in flow or load that would impact system performance. Process loading may be changed to accommodate different seasonal discharge permit requirements. The facility is normally designed in anticipation of future growth and, therefore, is typically underloaded early in the design period. All of these factors will affect the design of the aeration system and require that sufficient flexibility be provided to meet the variable oxygen demands that will occur. The components of the aeration system that must be designed to meet these changes include the blowers, air piping and appurtenances, and the diffusers. Air piping and blowers are addressed in later sections.

All diffusers have an allowable range of airflow rates that can be applied per unit. The range depends on size, shape, orifice diameter, and other characteristics of the device. The lower limits of this range are dictated by uniform airflow distribution from the system, and upper limits are those that cause diminishing improvements in oxygen transfer rate. To illustrate the constraints on airflow, consider the example of a typical ceramic disc diffuser. For this device, the allowable ratio of maximum to minimum airflow is about 5:1. Based on the change in OTE with airflow, the resulting ratio of maximum to minimum oxygen transfer rates would be approximately 4:1. It should be emphasized that diffuser density will play a significant role in this calculated turndown capacity. If turndown flexibility is dictated by growth over the design life of the facility, it is possible to provide only enough diffusers to meet initial diurnal and seasonal demands and to make provisions to add additional units over time to meet the ultimate demands of the system. In performing these calculations, it is important to consider mixing requirements as well as oxygen transfer rates. In systems operating under initial load conditions and in tapered aeration systems near the effluent end, mixing often controls airflow rate and may be an overriding consideration in diffuser layout and selection.

In the example above, the relationship between airflow rate and OTE was used to estimate oxygen transfer rate turndown. It is important to emphasize that this relationship may be different for different diffusers (see Table 3.7) and may change over time in process wastewater. When selecting a diffuser element, an examination of this relationship may be important. An example of this process is provided in the following. Figure 3.44 (Marx, 1998) provides data on the airflow rates and SOTE

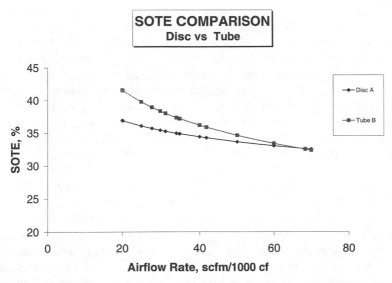

FIGURE 3.44 Comparison of performance of a disc and a tube vs. airflow (Marx, 1998).

values for two competing diffuser systems for two parallel grids in an aeration system. Disc A exhibits a lower sensitivity to airflow rate. Since the blower capacity is set, the maximum oxygen transfer rate is at the point where the two systems must provide the same SOTE. In this example, tube B will provide substantial benefit to the owner over diffuser B because the SOTE is much higher at average conditions where the system will normally operate. Note also that the turndown flexibility of disc A is significantly higher.

The operation of diffusers at their lowest allowable airflow rate has been shown to be the most efficient operating point for porous diffusers. It is tempting to operate a system at this low value but this practice can lead to operational problems. At low airflow, uniform air distribution across the diffuser may be difficult to obtain. Also at this low airflow, the head loss across the control orifice could also be low, requiring a change in orifice size to balance airflow throughout the entire system. If maldistribution occurs either along an individual diffuser header or within the entire grid of diffusers in the system, foulant deposition can begin, which may lead to premature fouling and poor performance of the entire system.

The reliability of a given diffuser system depends upon the mechanical integrity of the system and the maintenance required to ensure a high level of performance. Critical components to be considered in evaluating system integrity include the diffuser material, diffuser supports, diffuser connections, piping supports, and submerged air piping. Considerations for the diffuser material include physical and chemical resistance to the wastewater. Designers should incorporate mounting details that minimize build-up of stringy materials on diffuser piping. The supports and connections should be able to withstand stresses that will occur both during construction and operation. For example, tube-type diffusers will be subject to bending and relatively high stresses at the point of connection to the air header. Supports and air piping must be able to resist the dead weight of the equipment during installation as well as the buoyant forces of the system under normal operation. Gasket materials must be flexible and resistant to chemical or biological attack.

Required maintenance of diffused air systems has been described above. All systems need some preventative maintenance, but porous diffusers are typically more susceptible to wastewater components that may lead to fouling or deterioration. Routine maintenance is site specific depending on wastewater characteristics, process loads, flow regime, and system operation. Maintenance is performed to control fouling and to replace diffuser components when they deteriorate. To maximize OTE and minimize costs, fouling must be controlled. As fouling progresses, head loss across the diffuser increases thereby increasing blower energy costs. This gradual increase in pressure must be considered in the design of porous diffused air systems. Typical designs allow for head loss to increase by about 3.4 to 10.3 kPa (0.5 to 1.5 psi) before cleaning. Management of fouling at a given installation includes the provision of effective wastewater pretreatment to remove most of the fibrous material and heavy suspended solids. Air bumping is sometimes recommended to remove some deposits from the diffuser. The incorporation of in situ acid gas cleaning may serve to slow down fouling rate in some wastewaters. For systems that do not provide portable removal of diffuser headers for inspection, basins should be designed to allow isolation and rapid dewatering of the basin for appropriate cleaning and inspection of diffuser systems. Access to plant water that can deliver a high flow at about 415 kPa (60 psig) should be provided for diffuser cleaning.

All diffusers may be subject to gradual deterioration although those constructed from ceramic and stainless steel have demonstrated very long service lives. Deterioration may be due to buildup of inorganic materials within the diffuser that cannot be removed by ordinary cleaning methods or through breakdown of the diffuser material itself. The rate of deterioration depends on wastewater characteristics and diffuser type. The useful service life of a diffuser is generally considered to have been reached when the cost of replacement offsets the increased operating cost of the deteriorated element.

An important element in the design of the aeration system is the appropriate selection of the diffuser. Special testing of candidate diffusers using test headers or pilot plants is often justifiable when wastewater characteristics are suspected to have a significant influence on diffuser performance and/or service life. Present worth cost analyses are appropriate for both selecting diffusers and evaluating cost effectiveness of diffuser replacement.

3.5.1.8 Determine Aeration Rates

There are a number of different approaches to the design of diffused air systems. The procedure described below represents an iterative process where total airflow is calculated from the required transfer rate, OTR_f, and the estimated transfer efficiency, SOTE, for the diffuser system that was selected. The number of diffusers is ultimately determined based on the calculated total airflow rate. To start the process, the designer must determine the diffuser pattern (e.g., full floor grid, spiral roll) and whether tapering of airflow to meet demand will be implemented by varying diffuser density (if tapering is, in fact, selected as a design factor). If the flow regime is plug flow or basins-in-series, the aeration system may be laid out as a series of sectors or grids (typically three or four), each with a diffuser density that decreases from influent to effluent sectors. For completely mixed regimes, tapering is not practiced, whereas in dedicated step systems, the designer may or may not elect to provide some degree of diffuser tapering or may rely on adjustments in airflow rate for the distribution of oxygen along the tank length.

Once the oxygen requirements (AOR) have been calculated and the diffusers have been selected, it is possible to estimate the required airflow rate to meet the oxygen demand. Since the AOR will equal the OTR_f at steady state conditions, one may use Equation (2.53) to determine the standard oxygen transfer rate (SOTR) for a given grid within the tank. The designer can then determine the appropriate SOTE for the selected diffuser system. This value depends on the diffuser airflow rate, submergence, placement pattern, and diffuser density. It is often available from the equipment manufacturer. The calculation of total airflow rate for the given sector is then performed using Equation (2.51). An iterative process occurs whereby the designer selects an airflow rate per diffuser and estimates a diffuser density. Once a total airflow is calculated, the required number of diffusers for the preselected airflow rate per diffuser is determined. The diffuser density is subsequently calculated and compared with the estimated value. Either diffuser airflow rate or density can be readjusted until appropriate closure is achieved. It should be noted that diffuser density is used in its broadest definition to identify numbers of diffusers per sector whether in a full floor grid, located along one or two longitudinal walls, or placed in some other pattern. The design procedure described above should be effective for any diffuser type or configuration.

In these calculations, it is necessary for the designer to have information on field conditions (process water temperature, atmospheric pressure), beta, alpha and its spatial distribution, the target process water DO, and the steady-state DO saturation concentration at 20°C and 101.4 kPa (1 atm). One issue that the designer often faces is identifying the source for information on clean water performance data for the diffusers and on the appropriate values of alpha to use. This source should be the manufacturer of the equipment that was selected, although the information is sometimes unavailable or has been collected using nonstandard methodology. Today, most reputable manufacturers test their equipment in clean water using approved standard methods, but the information may be limited to a range of airflow, submergence, diffuser density, and pattern outside the actual system that is being designed. In those cases, the designer needs to estimate values of SOTE, preferably with the guidance of the manufacturer who knows the equipment.

The selection of alpha is often more difficult. If the manufacturer is unable to provide documented evidence of typical values for the facility being designed, it will be necessary to estimate values from the literature. Typical values of alpha for municipal wastewater have been presented in this text, but values for industrial or combined industrial/municipal wastewater are more difficult to obtain. Often the designer must ask for pilot studies with the wastewater and the selected diffusers to determine realistic alpha values. Since alpha varies with time of treatment (distance along a plug flow basin), the designer must also estimate appropriate values of alpha for each design sector if a plug flow regime has been selected. It is a good design strategy to be conservative in the estimate of alpha, especially for porous diffusers, and to provide sufficient flexibility in the aeration system because of the uncertainty of this value.

3.5.1.9   Check for Proper Mixing

Once airflow rates have been calculated, it is important to determine whether the diffused air system will provide sufficient mixing in each design sector. Details on mixing requirements are described in Section 3.4.4. As described in that section, mixing requirements are based on experience, and the designer must rely on the experience of the manufacturer (if any) and reported data in the literature.

3.5.1.10   Configure Diffuser System

After the number of diffusers has been selected, the diffuser system may be configured. Several iterations may be required to ensure that the entire range of oxygen demands can be met without exceeding the recommended airflow rate per diffuser. Important design considerations include basin inlet conditions, wastewater and airflow patterns within the basin, ability to isolate and dewater individual basins, access to diffusers within the basin and availability of plant water.

The distribution of influent wastewater and return sludge flows to the inlet end of the basin (or along the basin where step feed alternatives are selected) should be carefully considered. Depending upon basin size and configuration, it may be advisable to distribute these flows across the entire width of the basin. This distribution may minimize localized high velocity gradients and poor initial mixing in the inlet zone.

Provisions should be made for partially filling the basin without allowing the incoming flow to cascade directly onto the diffusers and in-basin piping. A drain system that permits each basin to be dewatered in a reasonable period of time (normally 8 to 24 hours) should be provided if diffusers are floor mounted and inaccessible for servicing at tank-side. The basin floor should be sloped to allow complete drainage to occur without ponding and to facilitate easy removal of residual solids. One arrangement that has been effectively used is the construction of a drain trough along the longitudinal wall of the basin, with the basin floor sloped to the trough and the trough sloped to drain to a collection sump or dewatering manhole.

Diffusers should be arranged in the tank to allow space for walking and access. Access is necessary both for installation and maintenance. Spacing between diffusers on adjacent laterals, between grids, and between each basin wall and adjacent diffusers should be examined. A minimum clear walkway space of about 50 cm (20 in) is usually adequate. Basin and diffuser cleaning require water at moderate pressures (approximately 400 to 700 kPa [60 to 100 psi]) at the nozzles. Hydrants with appropriate hose connections should be placed at frequent intervals (typically about 60 m [200 ft]).

3.5.1.11   Blower System Design

The description and design of the blower system are found in Chapter 4. Temporal variations in oxygen demand should be considered in selecting the appropriate number of blowers. Typically, the blowers are sized to allow one blower to meet minimum oxygen requirements, one or more blowers operating at full capacity to meet annual average requirements, and two or more blowers operating at full capacity to meet peak hour requirements.

3.5.1.12 System Flexibility

Sufficient flexibility should be provided to enable the system to be operated cost effectively over the entire life of the facility. The review should consider how the system will be operated at start-up and at the design loading. Over that period, the system must have sufficient flexibility to handle temporal variations in loading and oxygen demand, including hour-to-hour, day-to-day, and year-to-year variations.

Providing flexibility for year-to-year variations can be accomplished in several ways. Where the design period is relatively long and steady growth is expected, the designer/owner could choose to build a facility in phases. Another option is to construct all facilities in the first phase, with provisions for operating only a portion of the plant in the early design period. An additional alternative is to construct all of the basins, buildings, and major yard piping in the first phase and stage construction of the mechanical equipment (blowers, in-basin piping, and diffusers), as necessary. The decision on these alternatives depends on funding, projected growth patterns, and owner preference. A cost-effectiveness analysis of the alternatives is helpful in selecting the appropriate plan.

In any event, the final design must provide sufficient flexibility to allow economical operation over the design life. For example, if more basins and blowers are installed than are required to handle initial loads, capability should be provided to operate only as many basins and blowers as needed while holding the others in reserve. Similarly, if the number of diffusers required in a given basin or sector for the design year is significantly greater than required during start-up, space may be provided in the laterals to accommodate the maximum number of diffusers required. Not all holders need be filled with diffusers early in the design life.

Flexibility for handling seasonal, hour-to-hour and day-to-day variations in demand or changes in flow regime must also be provided in the system design. This is most often accomplished by providing the capability to adjust airflow to various sectors or basins in response to spatial and temporal changes in demand.

3.5.1.13 Air Piping Design

The air supply system delivers atmospheric air or high purity oxygen to the air diffusion system. It consists of three basic components: air piping, blowers, and air filters along with other conditioning equipment including gas injection diffuser cleaning systems. The air piping delivers air from the blowers to the diffusers. The blowers are designed to develop sufficient pressure to overcome the static head and line losses and deliver the required airflow to the diffusion system. Air filters are used to remove particulates from the inlet air stream to the blowers and may also be used to protect porous diffusers from air-side foulants.

The air piping should be designed to permit cost-effective installation and operation. Piping materials should be selected to provide the degree of durability (including resistance to mechanical damage, corrosion, and sunlight degradation) appropriate for the facility. Commonly used piping materials include carbon steel, stainless steel, ductile iron, fiberglass reinforced plastic (FRP), high-density polyethylene (HDPE), and polyvinyl chloride (PVC). Carbon steel, ductile iron, and FRP are the materials most often used for delivering air from the blowers to the basins because of their strength. Within the basin, stainless steel, HDPE, and PVC are often used because of their resistance to corrosion. The change is typically made at the droplegs into the basin. The choice between stainless steel and PVC for the air headers depends on the structural requirements of the diffuser connection. Stainless steel is often used for tube diffusers because of the cantilevered load applied to the lateral piping. However, PVC has been successfully used in tube installations where the connection between tube and lateral pipe has been designed for this force.

Both permanent flow meters and flow points for portable meter installation need to be properly located to allow accurate airflow determinations. An adequate number of flow points should be provided as required by the control requirements of the facility.

Piping should be sized to provide acceptable head loss at maximum airflow, including a head loss between the last positive flow split and the farthest diffuser of less than 10 percent of the loss through the diffuser. Losses through the blower inlet filter, control valves, and fittings all need to be considered in establishing total blower discharge pressure requirements. Basic principals of fluid mechanics can be used to determine head loss in air piping systems. At the rates of flow and velocities found in these systems, air can be treated as an incompressible fluid within the pipe and the Darcy–Weisbach equation can be used to determine head loss. An excellent source for the details of air piping design can be found in the *Design Manual, Fine Pore Aeration Systems* (EPA, 1989).

3.5.1.14 Control System Design

The control system is selected to meet the objectives of the wastewater treatment facility. A description of aeration control systems is found in Chapter 9. The design of this system is beyond the scope of this text but can be found elsewhere (EPA, 1989).

3.5.1.15 Retrofit Considerations

The retrofit of an aeration system is site specific. Many of the same considerations that apply to new systems apply to retrofit installations. These considerations include process oxygen requirements, diffuser selection, and configuration of the aeration system. There are some factors, however, that the designer cannot control such as basin configuration and flow regime.

In most instances where diffused air systems are being retrofitted, the existing air piping sizes are adequate for upgrading the system. Because the total airflow rate may decrease due to the higher efficiency diffusers, the size of the existing blower discharge headers and air mains that deliver air to the basins will usually be sufficient. The drop pipes into the basin may also be large enough. Replacement and recalibration of air metering devices must be considered at this time. The designer must also carefully check to determine if air piping is properly located to provide the air distribution and flow control capabilities required. Existing air distribution piping should be inspected for leaks, corrosion, and other conditions that may lead to premature failure.

Air filters will protect blowers from particulate intrusions but will not protect diffusers from air contaminants already in the downstream piping such as dirt, rust, or scale that were produced due to internal pipe corrosion, leaks or physical damage. Thorough cleaning of the air piping system may be required in some situations. Some designers prefer to provide air filters downstream of the blower discharge or in the drop pipes to protect new piping placed within the basin from debris accumulated in the older air distribution mains.

3.5.2 DESIGN EXAMPLE

The following example has been developed to illustrate one method for the design of a municipal wastewater activated sludge aeration system using diffused air aeration. The system will be a new design for 20 years into the future. The projected flow for this municipality is 0.232 m³/s (5.3 MGD). The current average flow is 0.114 m³/s (2.6 MGD). The loading and process conditions are presented below.

Process Loading Conditions for Municipality — 20 Year Design (lb/d = 2.205 × kg/d)

Variable	Min Month	Average Nonnitrifying Month	Average Nitrifying Month	Maximum Month Nitrifying	Peak Day Nonnitrifying
AOR, kg/d	1621	2454	4392	5255	5515
BOD₅, kg/d	2494	2993	2993	3492	5805
Temp, °C	10	15	20	25	25
Nitrifying	No	No	Yes	Yes	No
NOD, kg/d	—	—	1924	1924	—
Design DO, mg/l	2.0	1.0	2.0	2.0	0.5
Flow condition	Sustained	Sustained	Sustained	Sustained	Short term

Secondary treatment is to be provided to meet discharge requirements. Nitrification is required in summer months. The design requires an average hydraulic residence time of six hours with an average MCRT during the winter of four days and six days during the summer, when nitrification is required. The selected flow regime for this municipality is a plug flow activated sludge process consisting of four parallel aeration basins, each 7.0 m (23 ft) wide by 40 m (132 ft) long with a sidewater depth of 4.6 m (15 ft) (Figure 3.45). Diffuser submergence is 4.3 m (14 ft). Four basins may appear to be a large number for this small plant but were selected because of the wide variation in the process loading from start-up conditions to the 20-year design value (a doubling in flow and load over the 20 years). This variation is an economic issue. Initial construction costs will be higher but additional basins are needed for maintenance of the diffusers. Furthermore, operating costs may be

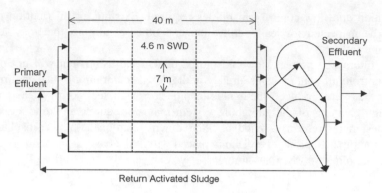

FIGURE 3.45 Design problem — aeration tank layout.

reduced since only the number of basins needed to satisfy maximum process oxygen requirements must be in service at any point in the life of the facility.

The next step in the aeration system design process is the estimation of spatial variations in process oxygen requirements along the plug flow basins. For the dimensions selected for these four parallel basins, it can be calculated that the hydraulic flow pattern for each basin would be approximated by three equal-sized basins in series. Therefore, it was decided that the air diffusion system would be segmented into three equal sized aeration zones. It was also determined that oxygen distribution would be achieved by tapering the diffusers in proportion to the oxygen demand in each of the three zones. The estimation of spatial oxygen demand was briefly described above and can be evaluated by appropriate biotreatment modeling or by the use of distribution factors obtained from practice (EPA, 1989). The actual oxygen requirements of each zone for one of the four parallel basins were calculated by oxygen demand distribution factors and appear below.

Actual Oxygen Requirements for One Basin — 20 Year Design (kg/d)
(lb/d = 2.205 × kg/d)

Zone	Minimum Month	Average Nonnitrifying Month	Average Nitrifying Month	Maximum Month Nitrifying	Peak Day Nonnitrifying
1	239	329	523	616	702
2	135	205	398	470	459
3	31	80	177	228	218
Total	405	614	1098	1314	1379

Following the estimation of AORs for each condition, the standard oxygen transfer rates (SOTRs) for each of the zones are calculated. The actual oxygen requirements (AOR) are equated to the field transfer rates (OTR$_f$) since the OTR$_f$s must satisfy the corresponding AORs. Equation (2.53) may then be used to estimate the individual SOTR values for each zone and flow condition. For this calculation, it is necessary to identify all of the parameters in the equation. These values are identified as follows:

- alpha values for each zone and flow condition were determined as follows

---

Alpha Values for Zone and Flow Condition

Zone	Minimum Month	Average Nonnitrifying Month	Average Nitrifying Month	Maximum Month Nitrifying	Peak Day Nonnitrifying
1	0.30	0.20	0.25	0.25	0.20
2	0.50	0.30	0.40	0.40	0.30
3	0.80	0.60	0.70	0.70	0.60

- Theta is 1.024; the values of wastewater temperature for each flow condition are given above.
- Omega, the pressure correction, is estimated as P_b/P_s; the elevation of the plant is 305 m (1007 ft); the value of P_b at 305 m is 98.6 kPa (14.3 psi). Omega = 0.97.
- Tau, the temperature correction, is estimated from DO surface saturation values at the given wastewater temperature and is given as $Tau = C^*_{st}/9.09$.
- Beta is estimated to be 0.98.
- The value of $C^*_{\infty 20} = 10.5$ mg/L from clean water testing of the selected aeration device at a submergence of 4.3 m (14 ft).
- The value of C_L for each zone is given above.

Using Equation (2.53), the following values of SOTR were calculated for each zone and flow condition.

Standard Oxygen Transfer Rates for Each Basin — 20 Year Design (kg/d) (lb/d = 2.205 × kg/d)

Zone	Minimum Month	Average Nonnitrifying Month	Average Nitrifying Month	Maximum Month Nitrifying	Peak Day Nonnitrifying
1	1039	1937	2752	3241	3898
2	347	787	1284	1568	1702
3	50	156	327	430	396
Total	1436	2880	4364	5239	5996

At this point, the designer must determine the performance characteristics for the diffused air device that was selected for this facility. If the design is preliminary, this information may be obtained from estimates in the literature such as the values provided in this text, the *Design Manual, Fine Pore Aeration Systems* (EPA, 1989), or the open literature. Final designs dictate that this information should be obtained from the manufacturer(s) of the device under consideration. For this example, a hypothetical set of performance data is used for a 23 cm (9 in) perforated membrane disc in a full floor grid configuration as given below.

Clean Water Test Performance Data-Perforated Membrane Disc (23 cm) Submergence — 4.3 m (14 ft)

AIRFLOW (m^3_N/h)	AIRFLOW (scfm)	SOTE @ Density-7.4%	SOTE@ Density-9.9%	SOTE@ Density-12.4%	SOTE@ Density-18.5%
0.78	0.5	30	33	36	38
1.57	1.0	28	30	32	34
2.35	1.5	27	29	31	32
3.14	2.0	26.5	28	28.5	31
3.93	2.5	26.3	27	28	30.5

The following design steps will use (Equation 2.51) in conjunction with the data in the table above. It is an iterative process whereby a value of SOTE is selected based on an estimate of diffuser density and diffuser airflow rate. A total airflow rate, G_s, is then calculated from Equation (2.51) and, for the selected airflow rate per diffuser, a total number of diffusers are calculated. The actual diffuser density is calculated and compared with the estimated value. A series of iterations follows until airflow per diffuser, diffuser density and SOTE are appropriate. Then, a calculation is performed to determine the SOTR at minimum allowable diffuser airflow rate, and this value is compared with the minimum oxygen requirement to determine whether more oxygen is provided than is required at this lower level of airflow (resulting in wasted energy at minimum turndown). At this point, adjustments may be made in diffuser density and airflow rate per diffuser to provide a more efficient design. Finally, a check must be made to determine whether sufficient mixing will be provided at minimum airflow rate per diffuser.

Zone 1

The first zone will need to satisfy the highest oxygen demands. It will, therefore, require the highest diffuser densities and airflow rates per diffuser. This zone is one-third of the basin length, 13.2 m (43.3 ft) and is 7.0 m (23 ft) wide. For this area, an 18.5 percent diffuser density was selected with airflow per diffuser of 3.93 m^3_N/h (2.5 scfm), providing an SOTE of 30.5 percent. Peak day will control the design.

$$Gs = \frac{3898(\text{kg/d})}{0.3 \times 24 \times 0.305 \dfrac{\text{kg/d}}{m^3_N/h}} = 0.139 \times 3898/0.305 = 1776 \ m^3_N/h \ (1127 \text{ scfm})$$

Using Equation (2.51)

Number of diffusers = 1776 m^3_N/h/3.93 m^3_N/h-diffuser = 452

$$\text{Check density: Density} = \frac{452 \text{ diffusers} \times 0.038 \text{ m}^2/\text{diffuser}}{(7 \text{ m} \times 13.2 \text{ m})} = 0.187 \text{ or } 18.7$$

percent (vs. 18.5 percent selected). This figure is acceptable and conservative.

Check SOTR at minimum acceptable airflow/diffuser: [Minimum airflow = 0.78 m^3_N/h/diffuser (0.5 scfm); SOTE = 38 percent]: SOTR = 452 diffusers × 0.78 m^3_N/h/diffuser × 0.38/0.139 = 964 kg/d (2142 lb/d). This figure compares with 1039 kg/d (2291 lb/d) at minimum flow; thus, demand controls airflow rate, not minimum allowable airflow, and excessive energy will not be consumed at minimum turndown.

Check mixing: Select G = 60 sec^{-1}, and minimum airflow rate is calculated at 1.52 m^3_N/h/m² (0.09 scfm/ft²) by Equation (3.6). Minimum mixing airflow required will be 1.52 m^3_N/h/m² × 7 m × 13.2 m = 140 m^3_N/h (90 scfm). At minimum allowable airflow rate per diffuser, minimum airflow will be 0.78 m^3_N/h × 452 diffusers = 353 m^3_N/h (226 scfm). This rate exceeds minimum mixing requirement; therefore, mixing requirement does not control airflow rate, and sufficient mixing will occur at minimum turndown.

Zone 2

In Zone 2, the peak day SOTR requirements control the design. Several alternative diffuser density/airflow rate combinations are possible. Select a diffuser density of 12.4 percent and airflow rate of 3.14 m^3_N/h (2.0 scfm), which would yield an SOTE of 28.5 percent. Using the same calculation procedure illustrated above, the following design information is obtained.

(1) G_s = 830 m^3_N/h (488 scfm)
(2) Number of diffusers = 264.
(3) Calculated density = 10.9 percent; this figure is significantly lower than estimated (12.4 percent).
     Try 9.9 percent at an airflow of 3.14 m^3_N/h producing an SOTE of 28 percent.
(4) New G_s = 845 m^3_N/h (536 scfm).
(5) New number of diffusers = 269.
(6) New density = 11 percent; this is a little better and conservative. Additional iterations will not be necessary.
(7) Check SOTR at minimum acceptable airflow/diffuser: At allowable minimum airflow of 0.78 m^3_N/h/diffuser, SOTE = 33 percent; SOTR = 498 kg/d (1105 lb/d) which compares with an oxygen demand (SOTR) of 347 kg/d (764 lb/d) at minimum flow. Since the allowable minimum airflow controls airflow to Zone 2 during minimum wastewater flow, the target DO will be exceeded during this period, and some energy will be wasted.
(8) Check minimum mixing requirements. The required airflow for adequate mixing of Zone 2 would be 140 m^3_N/h (90 scfm), the same as Zone 1 (step 5). At allowable minimum airflow per diffuser, the total airflow would be 210 m^3_N/h (130 scfm). Therefore, mixing requirement does not control airflow rate in this zone.

Zone 3

The maximum month SOTR controls oxygen requirements in Zone 3. Estimating a diffuser density of 7.4 percent and airflow rate per diffuser of 1.57 m^3_N/h (1.0 scfm), the SOTE would be 28 percent.The calculations follow.

(1) $G_s = 213$ m^3_N/h (136 scfm)
(2) Number of diffusers = 136.
(3) Calculated density is 5.6 percent. This calculation compares with estimated value of 7.4 percent. By linear extrapolation, estimate a value of SOTE = 25.5 percent for a density of 5.6 percent and airflow of 1.57 m^3_N/per diffuser.
(4) New $G_s = 226$ m^3_N/h (143 scfm).
(5) New number of diffusers = 144.
(6) New calculated density = 5.9 percent; this estimate is acceptable.
(7) Check SOTR at minimum allowable airflow rate per diffuser. At allowable airflow of 0.78 m^3_N/h/diffuser (0.5 scfm/diff), the estimated SOTE will be 28 percent by linear extrapolation; SOTR = 226 kg/d (504 lb/d) compared with an SOTR required at minimum flow of 50 kg/d (111 lb/d). As in Zone 2, the minimum allowable airflow rate per diffuser controls airflow in this zone during minimum wastewater flow conditions resulting in higher DO values and wasted energy.
(8) Check minimum mixing requirements. The required airflow is again 140 m^3_N/h (90 scfm) for adequate mixing of Zone 3, the same as Zones 1 and 2. At minimum allowable airflow rate per diffuser, the total airflow rate in this zone = 112 m^3_N/h (72 scfm), which indicates that mixing will control airflow in Zone 3. The minimum airflow rate allowable due to mixing considerations would be 0.97 m^3_N/h/diffuser (0.6 scfm/diffuser). Note that this exacerbates the already excessive oxygen transfer in this zone as calculated in (7) above.

Summary

Aeration rates were calculated for each flow condition and zone for the diffuser densities selected above. They are tabulated below.

Summary of Airflow Rates for Flow Condition and Zone — 20 Year Design Airflow — m^3_N/h (scfm = 0.637 × m^3_N/h)

Zone	Number of Diffusers	Minimum Month	Average Nonnitrifying Month	Average Nitrifying Month	Maximum Month Nitrifying	Peak Day Nonnitrifying
1	452	380	816	1195	1453	1776
2	269	210*	353	615	765	845
3	144	140**	140**	165	226	204
Basin Total	865	730	1309	1975	2444	2825
Syst. Total	3460	2920	5236	7900	9776	11300

* Controlled by minimum allowable airflow rate/diffuser; ** mixing controlled.

Once these calculations are performed, the designer should review the system design and identify any drawbacks that may affect the construction or operation of the system. A calculation of the system capacity at start-up and one-half way through the design life is instructive assuming a linear increase in load over the 20-year life. At start-up, it appears that Zone 1 will not be significantly inefficient with respect to excess aeration capacity except during minimum month flow conditions (i.e., the airflow rate per diffuser will be greater than the minimum allowable for all flow conditions except minimum month). In Zone 2, the aeration system will need to be operated at minimum allowable airflow per diffuser during average, nonnitrifying periods and minimum month periods during the start-up years. Observation of the data in the table above indicates that Zone 3 is mixing limited in the design year for low flow and average winter months. It is also mixing limited for most other flow conditions early in the design period. As previously mentioned, this results in higher operating costs than would occur if all zones were operated to avoid mixing limitations.

Zone 1 has been designed for a diffuser density that may create construction and operational difficulties. These characteristics are described more fully in the calculations that follow.

Finally, it is normally desirable that the airflow rate per diffuser in each zone be about the same to minimize head loss and difficulties with airflow control that may lead to poor airflow distribution and premature fouling. For average flow-nitrification conditions, the airflow is 2.64, 2.29 and 1.15 m^3_N/h/diffuser (1.68, 1.46 and 0.73 scfm/diffuser) for Zones 1, 2, and 3 respectively.

Several options are available to address these concerns. One design option is to place fewer diffusers in Zone 1 without changing the allowable airflow rate per diffuser. This would allow greater spacing between diffusers but would result in low to zero DO in that zone, thereby passing system oxygen demand downstream to Zones 2 and 3. The design could be modified so that Zone 3 could be operated to avoid mixing limited conditions some, or all of the time. This modification would also help to balance unit airflow rates in the three zones. A drawback to this strategy is that operation at low DO in Zone 1 may cause sludge bulking some of the time. As an alternative to removing diffusers from Zone 1, this zone could be deliberately operated at low airflow rates, and therefore, low DO forcing a greater load downstream as described above. This strategy is tempting during the earlier years of design life when there is excess capacity in the system. During the later periods in the design, when oxygen demands increase and nitrification becomes more critical, the operation can revert to the original design airflows.

A second design option would be to operate the basins in a step-feed mode. This option would allow part of the influent load to be introduced into Zone 2 and, perhaps, Zone 3. If this option is selected, it will be necessary to reevaluate the proper values of alpha and AOR distribution in the zones. Step-feed offers an advantage of superior sludge management during qualitative or quantitative shock loads to the plant but may produce lower treatment efficiency during some periods.

Once the diffuser number and airflow rates are determined, the designer may configure the diffuser system. A full floor grid was selected. Assume that one drop-leg will furnish air to each of the three zones. Each zone has a floor area of 7 m × 13.2 m,

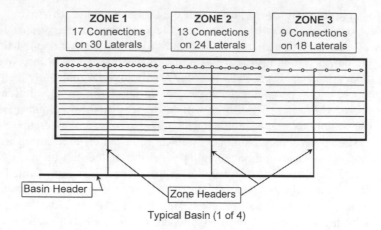

Typical Basin (1 of 4)

FIGURE 3.46 Design problem — diffuser system layout.

or 92.4 m² (23 ft × 43.6 ft = 996 ft²). Designers often provide extra baseplates in each zone for contingency. The calculations for each zone follow.

Zone 1

There are 452 diffusers in Zone 1, or 0.20 m²/diffuser (2.20 ft²/diffuser). This would require a 0.45 m (1.5 ft) spacing, center-to-center. Dividing the tank width by this spacing results in 7 m/0.45 = 15.5, say, 15 laterals placed on each side of the drop-leg main header. Note that the equal spacing between laterals will be about 46 cm (18 in), which is the minimum desirable spacing between laterals containing 23 cm (9 in) disc diffusers. Typically, the designer will leave approximately a 60 cm (24 in) clearance between the end of the headers and the wall, approximately 30 cm (12 in) spacing at the end of the zone, and will allow about 60 cm clearance at the central main header. This would leave 13.2 m – 0.6 m – 0.3 m – 0.6 m = 11.7 m for diffuser baseplates (about 38.3 ft). At a minimum spacing between discs of 33 cm (13 in) center-to-center, each lateral could accommodate 11.7 m/0.33 m = 35.5, say 34 diffusers for a total of 15 × 34 = 510 diffusers, or a 13 percent contingency. Leave four baseplates empty per lateral, uniformly distributed along the longitudinal axis of the zone. See Figure 3.46 for the layout of this system.

Zone 2

There are 269 diffusers in Zone 2, or 0.34 m²/diffuser (3.7 ft²/diffuser) with a spacing of 0.58 m (1.9 ft) center-to-center. Use 7 m/0.58 = 12 laterals in this zone on each side of the main header. Each lateral should accommodate a minimum of 269/12 = 22 diffusers. Adding a 20 percent contingency will place 26 baseplates on each lateral spaced at 45 cm (17.5 in) centers. Leave 4 baseplates empty per lateral.

Zone 3

By the same type of calculations, there will be nine laterals in Zone 3. Each lateral will contain 18 diffuser pods, of which, two will be blank, providing a contingency of about 12 percent in this zone.

The next step of the design will be the selection and sizing of the blowers, followed by the final piping design, filter selection and control layout. An example of blower calculations is found in Chapter 4. Details of piping design and layout along with control system selection and design may be found in the EPA *Design Manual, Fine Pore Aeration Systems* (1989).

3.6 NOMENCLATURE

A_d	m^2	total projected area of diffuser media
A_t	m^2	total surface area of aeration basin
AE_f	kg/kWh, lb/hp-h	aeration efficiency under process conditions
C	mg/L	surfactant concentration
C_1		empirical coefficient
C_L	mg/L	bulk liquid phase oxygen concentration
C_{20}^*	mg/l	clean water oxygen saturation concentration at diffuser depth and 20°C
$C_{\infty 20}^*$	mg/l	clean water oxygen saturation concentration at diffuser depth and 20°C
DWP	cm of water	dynamic wet pressure
d_B	cm	bubble diameter
E	W	power transferred to the fluid
F/M	lb BOD_5/d-lb MLSS	food to microorganism ratio
G	s^{-1}	root mean square velocity gradient
G_s	m_N^3/h, scfm	airflow rate at standard conditions
G_{sd}	m_N^3/h-diff	airflow rate per diffuser at standard conditions
H	m	sidewater depth
H_s	m	diffuser submergence
K_L	cm/h	overall liquid film coefficient
$K_L a$	h^{-1}	oxygen transfer coefficient
$K_L a_{20}$	h^{-1}	clean water oxygen transfer coefficient at 20°C
m		alpha factor for surfactant data
m		empirical constant
n		empirical coefficient
OTE		oxygen transfer efficiency
OTE_f	–, %	oxygen transfer efficiency under process conditions
OTR_f	kg/h, lb/h	oxygen transfer rate under process conditions
P_1	kPa, psia	absolute pressure at the surface
P_2	kPa, psia	absolute pressure at the depth of injection
SAE	kg/kWh, lb/hp-h	standard aeration efficiency
SOTE	–, %	standard oxygen transfer efficiency
$SOTE_a$	–, %	standard oxygen transfer efficiency at gas flow G_{sa}
$SOTE_b$	–, %	standard oxygen transfer efficiency at gas flow G_{sb}
SOTR	kg/h, lb/h	standard oxygen transfer rate

SRT   d   solids retention time
t   °C   temperature
V   m³   tank volume
W   W/m³   power dissipation
α   wastewater correction factor for oxygen transfer coefficient
β   wastewater correction factor for oxygen saturation
δ   depth correction factor for oxygen saturation
μ   N-s/m²   absolute viscosity
θ   temperature correction factor for oxygen transfer coefficient
τ   temperature correction factor for oxygen saturation
Ω   pressure correction factor for oxygen saturation

3.7 BIBLIOGRAPHY

Aeration Technologies, Inc. (1994). *Off-Gas Analyses Results and Fine Pore Retrofit Case History for Hartford, CN.*, EPA 600/R-94/105, NTIS No. PB94-200938, EPA, Cincinnati, OH.

Aiba, S. and Toda, K. (1963). "Effect of Surface Active Agents on Oxygen Absorption in Bubble Aeration." *J. Applied Microbiology,* 9, 443.

APHA (1995). *Standard Methods for the Examination of Water and Wastewater, 19th Edition,* APHA, AWWA, WEF, Washington, DC.

ASCE (1996). *Standard Guidelines for In-Process Oxygen Transfer Testing, ASCE-18-96,* American Society of Civil Engineers, Reston, VA.

ATV-Regelwerk (1996). *Messung der Sauerstoffzofuhr von Beluftungseinrichtungen in Belebungsanlagen in Reinwasser und in belebten Schlamm-Merkblatt ATV-M209,* Gesellschaft zur Forderung der Abwassertechnik E.V., Hennef, Germany.

Babbitt, H.E. (1925). *Sewerage and Sewage Treatment, Second Edition*, John Wiley and Sons, New York.

Baillod, C.R. and Hopkins, K. (1994). *Fouling of Fine Pore Diffused Aerators: An Interplant Comparison Study,* EPA 600/R94/103, NTIS No. PB94-200912, EPA, Cincinnati, OH.

Barnhart, E.L. (1966). "Factors Affecting the Transfer of Oxygen in Aqueous Solutions." Masters of Engineering (Sanitary Engineering) Thesis, Manhattan College.

Barnhart, E.L. (1969). "Transfer of Oxygen in Aqueous Solutions." *J. San. Engr. Div., ASCE,* 95, 645.

BBS Corp. (1990). *Off-Gas Analyses of Parkson Messner Aeration System at DuPage Co., IL.* BBS Corp., Consulting Engineers, Columbus, OH.

Bewtra, J.K. and Nicholas, W.R. (1964). "Oxygenation From Diffused Air in Aeration Tanks." *J.WPCF,* 36, 1195.

Boyle, W.C. and Redmon D.T. (1983). "Biological Fouling of Fine Bubble Diffusers- State-Of-Art." *J. Environ. Engr. Div., ASCE,* 109, 991.

Boyle, W.C. et.al. (1994). *Oxygen Transfer Studies at the Madison Metropolitan Sewerage District Facilities,* EPA 600/R-94/096, NTIS No. PB94-200847, EPA, Cincinnati, OH.

Brochtrup, J.A. (1983). "A Study of the Steady-State and Off-Gas Methods of Determining Oxygen Transfer in Mixed Liquor." Masters of Science Thesis, Dept. of Civil and Environmental Engineering, University of Wisconsin, Madison, WI.

Bushee, R.G. and Zack, S.I. (1924). 'Tests of Air Pressure Losses in Activated Sludge Plants." *Engineering News Record,* 93, 823.

Committee on Sewage and Industrial Wastes Practice (1952). *Air Diffusion in Sewage Works, MOP 5,* Federation of Sewage and Industrial Waste Associations, Champaign, IL.

Costerton, J.W. (1994*). Investigations into Biofouling Phenomena in Fine Pore Aeration Devices,* EPA 600/R-94/107, NTIS No. PB94-200953, EPA, Cincinnati, OH.

Currie, R.B. and Stenstrom, M.K. (1994). "Full Scale Field Testing of Aeration Diffuser Systems at Union Sanitary District." *Proc. 67th Annual Conference,* WEF, Chicago, IL.

DaSilva-Deronzier, G. et al. (1994). "Influence of a Horizontal Flow in the Performance of a Fine Bubble Diffused Air System." *Water Science Tech.* 30, 4, 89.

Dezham, P. et al. (1992). "Full Scale Process Water Testing of Membrane Aeration Panels." *Proc. 65th Annual Conference*, WEF, New Orleans, LA.

Donohue and Assoc. (1987). *Oxygen Transfer Testing of Counter-Current Aeration System,* Donohue and Assoc., Sheboygan, WI.

Donohue and Assoc. (1989). *Oxygen Transfer Testing of a Counter-Current Aeration System Plant,* Donohue and Assoc., Sheboygan, WI.

Donohue and Assoc. (1994). *Fine Pore Diffuser System Evaluation for Green Bay Metropolitan Sewerage District,* EPA 600/R94/093, NTIS No. PB94-200813, EPA, Cincinnati, OH.

Downing, A.L. and Bayley, R.W. (1961). "Aeration Processes for the Biological Oxidation of Wastewaters." *Chemical Engineering,* 157, A53.

Downing, A.L. et al. (1961). "Aeration and Biological Oxidation in the Activated Sludge Process." *The Institute of Sewage Purification,* Conf. Paper No. 2, Brighton, UK.

Doyle, M.L. and Boyle, W.C. (1985). "Translation of Clean to Dirty Water Oxygen Transfer Rates." *In: Proc. Seminar-Workshop on Aeration Systems-Design, Testing, Operation, and Control,* EPA 600/9-85/005, NTIS No. PB85-173896, EPA, Cincinnati, OH.

Eckenfelder Jr., W.W. (1959). "Factors Affecting Aeration Efficiency of Sewage and Industrial Wastes." *J.WPCF,* 31, 60.

Egan-Benck, K. et al. (1992). "Experiences with Three Types of Diffusers at an Energy Savings, Award Winning Plant." *65th Annual Meeting of the Central States Water Pollution Control Association,* Fontana, WI.

Eimco (1986). *Evaluation of the Oxygen Transfer Capabilities of the Eimco Elastox-D Fine Bubble Rubber Diffuser,* Eimco Process Equipment Co., Salt Lake City, UT.

Environmental Leasing Corp. (1987). *Measurement of Oxygen Transfer in Clean Water-Counter-Current Aeration, Cleveland, TX,* ELC, Houston, TX.

EPA (1985) Summary Report — Fine Pore Aeration Systems, USEPA, EPA/625/8-85/010, Oct. 1985, Water Engineering Research Laboratory, Cincinnati, OH.

EPA (1989). *Design Manual, Fine Pore Aeration Systems,* EPA 625/1-89/023, Risk Reduction Research Labs, USEPA, Cincinnati, OH.

Ernest, L.A. (1994). *Case History Report on Milwaukee Ceramic Plate Aeration Facilities,* EPA 600/R-94/106, NTIS No. PB94-200946, EPA, Cincinnati, OH.

Ewing Engineering Co. (1994). *Characterization of Clean and Fouled Perforated Membrane Diffusers,* EPA 600/R-94/108, NTIS No. PB94-200961, EPA, Cincinnati, OH.

Fair, G.M., Geyer, J.C., and Okun, D.A. (1966). *Water and Wastewater Engineering- Vol 2,* John Wiley and Sons, New York.

Fisher, M.J. and Boyle, W.C. (1999). "The Effect of Anaerobic and Anoxic Selectors on Oxygen Transfer in Wastewater." *Water Environment Research,* in press.

Gillot, S. et al. (1997). "Oxygen Transfer Under Process Conditions in an Oxidation Ditch Equipped with Fine Bubble Diffusers and Slow Speed Mixers." *Proc. 70th Annual Conference,* WEF, Chicago, IL.

Groves, K. et al. (1992). "Evaluation of Oxygen Transfer Efficiency and Alpha-Factor on a Variety of Diffused Aeration Systems." *Water Environment Research,* 64, 691.

GSEE, Inc. (1986). *Comparison of Oxygen Transfer Capabilities of Messner Panels and Other Fine Bubble Diffusers-McMurray, Pa.*, GSEE, Inc., Lavergne, TN.

GSEE, Inc. (1998). *Evaluation of the Oxygen Transfer Capabilities of the O_2-Okonom Magnum Membrane Diffuser for Dry Creek WWTP, KY,* GSSE, Inc., Lavergne, TN.

Guard, S. et al. (1990). *Full Scale Comparisons of Changes in Oxygen Transfer of Membrane Diffusers,* Eimco Process Equipment Co., Salt Lake City, UT.

Hantz, P.J. (1980). "Effect of the Chemical Constituents in Water on Oxygen Transfer." Masters of Science Thesis, Dept. of Civil and Environmental Engineering, The University of Wisconsin, Madison, WI.

Hartley, K.J. (1990). "Fouling and Cleaning of Fine Bubble Ceramic Diffusers." *Report 14,* Urban Water Research Association of Australia, Brisbane, Queensland, Australia.

Houck, D.H. and Boon, A.G. (1981). *Survey and Evaluation of Fine Bubble Dome Diffuser Aeration Systems,* EPA 600/2-81/222, EPA, Cincinnati, OH.

Huibregtse, G.L. (1987). "Evaluation of the IFU Fine Bubble Membrane Disc Diffuser." *Internal Report,* Envirex, Inc. Waukesha, WI.

Huibregtse, G.L. et al. (1982). "Factors Affecting Fine Bubble Diffused Aeration." unpublished paper presented at Central States Water Pollution Control Association Annual Meeting, Bloomingdale, IL, May 19–21.

Hung, J. (1998). *Ceramic Diffuser Fouling Studies- A Progress Report,* PhD candidate, Dept. of Civil and Environmental Engineering, University of Wisconsin, Madison, WI.

Hurd, C.H. (1923). "Design Features of the Indianapolis Activated Sludge Plant." *Engineering News Record,* 91, 259.

Hwang, H.J. and Stenstrom, M.K. (1985). "Evaluation of Fine Bubble Alpha Factors in Near-Full Scale Equipment." *J.WPCF,* 57,1142.

Jackson, M.L. (1982). "Deep Tank Aeration/Flotation for Fermentation Wastewater Treatment." *36th Purdue Industrial Waste Conference,* Lafayette, IN, 363.

Jackson, M.L. and Shen, C.C. (1978). "Scale Up and Design for Aeration and Mixing in Deep Tanks." *AIChE J.,* 24, 63.

Johnson, T.L. (1993). "Design Concepts for Activated Sludge Aeration Systems." Ph.D Thesis, Dept. of Civil Engineering, University of Kansas, Lawrence, KS.

Kim, Y.K., Mechanisms and Effects of Fouling in Fine Pore Ceramic Diffuser Aeration, PhD Thesis, University of Wisconsin, Madison, WI, 1990.

Kim, Y.K. and Boyle, W.C. (1993). "Mechanisms of Fouling of Fine Pore Diffusers." *J. Env. Engr. Div.,* ASCE, 119, 1119.

Leary, R.D. et al. (1969). "Full Scale Oxygen Transfer Studies of Seven Diffuser Systems." J. WPCF, 41, 459.

Marrucci, G. and Nicodemo, L. (1967). "Coalescence of Gas Bubbles in Aqueous Solutions of Inorganic Electrolytes." *Chemical Engr. Sci.* 22, 1257.

Martin, A.J. (1927). *The Activated Sludge Process,* MacDonald and Evans Publ., London, UK.

Marx, J. (1998). Personal communication, RUST E&I, Sheboygan, WI.

Marx, J. and Redmon, D. (1991). "Oxygen Transfer Performance of Rotating Bridge Aerators." *64th Annual Conference, WPCF*, Toronto, CN.

Masutani, G. and Stenstrom, M. K. (1984). "A Review of Surface Tension Measuring Techniques, Surfactants, and Their Implications for Oxygen Transfer in Wastewater Treatment Plants." Water Resources Program, School of Engineering and Applied Sciences, UCLA, Los Angeles, CA.

Mueller, J.A. et al. (1996). "Impact of Selectors on Oxygen Transfer- A Full Scale Demonstration." *Proc. 69th Annual Conference,* WEF, Dallas, TX, 427.

Mueller, J.A., Kim, Y-K, Krupa, J.J., Shkreli, F., Nasr, S., and Fitzpatrick, B. (2000). "Full-Scale Demonstration of Improvement in Aeration Efficiency." *ASCE J. Environ. Engr.,* 126(6), 549–555.

O'Connor, D.J. (1963). "Effects of Surface Active Agents on Reaeration." *Intl. J. Air and Water Poll.,* 5, 123.

Parkson Corp. (1991). *Oxygen Transfer Evaluation of Parkson Aeration Panel Diffuser System for City of Woonsocket, RI,* Parkson Corp., Ft. Lauderdale, FL.

Pasveer, A. and Sweeris, S. (1965). "A New Development in Diffused Air Aeration." *J. WPCF,* 37, 1267.

Paulson, W.L. (1976). "Oxygen Absorption Efficiency Study- Norton Co. Dome Diffusers." *Report to Norton Co.,* Worcester, MA.

Pöpel, H.J. and Wagner, M. (1991). "Welche Sauerstoffeinstrags- und Ertragswerte sind mit Druckluftsbeluftungssytemen Erreichbar?" *Design for Nitrogen Removal and Guarantees for Aeration, Proc. Of Workshop, Vol 50E,* Technical University Braunschweig, Braunschweig, Germany.

Pöpel, H.J. and Wagner, M. (1994). "Modeling and Simulation of Oxygen Transfer in Deep Aeration Tanks and Comparison with Full Scale Data." *Proc. 17th International Biennial Conference,* IAWQ, Budapest, Hungary.

Pöpel, H.J. et al. (1993). *Oxygen Transfer Rate and Aeration Efficiency of Sanitaire Membrane Disc Aerators,* Institute for Water Supply, Wastewater Technology, and Regional Planning, University of Darmstadt, Darmstadt, Germany.

Pöpel, H.J. et al. (1991). *Oxygen Transfer Rate and Aeration Efficiency of the O_2-Okonom Membrane/Flexible Tube Diffuser,* Institute for Water Supply, Wastewater Technology, and Regional Planning, University of Darmstadt, Darmstadt, Germany.

Redmon, D.T. (1998). Personal communication. Redmon Engineering Co., Milwaukee, WI.

Redmon, D.T. et al. (1983). "Oxygen Transfer Efficiency Measurements on Mixed Liquor Using Off-Gas Technique." *J. WPCF,* 55, 1347.

Reith, M.G. et al. (1995). "Effects of Operational Variables on the Oxygen Transfer Performance of Ceramic Diffusers." *Water Environment Research,* 67, 781.

Roe, F.C. (1934). "The Installation and Servicing of Air Diffuser Mediums." *Water Works and Sewerage,* 81, 115.

Rooney, T.C. and Huibregtse, G.L. (1980). "Increased Oxygen Transfer Efficiency with Coarse Bubble Diffusers." *J.WPCF,* 52, 2315.

Sanitaire (1998). "Report on EPDM Silver Series Diffusers." *Internal Report,* Sanitaire-Water Pollution Control Corp., Brown Deer, WI.

Sanitaire (1976-1986). "Oxygen Transfer." *Ceramic Disc Diffuser System Reports,* Sanitaire-Water Pollution Control Corp., Brown Deer, WI.

Sanitaire (1993). *Side by Side Evaluation of Sanitaire S-T Membrane Disc Grid Systems and Parkson Panels at Carmel, IN,* Sanitaire-Water Pollution Control Corp., Brown Deer, WI.

Schmidt-Holthausen, H.J. and Zievers, E.C. (1980). "50 Years of Experience in Europe with Fine Bubble Aeration." *53rd Annual Conference WPCF,* Las Vegas, NV.

Schmit, F.L. et al. (1978). "The Effect of Tank Dimensions and Diffuser Placement on Oxygen Transfer." *J.WPCF,* 50, 1750.

Semblex (1987). *Static Tube Aerator Tests,* Semblex, Springfield, MO.

Stenstrom, M.K. (1996). Personal communication, Dept. of Civil Engineering, UCLA, Los Angeles, CA.

Stenstrom, M.K. (1997). *Off-Gas Test Report for Orange County Water Reclamation Plant No. 1,* M.K. Stenstrom, Los Angeles, CA.

Stenstrom, M.K. and Gilbert, R.G. (1981). "Effects of Alpha, Beta, and Theta Factors on the Design, Specification, and Operation of Aeration Systems." *Water Research*, 15, 643.

Stenstrom, M.K. and Masutani, G. (1994). *Fine Pore Diffuser Fouling- The Los Angeles Studies*, EPA 600/R94/095, NTIS No. PB94-200839, EPA, Cincinnati, OH.

Waddington, R. (1995). "A Study of Ceramic Disc Diffuser Performance Problems at Madison Metropolitan Sewerage District Plant." Masters of Science Thesis, Dept. of Civil and Environmental Engineering, University of Wisconsin, Madison, WI.

WEF (1991). *Design of Municipal Wastewater Treatment Plants, Vol. 1, Manual of Practice 8*, Water Environment Federation, Alexandria, VA.

Wilfey-Weber, Inc. (1987). *Oxygen Transfer Efficiency of Wilfey-Weber Diffusers*, Wilfey-Weber, Inc., Englewood, CO.

Wilfey-Weber, Inc. (1998). *Clean Water Performance of Dura-Disc Plus Membrane Diffusers*, Wilfey-Weber, Inc., Englewood, CO.

WPCF (1988). *Aeration- Manual of Practice FD-13*, WEF, Alexandria, VA.

Yunt, F.W. (1980). *Results of Mixing Efficiency Tests with Norton Dome Aeration System at LA Glendale Treatment Plant*, Los Angeles County Sanitation Districts, Whittier, CA.

Yunt, F.W. and Hancuff, T.O. (1979). "Relative Number of Diffusers for the Norton Dome and Sanitaire Aeration Systems to Achieve Equivalent Oxygen Transfer Performance." *Report to Los Angeles County Sanitation Districts*, Whittier, CA.

Yunt, F.W. and Hancuff, T.O. (1988). *Aeration Equipment Evaluation-Phase I- Clean Water Test Results*, EPA 600/2-88/022, NTIS No. PB 88-180351, USEPA, Cincinnati, OH.

Yunt, F.W. and Stenstrom, M.K. (1990). *Aeration Equipment Evaluation-Phase II, Process Water Test Results*, EPA Contract No. 68-03-2906, EPA, Cincinnati, OH.

4 Deep Tank Aeration with Blower and Compressor Considerations

4.1 INTRODUCTION

Typical depths of diffused aeration tanks vary over a range from 3.50 to 6.00 m. This range is illustrated by an evaluation of 98 published performance tests in Germany (Pöpel and Wagner, 1989) showing the following tank depth distribution:

- tank depths greater than 6.00 m: 10 percent
- tank depths 4.00 to 6.00 m: 50 percent
- tank depths less than 4.00 m: 40 percent

Greater tank depths, 20 to 30 m, equipped with special ejector systems for oxygenation, have been used for treating industrial effluents only by applying the so-called "tower-biology" (Bayer company; Diesterweg et al., 1978) and bio-high-reactor (Hoechst company; Leistner et al., 1979). These systems produce very small bubbles (micrometer range), which remain stable at the high salinity (some 20 g/l) of the wastewater. However, at municipal wastewater conditions, these bubbles would coalesce and lead to poor oxygen transfer performance.

There is, however, a strong tendency towards greater tank depths, probably due to the following reasons:

- when upgrading wastewater treatment plants for biological nutrient removal, especially for biological nitrogen removal, the required increase of tank volume leads to much less area usage at greater depth;
- due to the higher oxygen transfer efficiency at greater tank depth, less air is required, producing less off-gas and odor problems and leading to less extensive gas cleaning equipment;
- in addition to the rise of the oxygen transfer efficiency, also an increase of the aeration efficiency is expected, which would lead to energy savings.

Consequently, a number of activated sludge plants in Europe have been upgraded for nutrient removal using significantly greater tank depths than stated above. Table 4.1. (Wagner, 1998) gives more detailed information on this development. In this context, deep diffused aeration tanks can be defined by having a depth of (significantly) greater than 6.00 m.

TABLE 4.1
Examples of Deep Aeration Tanks at
European Municipal Wastewater Treatment Plants

City	Water Depth m	Aeration Tank Volume m³	Diffuser Material	Type of Blower
Bonn, D	12.90	135,100	di-m	C + S
Bottropp, D	10.00	31,300	pl-m + do-c	C
Frankfurt, D	8.00	57,600	di-rpp	C
Heilbronn, D	7.80	45,000	di-m	C
Helsinki, SF	12.00	60,000*	di-m	C
Stockholm, S	12.00	110,000*	di-m	C

diffuser submergence ≈ water depth – 0.25 m
* = average of variable volume allotted to nitrification, i.e., under aeration
C = centrifugal blower   pl = plate
S = crew compressor   c = ceramic
di = disc   m = membrane
do = dome   rpp = rigid porous plastic

Possible disadvantages of deep aeration tanks have also been envisaged immediately with the advent of greater tank depth (ATV-Arbeitsbericht, 1989). In each
case, these have to be carefully considered, and measures need to be taken to prevent
any process impairment, if required. The potential drawbacks are:

- decreased CO_2 stripping from the wastewater due to the required smaller
airflow rates, giving rise to a more intensive lowering of the pH-value,
especially at low alkalinity. This occurrence may impair or even terminate
nitrification unless countermeasures like addition of lime (pH) or soda
ash (pH and alkalinity) are taken;
- supersaturation of mixed liquor, with respect to all gases, due to the
high(er) water pressure. Whereas the oxygen is generally utilized, a serious supersaturation with respect to nitrogen may remain in the tank
effluent and lead to (partial) solids flotation in the secondary clarifier. This
problem can be solved by either limiting the tank depths to (not yet
precisely known) values to avoid excessive nitrogen supersaturation or by
installing special constructions for gas release between aeration tank and
secondary clarifier;
- the process of aeration and gas transfer in deeper tanks has been thoroughly investigated and modeled only recently (Pöpel and Wagner, 1994;
Pöpel et al., 1998). Hence, there was (is) much uncertainty with respect
to design of diffused aeration systems in deep tanks.

In this chapter, the process of oxygen transfer in deep tanks is characterized and
modeled, based on the involved physical mechanisms. Although these hold, obviously,

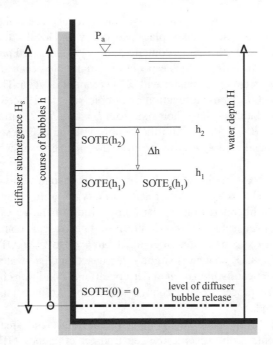

FIGURE 4.1 Schematic of deep tank.

for any water depth, some of them can be neglected for more shallow tanks without greater inaccuracies. The model is then verified by an extensive investigation and evaluation program leading to useful empirical relations for design. The application of the model is outlined at the end of the first section.

The question of (higher) aeration efficiency in deep aeration tanks is covered in the following section. First, the components of the air supply system and their energy requirements are discussed, followed by an outline of different types of blowers and their energy consumption as a function of diffuser submergence. The above model is then applied to develop principles of blower selection for optimum aeration efficiency and hence maximum energy savings.

4.2 OXYGEN TRANSFER IN DEEP TANKS

4.2.1 CHARACTERIZATION OF THE PROCESS OF OXYGEN TRANSFER IN DEEP TANKS

In an aeration tank of H (m) of water depth, the bubbles are released at the depth of diffuser submergence of H_S (m), generally 0.20 to 0.30 m less than the wastewater depth H. The actual difference depends upon the height of the specific diffuser system construction (see Figure 4.1). The water level is exposed to the atmospheric pressure, P_a. The total pressure, P_t, at the bubble release level ($h = 0$) is given as follows.

$$P_t = P_a + \rho \cdot g \cdot H_S \qquad (4.1)$$

Because of this pressure, the bubble volume is reduced as is the interfacial area, A, through which gas transfer takes place. Secondly, the local saturation concentration of oxygen, $c_{s,}$ (and other gases contained in air) is increased proportional to this pressure growth. This c_s-increase is especially remarkable because the air composition is still unchanged by gas transfer with 21 percent of oxygen. Thirdly, the oxygen transfer coefficient, k_L, being a function of bubble size, is reduced accordingly.

Following the bubbles along their rise from $h = 0$ to $h = H_S$ after bubble release, the total pressure P_t is reduced, and the bubble volume expands. This occurrence causes the interfacial area A to grow again and k_L to increase, eventually attaining its "normal value".

Also, by this pressure decrease, the saturation concentrations of all gases contained in air are reduced again. With respect to oxygen utilized by activated sludge or carbon dioxide liberated from it, the composition of the air is changed, which also affects the local saturation concentration. The oxygen content of the air is reduced due to the oxygen transfer efficiency from $h = 0$ to $h = h$ (OTE(h) as indicated in Figure 4.1). The CO_2 content is slightly decreased in clean water (tests) by some stripping and significantly *increased* under operational conditions by biological CO_2 production. These processes also change the bubble volume (slightly), which is normally neglected.

Consequently, despite the enlargement of the interfacial area, A, and the gas transfer coefficient, k_L, the specific oxygen transfer efficiency OTE$_s$ is continually decreasing (see Figure 4.1). This decrease is mainly due to the reduction of c_s by the changes of pressure and air composition.

When approaching the water level ($h \approx H_S$), the bubbles reach characteristics (with the exception of gas composition) they would have without any additional water pressure, hypothetically at a tank depth of zero or in very shallow tanks. These conditions of an aeration system of zero (or very small) depth and unchanged air composition are indicated by a subscript of zero:

- bubble volume V_B: V_{B_0} (m^3)
- bubble diameter d_B: d_{B_0} (m)
- interfacial area A: A_0 (m^2)
- specific interfacial area a: a_0 (m^{-1})
- gas transfer coefficient k_L: k_{L_0} (m/h)
- saturation concentration c_s: c_{s_0} (g/m^3), if air composition is not changed

These "standard values" are used as references in modeling the described mechanisms later.

Again, it is pointed out, that the above processes and changes of bubble and transfer characteristics occur in aeration tanks of conventional or even shallow depth. However, the consequences for the rate and efficiency of gas transfer are so small that they can be neglected, and it is only in tanks of greater depth that they have to be taken into account quantitatively.

With respect to oxygen transfer to the water, it should be noted that there is an important oxygen concentration gradient in the rising bubbles. The highest oxygen content is present immediately after bubble release and the lowest when the bubbles leave the water at the surface. In the technique of off-gas measurement, use is made of this phenomenon. On the other hand, the (waste) water content of an aeration tank is fully mixed in the vertical direction. This difference has been shown in the multitude of oxygen transfer tests under clean and dirty water conditions with oxygen probes placed at different depths within a tank. In other words, there is no oxygen gradient present in the (waste) water. Finally, this means that transfer of oxygen takes place only during the bubble rise from $h = 0$ to $h = H_S$, and this transferred oxygen is then distributed over the full body of water or over the complete water depth H. In modeling oxygen transfer, this has to be taken into account quantitatively. This influence is strong in shallow tanks, where the difference between water depth and depth of diffuser submergence is relatively large. It diminishes as the water depth increases.

4.2.2 MODELING OF THE PROCESS OF OXYGEN AND GAS TRANSFER IN DEEP TANKS

4.2.2.1 Influence of Depth and Water Pressure on the Transfer Parameters

To quantify the influence of atmospheric plus water pressure on the transfer of oxygen, the pressure situation within the tank has to be thoroughly defined and quantified. To this end, the hydraulic pressure (m water column, WC) within the tank at depth h (see Figure 4.1) is converted into the standard unit P (Pa; N/m^2) and then related to the atmospheric standard pressure of $P_a = 101\ 325$ Pa $= 101.325$ kPa. A bubble at depth h is exposed to an additional water pressure of ΔP (m WC) $= (H_S - h)$, or ΔP (Pa) $= 9,810 \cdot (H_S - h)$, and hence, to a total pressure of $P_a + \Delta P$. Relating this total pressure to the atmospheric standard pressure of P_a yields the relative pressure π.

$$\pi = 1 + \frac{\Delta P}{P_a} = 1 + \frac{9,810 \cdot (H_S - h)}{101,325}$$

$$= 1 + z \cdot (H_S - h) = 1 + 0.0968 \cdot (H_S - h) \approx 1 + 0.1 \cdot (H_S - h)$$

(4.2)

the conversion factor, z, being $z = 9,810/101,325 = 0.0968 \approx 0.1$.

The rounded value of 0.1 reflects the rule of thumb, that 10 m of water column will double the standard pressure. In the following, the relative pressure π is the relevant pressure parameter for quantifying the influence of tank depth on oxygen transfer via the influenced parameters k_L, a, and c_s. These parameters, together with the water volume of the aeration tank, V, define the standard oxygen transfer rate SOTR (kg/h).

$$SOTR = \frac{k_L \cdot a \cdot c_s \cdot V}{1000}$$

(4.3)

The following definitions apply.

V	water volume of aeration tank	[m³]
A	total interfacial area	[m²]
a	specific interfacial area = A/V	[m⁻¹]
A_{at}	bottom area of aeration tank	[m²]
k_L	liquid film coefficient	[m/h] where $k_L \cdot a$ is similar to $K_L a_{20}$ in Equation (2.42)
c_s	oxygen saturation concentration	[mg/l] similar to $C^*_{\infty 20}$ in Equation (2.42)
G_s	standard airflow rate	[m$_N$³/h at STP]

As pointed out when characterizing the process of oxygen transfer in deep tanks, the first three parameters of Equation (4.3), k_L, a, and c_s, depend on water pressure and c_s, additionally on oxygen reduction within the bubble air. Since these effects are normally neglected, this equation is actually applicable for very shallow tanks ($H \rightarrow 0$), only and should be written for these conditions with a subscript of zero.

$$SOTR_o = \frac{k_{Lo} \cdot a_o \cdot c_{so} \cdot V}{1000} \tag{4.4}$$

This approach holds also for the standard oxygen transfer efficiency SOTE (–, %) and its specific value SOTE$_s$ (m⁻¹, %/m), based on the fraction or percent of oxygen absorbed per meter water depth, H. It differs slightly from per meter of bubble rise H_S, although generally reported in this latter way. Both SOTE parameters will be extensively applied in modeling. With an oxygen content of ambient air of 300 g/m$_N$³, the result is similar to Equation (2.51).

$$SOTE = \frac{\text{mass of O}_2 \text{ transferred}}{\text{mass of O}_2 \text{ supplied}} = \frac{k_L \cdot a \cdot c_s \cdot V}{300 \cdot G_s} = \frac{SOTR}{0.3 \cdot G_s} \tag{4.5}$$

More accurately for shallow tanks ($H \rightarrow 0$), the SOTE$_0$ is defined as follows

$$SOTE_o = \frac{k_{Lo} \cdot a_o \cdot c_{so} \cdot V}{300 \cdot G_s} = \frac{SOTR_o}{0.3 \cdot G_s} \tag{4.6}$$

Similarly, the specific oxygen transfer efficiency SOTE$_s$ can be formulated. It has to be noticed, however, that SOTE$_s$ is reduced during the bubble rise due to pressure changes and oxygen reduction in the air, as will be shown quantitatively later. Hence, the average value SOTE$_{sa}$ over the full bubble rise is calculated by dividing SOTE by the water depth H (not by the depth of diffuser submergence H_S).

$$SOTE_{sa} = \frac{\text{average mass of O}_2 \text{ transferred}}{(\text{mass of O}_2 \text{ supplied}) \cdot (\text{water depth H of aeration tank})}$$

$$= \frac{k_L \cdot a \cdot c_s \cdot V}{300 \cdot G_s \cdot H} = \frac{SOTR}{0.3 \cdot G_s \cdot H} \tag{4.7}$$

Again, this equation can be expressed for very shallow tanks ($H \to 0$).

$$SOTE_{so} = \frac{k_{Lo} \cdot a_o \cdot c_{so} \cdot V}{300 \cdot G_s \cdot H} = \frac{SOTR_o}{0.3 \cdot G_s \cdot H} \quad (4.8)$$

The process of oxygen transfer in deep tanks is modeled by expressing the parameters varying with depth (k_L, a, and c_s) as functions of their value for shallow tanks (k_{Lo}, a_0, and c_{so}). These functions are derived based on the physical laws governing the depths dependent processes as characterized in Section 4.2.1.

The pressure influence on the bubble size is modeled by the universal gas law ($P \cdot V = m \cdot R \cdot T$), to which the relative pressure π (Equation 4.2) is applied ($\pi \cdot V = m \cdot R \cdot T / P_a =$ constant). Hence, the product of the relative pressure π and the bubble volume V_B is constant, and the bubble volume V_{B_0} is reduced inversely proportional to the relative pressure π as defined in Equation 4.2.

$$V_B = \frac{V_{Bo}}{\pi} = \frac{V_{Bo}}{1 + z \cdot \left(H_S - h\right)} \quad (4.9)$$

Assuming geometrically similar deformation of the bubble by compression, the bubble diameter d_{B_0} is changed by the 1/3-power of the volume change.

$$d_B = \frac{d_{Bo}}{\pi^{(1/3)}} = \frac{d_{Bo}}{\left[1 + z \cdot \left(H_S - h\right)\right]^{(1/3)}} \quad (4.10)$$

Finally, the total area, A, and the specific area, a, are related by the second power of the diameter. This relationship leads to the dependence of the interfacial area on pressure and on depth $H_S - h$.

$$A = \frac{A_o}{\pi^{(2/3)}} = \frac{A_o}{\left[1 + z \cdot \left(H_S - h\right)\right]^{(2/3)}}$$

$$a = \frac{a_o}{\pi^{(2/3)}} = \frac{a_o}{\left[1 + z \cdot \left(H_S - h\right)\right]^{(2/3)}}$$

$$(4.11)$$

Next to the area parameters, the liquid film coefficient, k_L, is influenced by the pressure-dependent bubble diameter, d_B, as was shown by Mortarjemi and Jameson (1978) and Pasveer (1955). Their findings are plotted in Figure 4.2. Already in 1935, Higbie proposed the penetration theory for quantifying this interrelationship as given in Equation 2.21.

$$k_L = 2\sqrt{\frac{D \cdot v_B}{\pi \cdot d_B}} \quad (4.12)$$

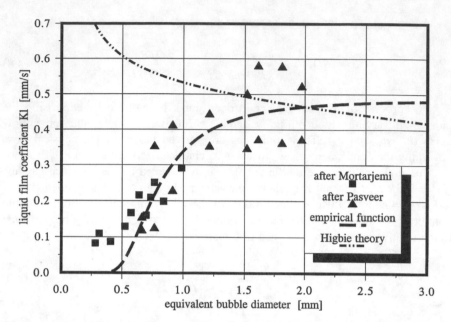

FIGURE 4.2 Liquid film coefficient as a function of the equivalent bubble diameter after Mortarjemi and Pasveer, Higbie theory and empirical function. (From Pöpel and Wagner, 1994, *Water Science and Technology*, 30, 4, 71–80. With permission of the publisher, Pergamon Press, and the copyright holders, IAWQ.)

Here, v_B (m/h) is the rise or slip velocity of the bubble with respect to water. As follows from Figure 4.2, this equation is valid only for bubbles greater than 2 mm. Generally, fine bubbles have an equivalent diameter of some 2 mm, so that the Higbie theory cannot yield correct results for compressed fine bubbles of smaller than 2 mm. By combining the results of Mortarjemi, Jameson, and Pasveer [$k_L = f(d_B)$] with Equation 4.10 [$d_B = f(d_{B_0}, H_S\text{-}h)$], an empirical relation is developed relating the liquid film coefficient to depth.

$$k_L = k_{Lo} \cdot \exp\left[-0.0013 \cdot \left(H_S - h\right)\right] \tag{4.13}$$

This function proceeds from a liquid film coefficient $k_{L_0} = 0.48$ mm/s, typical for an equivalent bubble diameter of $d_B = 3.0$ mm. Figure 4.2 shows that the k_L data are fitted very well by Equation 4.13. It should be noted, however, that a bubble diameter of 2 mm is reduced to only 1.55 mm in a 12 m deep tank. Hence, the liquid film coefficient is influenced only slightly under practical conditions.

The last parameter influenced by pressure is the oxygen saturation concentration. This effect is quantified by multiplication of c_{s_0}, the standard saturation concentration without water pressure, with the relative pressure π.

$$c_s = c_{so} \cdot \pi = c_{so} \cdot \left[1 + z \cdot \left(H_S - h\right)\right] \tag{4.14}$$

In this case, however, the parameter c_{s_0} is also affected by the oxygen transfer during bubble rise, decreasing the oxygen partial pressure in the bubble air. This influence is quantified via the standard oxygen transfer efficiency SOTE(h) during the bubble rise from $h = 0$ to $h = h$. In Figure 4.1, for instance, the SOTE-values for $h = h_1$ and $h = h_2$ are depicted for the purpose of illustration; quantities, which are yet unknown. With SOTE(h), as standard oxygen transfer efficiency from the level of bubble release until depth h, the saturation concentration is decreased correspondingly.

$$c_s = c_{so} \cdot \left[1 - SOTE(h)\right] \tag{4.15}$$

By combining Equations 4.14 and 4.15, the final expression for the saturation concentration at any height above the diffusers, h, is obtained.

$$c_s = c_{so} \cdot \left[1 + z \cdot \left(H_S - h\right)\right] \cdot \left[1 - SOTE(h)\right] \tag{4.16}$$

In summary, the influence of depth on the three basic transfer parameters, a, k_L, and c_s, can be expressed by simple mathematical functions found in Equations 4.11, 4.13, and 4.16, respectively. They include the respective values without water pressure, a_0, k_{L_0}, and c_{s_0}, and the standard oxygen transfer efficiency during bubble rise from the release level until h.

4.2.2.2 Development of the Model

To develop the transfer model for deep tanks, the pressure influenced transfer parameters, Equations 4.11, 4.13, and 4.16, are inserted into Equations 4.7 and 4.8 to define the specific standard oxygen transfer efficiency as a function of depth.

$$SOTE_s(h) = \frac{k_{Lo} \cdot a_o \cdot c_{so} \cdot V}{300 \cdot G_s \cdot H} \cdot \left[1 - SOTE(h)\right] \cdot \frac{\left[1 + z \cdot \left(H_S - h\right)\right]^{(1/3)}}{\exp\left[+0.0013 \cdot \left(H_S - h\right)\right]} \tag{4.17}$$

$$SOTE_s(h) = SOTE_{so} \cdot \left[1 - SOTE(h)\right] \cdot \frac{\left[1 + z \cdot \left(H_S - h\right)\right]^{(1/3)}}{\exp\left[+0.0013 \cdot \left(H_S - h\right)\right]} \tag{4.18}$$

$$= SOTE_{so} \cdot \left[1 - SOTE(h)\right] \cdot \Phi(h)$$

$$\Phi(h) = \frac{\left[1 + z \cdot \left(H_S - h\right)\right]^{(1/3)}}{\exp\left[+0.0013 \cdot \left(H_S - h\right)\right]} \tag{4.19}$$

Equations 4.18 and 4.19 state that the specific standard oxygen transfer efficiency $SOTE_s$ at any depth position, h, within the tank depends on

- the specific standard oxygen transfer efficiency of the aeration system in a very shallow tank, $SOTE_{so}$. This parameter is further applied as a characteristic for the effectiveness of the aeration system and is referred to as "basic specific oxygen transfer efficiency" $SOTE_{so}$;
- the standard oxygen transfer efficiency up to this position, and
- a (mathematical) function $\Phi(h)$ of this position h and the depth of submergence H_S of the diffuser system.

The differential equation for the deep tank model is derived on the basis of this approach and the transfer efficiencies depicted in Figure 4.1. The rise of the bubbles from the release level to the tank depths h_1 and h_2 yields the respective standard oxygen transfer efficiencies, $SOTE(h_1)$ and $SOTE(h_2)$. At depth h_1, the specific standard oxygen transfer efficiency amounts to $SOTE_s(h_1)$. The increase of SOTE over the reach from h_1 to h_2 is quantified by the product of the local specific standard oxygen transfer efficiency [$SOTE_s(h_1)$] and the bubble rise Δh.

$$SOTE(h_2) = SOTE(h_1) + SOTE_s(h_1) \cdot \Delta h \qquad (4.20)$$

with $\Delta h = h_2 - h_1$
Equation 4.20 can be rearranged into a difference equation.

$$SOTE_s(h) = \frac{SOTE(h_2) - SOTE(h_1)}{\Delta h} \qquad (4.21a)$$

Applying the limit of $\Delta h \to 0$ yields a differential equation.

$$SOTE_s(h) = \frac{d[SOTE(h)]}{dh}$$

$$= SOTE_{so} \cdot [1 - SOTE(h)] \cdot \frac{\left[1 + z \cdot (H_S - h)\right]^{(1/3)}}{\exp\left[+0.0013 \cdot (H_S - h)\right]} \qquad (4.21b)$$

$$= SOTE_{so} \cdot [1 - SOTE(h)] \cdot \Phi(h)$$

The last two lines of Equation 4.21 are obtained by inserting the derived Equation 4.18 for quantifying $SOTE_s(h)$ to give the final differential equation of the model. Equation 4.21 is a nonhomogeneous linear differential equation of the first order, which can only be solved numerically (e.g., by the Runge–Kutta Method) due to the structure of $\Phi(h)$. The solution can also found by means of a PC spreadsheet. The numerical integration has to proceed from $h = 0$ to $h = H_S$.

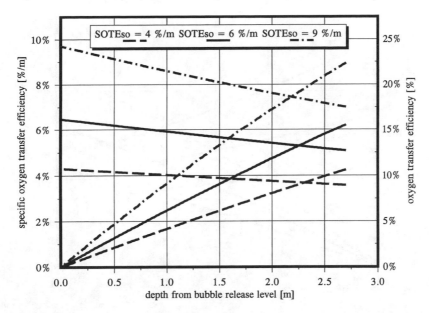

FIGURE 4.3 Specific (%/m) and standard (%) oxygen transfer efficiency in a tank of 3.00 m water depth and a depth of diffuser submergence of 2.70 m. (From Pöpel and Wagner, 1994, *Water Science and Technology*, 30, 4, 71–80. With permission of the publisher, Pergamon Press, and the copyright holders, IAWQ.)

4.2.2.3 Model Results

By integration of the model, the influence of depth on oxygen transfer can be shown for different conditions (depth H and $SOTE_{so}$) via graphical presentation. The progress of the standard oxygen transfer efficiency $SOTE(h)$, as a function of bubble rise, is the basic result of the integration. Additionally, the local specific standard oxygen transfer efficiency ($SOTE_s(h)$ in %/m) along this lift is obtained as an intermediate result. Due to interactions of pressure and oxygen uptake, as quantified by Equations 4.11, 4.13, and 4.16, $SOTE_s(h)$ has its maximum value at the bubble release level and is continuously decreasing thereafter. The standard oxygen transfer efficiency $SOTE(h)$, however, is increased correspondingly. These changes exhibit an almost linear relation to the bubble rise in shallow tanks (where the slight influence of pressure prevails). A more curved dependency exists in deeper tanks, where, along with the total pressure, the decrease in oxygen partial pressure of the bubbles due to the oxygen uptake becomes important.

This dependency is illustrated by the following examples for three different tank depths (3.00, 6.00, and 12.00 m with a bubble release level of 0.30 m above the tank bottom). These depths are combined with three different aeration systems, which are identified by their basic specific oxygen transfer efficiency $SOTE_{so}$ (4, 6, and 9 %/m). For each tank depth, the specific oxygen transfer efficiency $SOTE_s(h)$ and the standard oxygen transfer efficiency $SOTE(h)$ are depicted as a function of the bubble rise from release ($h = 0$) until water level ($h = H_S = H - 0.3$ m) in Figures 4.3 to 4.5.

As can be read from the figures, the function lines are almost straight in Figure 4.3 ($H = 3.00$ m) and become increasingly curved when going to Figures 4.4

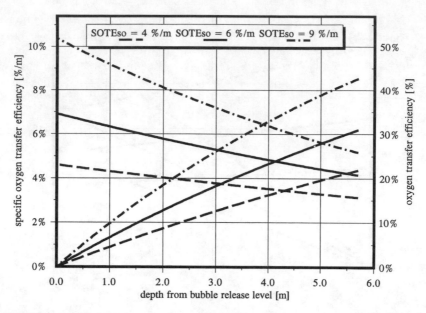

FIGURE 4.4 Specific (%/m) and standard (%) oxygen transfer efficiency in a tank of 6.00 m water depth and a depth of diffuser submergence of 5.70 m. (From Pöpel and Wagner, 1994, *Water Science and Technology*, 30, 4, 71–80. With permission of the publisher, Pergamon Press, and the copyright holders, IAWQ.)

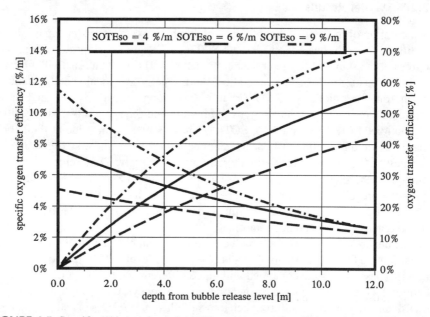

FIGURE 4.5 Specific (%/m) and standard (%) oxygen transfer efficiency in a tank of 12.00 m water depth and a depth of diffuser submergence of 11.70 m. (From Pöpel and Wagner, 1994, *Water Science and Technology*, 30, 4, 71–80. With permission of the publisher, Pergamon Press, and the copyright holders, IAWQ.)

($H = 6.00$ m) and 4.5 ($H = 12.00$ m). In this sequence, the standard oxygen transfer efficiency of the three aeration systems is strongly increasing from shallow (11, 16, and 22 percent) to greatest depth (41, 55, and 71 percent), and the local specific oxygen transfer efficiency $SOTE_s(h)$ is reduced due to oxygen depletion in the air bubble. In the deepest tank (Figure 4.5), the specific oxygen transfer efficiencies of all three aeration systems are attenuated from 5.1 to 11.4 %/m at bubble release to almost the same value, 2.3 to 2.7 %/m, near the water level.

The above information on $SOTE_s(h)$ and its characteristics illustrates very clearly the changes of this parameter, as well as oxygen transfer, during bubble rise in tanks of different depths. For practical application, however, the average value over the full tank depth H, $SOTE_{sa}$, as defined by Equation (4.7), is of more importance. It can be calculated from the obtained values for $SOTE(h = H_S) = SOTE$.

$$SOTE_{sa} = \frac{SOTE(h = H_S)}{H} = \frac{SOTE}{H} \qquad (4.22)$$

In the 12.00 m deep tank, for instance, $SOTE_{sa}$ is calculated from the above SOTE values (41, 55 and 71 percent) of the three different aeration system as 3.4, 4.6, and 5.9 %/m. This figure is much lower than the three basic specific oxygen transfer efficiencies of 4.0, 6.0 and 9.0 %/m, mainly due to oxygen depletion in the air during bubble rise. In generalizing this information, the SOTE and the $SOTE_{sa}$ values for tanks from $H = 0.00$ m to $H = 15.00$ m depth are calculated and plotted versus tank depth H in Figure 4.6. Six different aeration systems with basic specific oxygen transfer efficiencies from $SOTE_{so} = 4$ %/m to 9 %/m are used. The bubble release level is assumed 0.30 m above the tank bottom, important only for the specific oxygen transfer efficiency $SOTE_{sa}$.

The characteristics of the $SOTE_{sa}$ lines near the bubble release level differ considerably from the local $SOTE_s(h)$ lines in Figures 4.3 to 4.5 for the following reason: in a tank with a depth equal to the bubble release level, no oxygen can be transferred, and hence, $SOTE(h = 0) = 0$ and also $SOTE_{sa} = SOTE/H = 0$ (Equation 4.22). When increasing the tank depth, the bubble rise (H_S) is still very small as is the SOTE. This little quantity is divided by $H > H_S$, leading to an insignificant average specific oxygen transfer efficiency $SOTE_{sa}$. As can be seen from Figure 4.6, $SOTE_{sa}$ reaches maximum values at tank depths close to $H = 2.70$ m (system with $SOTE_{so} = 9$ %/m) until $H = 5.75$ m (system with $SOTE_{so} = 4$ %/m). Both depicted functions, $SOTE_{sa} = f(H)$ and $SOTE = f(h)$, will be applied later for designing aeration systems in deeper tanks.

4.2.3 MODEL VERIFICATION

The derived model is verified in two ways. First, 98 published performance tests in aeration tanks of different depth varying from 3.40 m to 12.00 m (Pöpel and Wagner, 1994) are evaluated, and the results verify the model qualitatively. Secondly, the results of an extensive full-scale experiment with water depths from $H = 2.50$ m to $H = 12.50$ are applied for a more rigorous certification of the model.

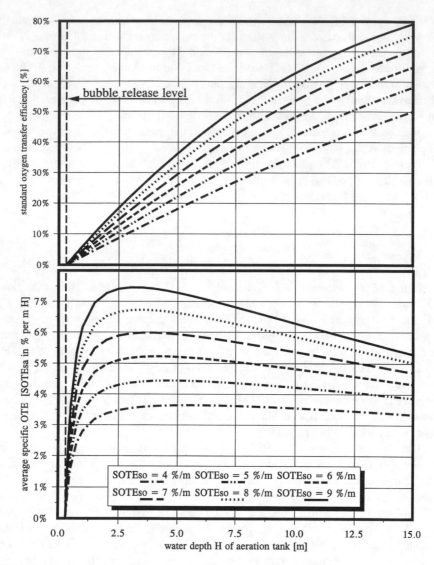

FIGURE 4.6 Standard oxygen transfer efficiency SOTE (%) and average specific oxygen transfer efficiency $SOTE_{sa}$ (%/m) as a function of water depth and of six aeration systems defined by their basic $SOTE_{so}$ (%/m). (From Pöpel and Wagner, 1994, *Water Science and Technology*, 30, 4, 71–80. With permission of the publisher, Pergamon Press, and the copyright holders, IAWQ.)

4.2.3.1 Qualitative Verification

The oxygen transfer results from 98 published performance tests are presented in two ways for comparison with the model. First, the data are depicted for six depth classes as a function of the specific airflow rate (m_N^3 of air per hour per m^3 of aerated water volume) in two figures (Figure 4.7 and 4.8). In Figure 4.7, the standard oxygen transfer efficiency SOTE (%) is plotted on the ordinate, whereas in Figure 4.8, the average specific oxygen transfer efficiency $SOTE_{sa}$ (%/m) is plotted. Secondly, the measured

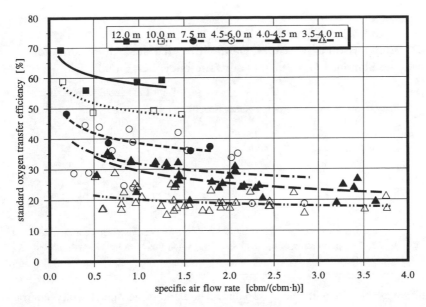

FIGURE 4.7 Standard oxygen transfer efficiency [%] as a function of the specific airflow rate [cbm/(cbm·h)] and of the water depth H [m]. (From Pöpel and Wagner, 1994, *Water Science and Technology*, 30, 4, 71–80. With permission of the publisher, Pergamon Press, and the copyright holders, IAWQ.)

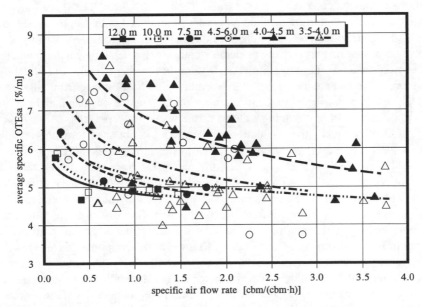

FIGURE 4.8 Average specific oxygen transfer efficiency [%/m] as a function of specific airflow rate [cbm/(cbm·h)] and of the water depth H [m]. (From Pöpel and Wagner, 1994, *Water Science and Technology*, 30, 4, 71–80. With permission of the publisher, Pergamon Press, and the copyright holders, IAWQ.)

TABLE 4.2
Comparison of Measured Data with Calculated Model Data for the Standard Oxygen Transfer Efficiency, SOTE (%)

Tank Depth Range	Data Range Measured	Data Calculated with SOTE$_{so}$ =		
		4 %/m	6 %/m	9 %/m
3.4–4.0	15–29	15	21	30
4.0–4.5	19–35	16	24	33
4.5–6.0	19–45	20	28	40
7.5	36–48	28	39	52
10.0	48–59	36	48	63
12.0	56–69	42	55	70

Reprinted from Pöpel and Wagner, 1994, *Water Science and Technology*, 30, 4, 71–80. With permission of the publisher, Pergamon Press, and the copyright holders, IAWQ.

data are compared with the model calculated for basic specific oxygen transfer efficiencies, SOTE$_{so}$, from 4 %/m to 9 %/m in two Tables (4.2 and 4.3), referring to the SOTE (%) and the SOTE$_{sa}$ (%/m), respectively.

With respect to SOTE, the significant increase of this parameter with increasing tank depth can be seen in Figure 4.7. A quantitative comparison is possible via Table 4.2 in which the measured SOTE data range for the six depth classes is given together with the model data calculated for 4 %/m, 6 %/m, and 9 %/m. The shaded areas of Table 4.2 indicate that the data variation is very pronounced in the depth ranges up to 6 m. This is due to the great differences in diffuser densities (diffusers per m²) of the investigated aeration tanks having moderate depths. In this depth range, the actual data are covered by an SOTE-range from 4 to 9 %/m. In the deeper tanks, the actual data are more stable and are theoretically represented by an SOTE-range from only 6 to 9 %/m. This can be attributed to the meagerness of data, on the one hand, and possibly also to the more stable streaming patterns of the water in deeper tanks.

An identical qualitative evaluation of the model is obtained from the test data with respect to the average specific oxygen transfer efficiencies, SOTE$_{sa}$ (%/m), in Figure 4.8 and Table 4.3. In Figure 4.8, the regression lines show lower values as the depth H increases, as predicted by the model in Figure 4.6 (bottom). This model does not hold for the lowest depth range 3.5 to 4.0 m, for which the regression line lies much lower than expected. Reasons for this behavior at very low depths could be more unstable streaming patterns in very shallow tanks or greater construction height of the air diffusion system leading to lower diffuser submergence. This data behaves as predicted for tanks below 2.5 m water depth by the model (see Figure 4.6, bottom, near left ordinate). This behavior is also shown by the lowest values of the data range in Figure 4.3, where the measured maximum values show a gradual decrease with increasing depth class as predicted by the model.

TABLE 4.3
Comparison of Measured Data with Calculated Model Data for the Average Specific Oxygen Transfer Efficiency SOTE$_{sa}$ (%/m)

		Data Calculated with SOTE$_{so}$ =		
Tank Depth Range	Data Range Measured	4 %/m	6 %/m	9 %/m
3.4–4.0	4.0–8.2	3.9	5.6	7.9
4.0–4.5	4.5–7.8	3.9	5.5	7.8
4.5–6.0	3.7–7.5	3.8	5.4	7.5
7.5	4.8–6.4	3.7	5.1	6.9
10.0	4.8–5.9	3.6	4.8	6.3
12.0	4.7–5.8	3.5	4.6	5.9

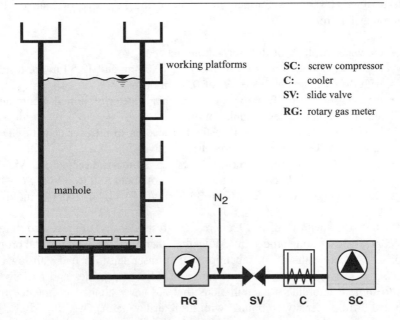

working platforms

SC: screw compressor
C: cooler
SV: slide valve
RG: rotary gas meter

manhole

N$_2$

RG   SV   C   SC

FIGURE 4.9 Schematic of the deep tank pilot plant.

The comparison of the shaded model data in Table 4.3 with the measured data range reveals the same information as concluded above for the SOTE.

4.2.3.2   Full-Scale Experimental Verification in Clean Water

A rigid quantitative verification of the deep tank model in clean water is carried out via a full-scale pilot program. The main parts of the pilot plant are the aeration tank, a screw compressor, the air piping system and the distribution frame with membrane disc diffusers (see Figure 4.9). Main element is the "deep tank," a stainless steel cylinder of 4.25 m diameter (area 14.2 m²) and a height of 13 m (volume 184.4 m³) with five working platforms at different elevations. Diffuser mounting is performed via a manhole near the tank bottom.

The water level is controlled by means of pneumatic valves for inlet and outlet and a pressure gauge at the tank bottom, ensuring that the preset water depth is also maintained at continuous through-flow of water or wastewater. The air supply is controlled by a screw compressor (Aerzener, type VM 137 D) into the distribution frame at two points. The diffuser frame allows different diffuser arrangements and densities to be investigated. The construction height of the diffuser system, including the necessary piping, amounts to 0.32 m. The disc diffusers are built from polypropylene and equipped with slotted membranes from the Gummi Jäger Company (Hanover). Altogether, four arrangements are investigated (9, 19, 36 and 55 discs), leading to diffuser densities of 4.5, 9.5, 17.9, and 27.4 percent respectively. Deoxygenation was performed with pure nitrogen gas during the clean water tests.

Experimental variables for determination of the influence of tank depth on oxygen transfer are

- the water depth H or diffuser submergence H_S;
  depths of $H = 2.50$ m, 5.00 m, 7.50 m, 10.00 m, and 12.50 m are tested
  with diffuser submergences H_S of 0.32 m or less.
- the diffuser density DD, expressed as square meter of slotted membrane
  area per square meter of tank bottom:
  9, 19, 36 and 55 discs are investigated leading to diffuser densities DD
  of 4.5, 9.5, 17.9, and 27.4 percent respectively.
- the airflow rate G_s is varied over three steps so that the second rate yields
  a volumetric standard oxygen transfer rate of about $SOTR_V = 100$ g/(m^{3}h)
  O_2, leading to airflow rates G_s of 35.5 m$_N^3$/h, 71 m$_N^3$/h, and 142 m$_N^3$/h.

The test series with 19 discs (9.5 percent diffuser density) are repeated to reveal the accuracy of the testing procedure. Altogether, therefore, the experimental program comprises 5 water depths, $4 + 1$ (repetition) = 5 diffuser densities, and 3 airflow rates, i.e., $5 \cdot 5 \cdot 3 = 75$ single tests. The wide range of diffuser densities and airflow rates leads to some extraordinary combinations that are never applied in practice (great depth and diffuser density combined with high airflow rate). They would also lead to operational problems in practice as well as in testing (great diffuser density combined with low airflow rates and consequently very low diffuser loading, especially at low water depth). The experimental results of these combinations were not included in the data evaluation. Altogether, 18 runs are not included in the evaluation due to this atypical behavior, leaving $75 - 18 = 57$ data sets for final evaluation.

Clean water testing is performed according to the nonsteady state method after deoxygenation with pure nitrogen gas N_2, according to the German standard (ATV, 1996) (see also Figure 4.9), leading to an oxygen content of 0.3 mg/l only. The increase of the oxygen content is measured on-line with seven probes (very accurate "Orbisphere probes", Giessen, Germany), arranged at different heights and positions with respect to the reactor cross section.

In addition to the oxygen concentration, a number of other parameters are determined: exact water depth at the start and end of each test; water temperature;

conductivity and pH of the water; applied amount of nitrogen; temperature and humidity of the applied air; airflow rate; temperature of the compressed air in the piping system ahead of and behind the rotary gas meter; pressure difference at the slide valve; pressure behind the slide valve and within the diffuser frame; and atmospheric pressure.

The data of each probe are evaluated with a computer program developed according to the U.S. standard (ASCE, 1991) with the aeration coefficient $k_L a_T$ and the saturation concentration $c_{s,T}$ as a result. An optimum fit to the data is accomplished by variation of the starting point and the number of data evaluated. Results with more than five percent deviation from the average of all probes are discarded (ATV, 1979). Finally, the aeration coefficients $k_L a$ and the saturation concentration are reduced to (former German) standard conditions ($T = 10°C$ and $P_a = 101.325$ kPa). The present standard (20°C) yields values some two percent higher ($OTR_{20}/OTR_{10} = \theta^{10}·c_{s,20}/c_{s,10} = 1.024^{10}·9.09/11.29 = 1.0206$). From both parameters, $k_L a$ and c_s, the standard oxygen transfer efficiency SOTE and the average specific oxygen transfer efficiency $SOTE_{sa}$, are calculated by means of Equations 4.5 and 4.7 respectively.

If the obtained $SOTE_{sa}$ values are converted to the "basic specific oxygen transfer efficiency" ($SOTE_{so}$-values), the tested aeration system would have at a diffuser submergence of zero. This conversion is facilitated by the computer program, "O₂-deep", developed on the basis of the derived model (Pöpel et al., 1997), as is explained in more detail in Section 4.2.4. Whereas the first set of data ($SOTE_{sa}$) is strongly influenced by water depth, the depth-corrected data ($SOTE_{so}$) cannot show any depth influence, if the model by which the data were corrected, precisely allows for all depth influences on SOTR and SOTE. A check on this property will be the final validation of the model. The remaining effects (diffuser density and airflow rate) are not affected by the depth correction.

A first impression of the results is given in Table 4.4, by presentation of the average specific oxygen transfer efficiency $SOTE_{sa}$ and the depth corrected basic specific oxygen transfer efficiency ($SOTE_{so}$), averaged over the different parameters tested, the diffuser density DD, the water depth H, and the airflow rate G_s. From Table 4.4, it is evident that both oxygen transfer efficiencies increase with increasing diffuser density. With respect to water depth, the generally experienced decrease of the average specific oxygen transfer efficiency ($SOTE_{sa}$) at depths greater than 4 to 5 m (compare with Figure 4.6; lower part) can be seen. In contrast, the depth corrected $SOTE_{so}$ values vary irregularly between 5.7 and 6.0 %/m, exhibiting a lower influence of depth than $SOTE_{sa}$. As usual, the highest specific oxygen transfer efficiency is obtained at the lowest airflow rate. This fact holds for the raw and for the depth corrected data.

A quantitative analysis of both specific oxygen transfer efficiencies ($SOTE_{sa}$ and $SOTE_{so}$) is performed by linear regression methods. The diffuser submergence H_S (m), the diffuser density DD (m²/m²), and the airflow rate G_s (m$_N$³/h) are independent variables. The dependent variable ($SOTE_{sa}$) is very difficult to treat with linear regression; hence, not $SOTE_{sa} = SOTE/H$ is applied but rather $SOTE/H_S$, which decreases almost linearly with depth. Due to the slight increase of the specific oxygen transfer efficiencies at high diffuser densities (see Table 4.4), the natural logarithm

TABLE 4.4
Average Values of the Average Specific Oxygen Transfer Efficiency
(SOTE$_{sa}$) and the Basic Specific Oxygen Transfer Efficiency (SOTE$_{so}$) at
Different Test Conditions (%/m)

Diffuser Density (%)			Water Depth H (m)			Airflow Rate G_s (m$_N$³/h)		
value	SOTE$_{sa}$	SOTE$_{so}$	value	SOTE$_{sa}$	SOTE$_{so}$	value	SOTE$_{sa}$	SOTE$_{so}$
4.5	4.22	4.94	2.5	4.81	5.65	35.5	4.96	6.03
9.5	4.98	6.05	5.0	5.18	5.98	71.0	4.88	5.94
17.9	5.24	6.53	7.5	4.99	5.93	142.0	4.68	5.62
			10.0	4.75	5.90			
			12.5	4.46	5.79			

of DD (ln DD) is applied as the variable for regression. The analysis results in the following equations:

original data as calculated from measurements

$$\frac{SOTE}{H_S} = 8.24 \cdot 10^{-2} - 1.171 \cdot 10^{-3} \cdot H_S + 8.28 \cdot 10^{-3} \cdot \ln(DD) - 2.77 \cdot 10^{-5} \cdot G_s \quad (4.23a)$$

correlation coefficient $r = 0.922$
standard deviation $s = 0.0024 \text{ m}^{-1} = 0.24 \text{ %/m}$

From Equation 4.23a, the average specific oxygen transfer efficiency can be calculated.

$$SOTE_{sa} = \frac{H_S}{H} \cdot \left(8.24 \cdot 10^{-2} - 1.171 \cdot 10^{-3} \cdot H_S + 8.28 \cdot 10^{-3} \cdot \ln(DD) - 2.77 \cdot 10^{-5} \cdot G_s\right)$$

$$(4.23b)$$

This equation has the same correlation coefficient, however, with a slightly smaller standard deviation ($H_S/H < 1$), and hence, a slightly higher accuracy. A graphical representation of the results is given in Figure 4.10. In the upper part, the influence of water depth on SOTE$_{sa}$ at different diffuser densities is plotted using the average airflow rate of the quoted values, 82.8 m$_N$³/h. The density of 27.4 percent has not been evaluated but is plotted nevertheless to show that the greatest influence of diffuser density occurs at low densities. The behavior of these lines is very similar to the model calculations depicted in Figure 4.6.

The bottom part of Figure 4.10 shows the same depth influence, while combined with the airflow rate, averaged over all applied diffuser densities, 10.6 percent. It is evident that the influence of the airflow rate G_s on the average specific oxygen transfer efficiency and hence on the standard oxygen transfer efficiency is small compared with the diffuser density effect.

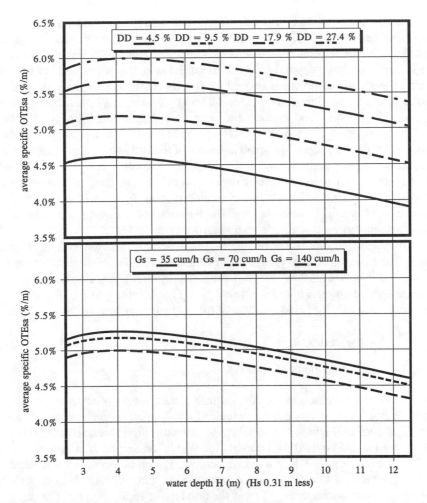

FIGURE 4.10 Influence on the average specific oxygen transfer efficiency of water depth H combined with diffuser density (top) and combined with airflow rate (bottom) according to verification data.

The final validation of the model is performed by analyzing the depth-corrected data $SOTE_{so}$ for any depth influences. If these are removed correctly from the data by the performed corrections with the program O_2-deep, then the $SOTE_{so}$-data should be altogether independent of depth. The regression with all parameters of Equation 4.23 showed no statistically significant influence of depth. Hence, only diffuser density DD and airflow rate are independent regression parameters.

$$SOTE_{so} = 9.00 \cdot 10^{-2} + 1.164 \cdot 10^{-2} \cdot \ln(DD) - 3.69 \cdot 10^{-5} \cdot G_s \qquad (4.24)$$

correlation coefficient $r = 0.904$
standard deviation $s = 0.0028 \text{ m}^{-1} = 0.28 \text{ %/m}$

The depth corrected $SOTE_{so}$ values (Equation 4.24) show good agreement with measured data (high correlation coefficient, low standard deviation) and no significant depth influence. This agreement shows that the model sufficiently corrects for the influence of water depth on oxygen transfer. For practical purposes, it is applicable to deep tanks using fine pore air diffusion with sufficient accuracy as indicated by the standard deviations of Equations 4.23 and 4.24, ranging from 0.2 to 0.3 %/m.

To visualize the trend of the depth corrected data $SOTE_{so}$, Equation 4.24 is depicted in Figure 4.11 by plotting $SOTE_{so}$ versus the diffuser density for the three applied airflow rates. Again, the small influence of the airflow rate is evident, whereas the diffuser density (extrapolated to 27.4 percent) controls $SOTE_{so}$ very effectively. This effect is similar to the results derived from 98 published performance tests (Pöpel and Wagner, 1989), which are summarized in Figure 4.12 by plotting the relative SOTR versus diffuser density. The intense data scattering is caused by the additional influences of water depth and airflow rate on SOTR.

Altogether, the model can be applied for designing aeration systems in deep tanks. The basic specific oxygen transfer efficiency $SOTE_{so}$ of an aeration system is influenced by the airflow rate and primarily by the diffuser density, as is the average specific oxygen transfer efficiency $SOTE_{sa}$. Contrary to $SOTE_{sa}$, however, the basic value $SOTE_{so}$ is independent of diffuser submergence and water depth.

4.2.4 Model Applications

The model can be applied in two ways:

(1) The main influences (depth, diffuser density, airflow rate) on oxygen transfer parameters can be visualized and applied for a rough parameter estimation (Figures 4.10 to 4.12). Additionally, this more qualitative information can be used for interpolation within the second application.

(2) The SOTR or SOTE of a known aeration system of a certain water depth can be used to calculate the corresponding parameters of this system at any other water depth. Whereas the first type of application must be based on sound engineering judgment of the applicant, the second use is elucidated in more detail as follows.

This main application of the model is to calculate oxygen transfer data of fine bubble air diffusion systems (to be) installed in deep tanks by applying the experience gained from similar aeration systems in tanks of conventional or lower depth. The similarity can be defined by quantifiable parameters, like airflow rate and diffuser density, and by less quantifiable parameters, like arrangement of the diffusers and hydraulic streaming patterns, both vertical and horizontal, within the tank. A diffuser layout of the full floor grid type with almost equal diffuser density will produce similar streaming patterns in the above sense and allow the model to be applied to different airflow rates.

For a model application of reasonable accuracy, Figure 4.6 can be applied. High accuracy is obtained when using the developed computer program, O_2-deep (Pöpel

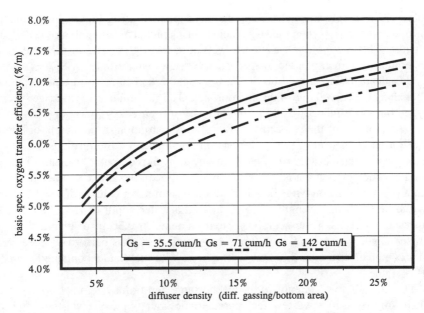

FIGURE 4.11 Basic specific oxygen transfer efficiency $SOTE_{so}$ as a function of diffuser density (%) and airflow rate (cum/h at STP).

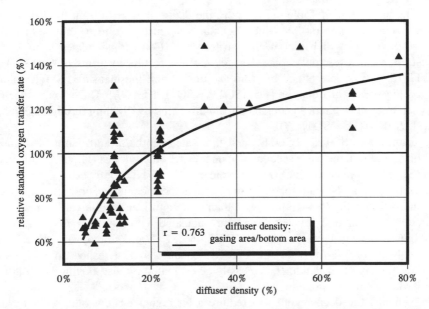

FIGURE 4.12 Influence of diffuser density on the standard oxygen transfer rate expressed as percentage of SOTR at 20% density. (Data from Pöpel and Wagner, 1989.)

et al., 1997). The rationale of the approach is explained using Figure 4.6. In the top figure, the standard oxygen transfer efficiency is depicted as a function of tank depth H (and height of bubble release level: 0.30 m in this figure) and of the efficacy of the aeration system expressed by its basic oxygen transfer efficiency $SOTE_{so}$. When the tank depth is increased, the SOTE is not increased linearly to tank depth but rather along the curved line of the appropriate $SOTE_{so}$. Similarly, the average specific oxygen transfer efficiency $SOTE_{sa}$ (bottom part of Figure 4.6) follows the declining line ($H > 3.50$ m) of the respective $SOTE_{so}$ line. A variation of the height of bubble release level of 0.30 m in Figure 4.6 has little influence on the result, especially at greater depths, but can accurately be taken care of by the computer program, O_2-deep.

The model application is illustrated by the following example. An aeration tank with a full floor coverage fine bubble aeration system has a volume of $V = 1,725$ m^3, a width of 15.00 m, a length of 25.00 m, and a water depth of $H = 4.60$ m. The construction height of the aeration system amounts to 0.30 m to give a depth of diffuser submergence of $H_S = 4.30$ m. The manufacturer has performed three clean water compliance tests at different airflow rates with the results contained in upper part of Table 4.5.

The manufacturer intends to install the same aeration system at another loca- tion having the same wastewater characteristics but twice the wastewater flow. Because of very limited space, the same tank area has to be applied with twice the tank depth, i.e., with $H = 9.20$ m. The depth of diffuser submergence amounts to $H_S = 8.90$ m. Because of the double plant loading, the required SOTR is twice that of the earlier performed tests, viz. 100, 250, and 460 kg/h. The required airflow rates have to be estimated.

The upper part of Table 4.5 refers to the depth of $H = 4.60$ m; the lower part to $H = 9.20$ m. The first line (line 1) contains the airflow rates G_s applied for the three tests, from which the specific airflow rate (G_s/V) is calculated (line 2) for illustration purposes, only. Line 3 states the test results in terms of SOTR. The SOTE (line 4) is determined by from G_s (line 1) and the measured SOTR values (line 3) by means of Equation (4.5) [SOTE = SOTR/($0.3 \cdot G_s$)]. The average specific oxygen transfer efficiency is obtained from this value by dividing through the water depth H ($SOTE_{sa}$ = SOTE/H).

From either SOTE or $SOTE_{sa}$ and the water depth H (and depth of diffuser submergence H_S), the basic specific oxygen transfer efficiency $SOTE_{so}$ is found either via Figure 4.6 (upper part for SOTE, bottom part for $SOTE_{sa}$) or by using the program O_2-deep. The results, valid for any water depth at the specified airflow rate, are given in line 6. From Figure 4.6, not more than two significant digits can be read; the stated results (three significant digits) are calculated with the program.

In test 1, for instance, a value of $SOTE_{so} = 7.87$ %/m is found, very close to the dotted lines for 8 %/m in Figure 4.6. The conditions with respect to SOTE and $SOTE_{sa}$ for any other depth, H, can easily be estimated by just moving along a line somewhat below the dotted one.

Although the deeper tank will require a bit higher airflow rate, reducing the $SOTE_{so}$ values insignificantly, the above results are transferred to a water depth of $H = 9.20$ m (lines 7 to 10) as a first estimate. In lines 7 and 8, the SOTE and the $SOTE_{sa}$ are estimated applying Figure 4.6 or the model as indicated. Then, the

TABLE 4.5
Example Data of a Full Floor Coverage Fine Bubble Aeration System of $H = 4.60$ m and of $H = 9.20$ m Water Depth

Line	Parameter	Unit	Test 1	Test 2	Test 3
	Conditions at $H = 4.60$ m water depth				
1	Airflow rate G_s	m_N^3/h	550	1,500	3,000
2	Specific airflow rate	$m_N^3/m^3/h$	0.32	0.87	1.74
3	SOTR	kg/h	50	125	230
4	SOTE	%	30.3	27.8	25.6
5	$SOTE_{sa}$	%/m	6.59	6.04	5.56
6	$SOTE_{so}$	%/m	7.87	7.09	6.44
	Conditions at $H = 9.20$ m water depth and at same airflow rate				
7	SOTE	%	54.5	50.8	47.5
8	$SOTE_{sa}$	%/m	5.92	5.52	5.16
9	SOTR (definition)	kg/h	100	250	460
10	required airflow rate	m_N^3/h	612	1,640	3,228
	Conditions at higher airflow rate				
11	Additional ΔG_s	m_N^3/h	62	140	228
12	Reduction of $SOTE_{so}$	%/m	0.05	0.08	0.10
13	Adjusted $SOTE_{so}$	%/m	7.82	7.01	6.34
14	Adjusted SOTE	%	54.3	50.4	47.0
15	Adjusted $SOTE_{sa}$	%/m	5.90	5.48	5.10
16	Required airflow rate	m_N^3/h	614	1,653	3,262
17	Add to first estimate	%	0.33	0.79	1.05
	Comparison of tank depth results				
18	Ratio of SOTR	—	2	2	2
19	Ratio of G_s	—	1.12	1.10	1.09

required airflow rate under these conditions (line 10) is calculated from the new standard oxygen transfer rates SOTR (line 9) and the obtained SOTE values (line 7), again by using Equation 4.5 [SOTR = SOTE·0.3·G_s]. The new airflow rates surpass the rates from line 1 by only small amounts (line 11), reducing the $SOTE_{so}$ values to a certain extent (compare Equation 4.24). This extent can be estimated from the test differences in line 1 (G_s) and line 6 ($SOTE_{so}$) as follows.

$$\Delta SOTE_{so} = \left(SOTE_{so,2} - SOTE_{so,1}\right) \cdot \frac{\Delta G_s}{G_{s,2} - G_{s,1}}$$

$$= (7.09 - 7.87) \cdot \frac{62}{1,500 - 550} = -0.051\%/m$$

The same approach is applied to calculate the $SOTE_{so}$ reduction for test 2 and test 3 conditions. The results are summarized in line 12. The adjusted $SOTE_{so}$ is obtained by subtracting $\Delta SOTE_{so}$ from the original value (line 13). The adjusted standard oxygen transfer efficiencies SOTE as well as $SOTE_{sa}$ are given in lines 14 and 15 after applying Figure 4.6 (upper part for SOTE, bottom part for $SOTE_{sa}$) or by using the program O_2-deep. The improved estimates of the airflow rate (line 16; computed by Equation 4.5) leading to a small reduction of $SOTE_{sa}$ differ from the first "rough" estimate by only 0.3 to 1.1 percent, although the depth has been doubled. The tiny improvement (lines 11 to 17) therefore seems unnecessary.

A comparison of the results for both tanks is performed in lines 18 and 19. The ratio of the SOTR values equals two (by example definition), whereas the required airflow rate ratios increase by only nine to 12 percent.

4.3 AERATION EFFICIENCY IN DEEP TANKS

4.3.1 THE AIR SUPPLY SYSTEM AND ITS COMPONENTS

4.3.1.1 Introduction

By the air supply system, atmospheric air or high purity oxygen is conveyed from their respective sources into the biological treatment units. The main components of the air supply system are discussed in this section, limited to the supply of atmospheric air. Special features required for the supply of high purity oxygen are dealt with in Chapter 6. Also, some of the important constituents of the air supply system have already been covered in Chapter 3, and reference is made to these sections to avoid repetition.

The main components of an air supply system for an activated sludge plant (Figure 4.13) and for artificially aerated attached growth reactors used for

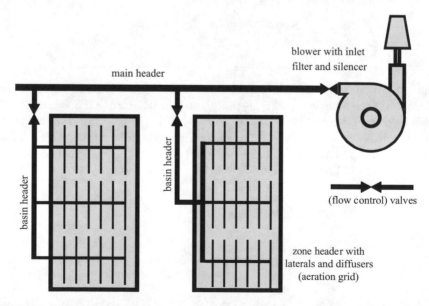

FIGURE 4.13 Schematic of the air supply system.

BOD_5-removal and/or nitrification (packed-bed reactors; aerated biological filters; and fluidized-bed reactors) is summarized as follows:

- air inlet with air filter, frequently combined with noise control (silencer);
- blower or compressor;
- outlet air filter and outlet noise control (silencer), if required;
- air supply piping, consisting of the following elements in the direction of air flow;
  - main header conveying the air to the basin headers each serving an activated sludge tank;
  - basin header transporting the air along a tank to several droplegs (or drop pipes) each serving a zone header, to which the laterals with the diffusers (grid) are connected, aerating one zone of an activated sludge tank;
  - alternately, the basin header directly feeding into a drop pipe for one activated sludge tank and this header continuing along the tank bottom serving the zone laterals with diffusers;
- diffusers, transferring air and oxygen into the activated sludge mixed liquor;
- necessary appurtenances like
  - isolation or shut off valves for disconnecting part(s) of the tanks;
  - airflow control valves;
  - airflow meters;
  - other measurement and control devices;
- control system (hardware and software) for automated control (or manual) of DO and aeration intensity (airflow rate) to all parts of the activated sludge plant.

The appropriate design and layout of the complete air supply system can ensure proper functioning of the biological conversions and reduce the energy expenditure to an economic minimum. The optimization of energy consumption is of especially great importance since the air supply requires roughly 70 percent of the total energy necessary for biological wastewater treatment. Optimization can essentially be achieved in four ways including: (a) by minimizing the frictional loss of head in the total air supply system; (b) by applying efficient fine pore diffusers; (c) by selecting a blower or compressor of high efficiency matching the operational requirements of airflow rate and back pressure; and (d) by implementing an optimum airflow and DO control system by adjusting the airflow rates as close as possible to the required variations with respect to space and time.

In the following section, the elements of the air supply system are described and discussed in more detail following the classification listed above. Because of their vital importance, blowers and compressors are not included below but handled separately in Section 4.3.2. Reference is made to more complete coverage in other sections.

4.3.1.2 Air Filtration and Noise Control

Fine particulate matter has to be removed from the atmospheric air prior to compression to protect blowers and compressors from abrasion and prevent airside clogging of fine pore diffusers. For blower protection, a 95 percent removal of particles of 10 μm or larger is sufficient. This requirement is also adequate for membrane diffusers, whereas former requirements, formulated mainly for ceramic diffusers, state a 90 percent removal for particles of 1 μm or larger (EPA, 1989; WPCF, 1988; see Section 3.5.5.1).

In wastewater aeration, fibrous media filters, renewable media filters, and electrostatic precipitators are in use. The most common type, fibrous media filters, can further be differentiated into the dry-type filter, built up of random mats of fibers. Size and type of the fiber material determines the degree of particulate removal. The second subgroup, viscous impingement filters, use high porosity filter media covered with viscous matter similar to oil. The viscous substance traps the dust particles impinged onto the filter media.

Renewable media filters require little space, are easy to maintain, but relatively costly to replace which limits their extensive application. The use of electrostatic precipitators is generally restricted to smoky areas.

Filtration is frequently combined with control of noise originating from mainly the blowers and compressor (motors, impellers of a dynamic compressor or drivers of a positive displacement blower, or PD-blower). Housing or sound insulated covering of the units is quite common, both containing wide openings with blind slats and (pre)filters for air intake. PD-blowers are generally equipped with silencers at the air intake and outlet side (ATV, 1997).

4.3.1.3 Air Supply Piping and Diffusers

The air supply piping system conveys the compressed air from the outlet of the blower or compressor to the diffusers and has to evenly distribute the air over all tanks (sections) in operation. Its main elements have been summarized within the introduction (4.3.1.1) and in Figure 4.13. Airflow and pressure meters as well as control valves are installed within this system to ensure the appropriate air distribution.

The important questions on pipe materials applied to prevent corrosion by moisture condensation on the inside and sunlight on the outside, as in the case of PVC, have been extensively discussed in Section 3.7.1.13. Additional information can be found in WPCF (1988) and EPA (1989).

With respect to sizing the air distribution pipes, it is important that the loss of head within the piping systems is small compared with the resistance of the diffusers to safeguard even air distribution. To this end, the total piping loss of head after the last split should not exceed 10 percent of the diffusers resistance (EPA, 1989). Another approach recommends airflow velocities in the range of 10 to 20 m/s at maximum airflow rates to solve this problem (ATV, 1997).

The last components of the air distribution system are the diffusers, together with blowers and compressors, the most important devices of the aeration system. Their importance is due to the energy demand caused by their relatively high resistance, by the requirement of producing small bubbles evenly over the full diffuser surface area, and by the possible problems of inside and outside fouling. Accordingly, all pertinent issues have extensively been covered already in Chapter 3.

4.3.2   BLOWERS AND COMPRESSORS AND THEIR ENERGY REQUIREMENTS

4.3.2.1   Introduction

The terms *blower* and *compressor* have never been rigidly defined, but normally, a blower is a device producing outlet air pressures of less than 100 kPa or 10 m WC. Compressors are able to generate air pressures (far) in excess of 100 kPa. Both types obviously "compress" the air. It is clear, therefore, from the respective pressure ranges, that "blowers" can be applied to the majority of activated sludge tanks with a water depth of 9 m (100 kPa outlet pressure) or lower. Deeper tanks require compressors.

Blowers and compressors are the dominant source of energy consumption of a wastewater treatment plant applying diffused aeration. More or less than 70 percent of the total energy demand of an activated sludge wastewater treatment plant is created by aeration. Appropriate selection of blowers and compressors can therefore lead to substantial energy and cost savings. Typical values of the specific energy consumption for aeration of an activated sludge treatment plant may range from < 15 to > 35 kWh per capita yearly.

4.3.2.2   Types of Blowers and Compressors and Their Characteristics

From the various types of blowers and compressors manufactured, basically only two groups are applied in wastewater treatment. These include (a) the positive displacement blower (PD-blower) and (b) the dynamic or centrifugal blower or compressor. PD-blowers successively compress a fixed volume of air in an enclosed space to a higher pressure. The two types applied in wastewater treatment are (a$_1$) the rotary-lobe blower and (a$_2$) the rotary helical screw compressor. Also, the dynamic type shows two subgroups (b$_1$), the multistage centrifugal blower and (b$_2$) the centrifugal turbine or turbo compressor.

The rotary-lobe blower (positive displacement blower) is equipped with either two-lobe (older type) or three-lobe rotors arranged in a closed casing (see Figure 4.14). The air displacement and compression is brought about by the revolution of the rotors in opposite directions to each other as shown in Figure 4.15. Hence, the compressed air does not flow continuously. Some air pulsation is produced which is less pronounced with the three-lobe rotor.

Rotary-lobe blowers are available from very small units (< 100 m$_N^3$/h) up to very large units approaching airflow rates of 100,000 m$_N^3$/h. Depending on the rotor length, the blowers applied in wastewater treatment can produce a pressure rise up to 100 kPa (10 m water column) and can be applied to water depths up to 9 m.

The inlet volumetric flow rate G (m^3_N/h) would be directly proportional to the rotational speed (rpm) of the lobes if there were no slippage through the clearances. The slippage depends upon the total clearance area and the differential pressure of the device. Hence, the operation characteristic of this blower shows a reduced inlet flow rate at higher outlet pressures. The volumetric capacity can easily be controlled by the rotational speed (Figure 4.16), e.g., via a variable-frequency drive. At a required pressure rise of 60 kPa (about 5 m water depth), for instance, the airflow

FIGURE 4.14 Two types of rotary-lobe blowers (positive displacement blowers). Left: Two-lobe PD-blower (older type). Right: Three-lobe PD-blower (modern type).

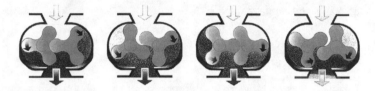

FIGURE 4.15 Schematic of a three-lobe PD-blower showing the progress of air displacement combined with compaction (from left to right) air intake at top, air delivery at bottom. (From Aerzener Maschinenfabrik GmbH, Germany. With permission.)

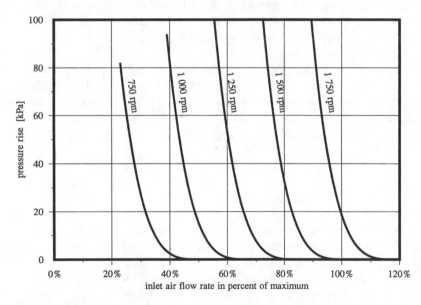

FIGURE 4.16 Inlet airflow rate of a rotary-lobe blower as a function of rotational speed (rpm) and pressure rise indicating the maximum capacity (top left) to prevent overheating. (From Aerzener Maschinenfabrik GmbH, Germany. With permission.)

FIGURE 4.17 Helical screw compressor. (From Aerzener Maschinenfabrik GmbH, Germany.) Left: 6-teeth female rotor (left) with 4-teeth male rotor (right). Right: open casing of helical screw compressor. (From Aerzener Maschinenfabrik GmbH, Germany. With permission.)

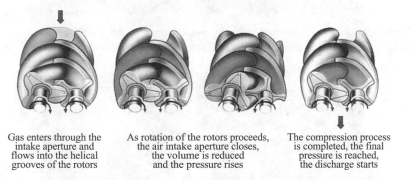

| Gas enters through the intake aperture and flows into the helical grooves of the rotors | As rotation of the rotors proceeds, the air intake aperture closes, the volume is reduced and the pressure rises | The compression process is completed, the final pressure is reached, the discharge starts |

FIGURE 4.18 Visualization of the process of air compression by a helical screw compressor. (From Aerzener Maschinenfabrik GmbH, Germany. With permission.)

rate of the blower depicted in Figure 4.16 can be controlled from 25 percent (750 rpm) to 93 percent (1750 rpm) of the maximum value, which is an airflow rate ratio of 1:3.7. At low airflow rates and high pressure rise, overheating of the blower can occur due to the reduced cooling action of the low airflow rate.

The rotary helical screw compressor (screw compressor) is applied in wastewater treatment only to a very limited extent. The positive air displacement is produced by two screws or rotors, the male and the female rotor (Figure 4.17, left), rotating at high speeds in opposite directions to each other within a closed casing (Figure 4.17, right). The process of compressing the air is shown in Figure 4.18. Due to the high rotational speed (< 10,000 rpm), pressure rises can reach 200 kPa and more in a single-stage unit, and therefore, the term *blower* would not be appropriate. The capacity range of intake airflow rates (300 to 60,000 m_N^3/h) matches, however, almost that of PD-blowers.

The operating characteristics are similar to the PD-blower (Figure 4.16). The intake airflow rate is (almost) proportional to the rotational speed in that higher pressure rises increase the air slippage. The main difference is the greater air compression.

Dynamic blowers or compressors very much resemble a centrifugal water pump in that the energy of the created streaming velocity is converted into the higher

FIGURE 4.19 Cutaway view of a turbo compressor showing the gear box, the impeller, and the discharge diffuser vanes in open position. (From HV-Turbo A/S, Denmark. With permission.)

FIGURE 4.20 Flow rate control elements of a turbo compressor. (From HV-Turbo A/S, Denmark.) Left: variable pre-rotation system controlling the inlet guide vanes. Right: the discharge diffuser system with almost closed vanes. (From HV-Turbo A/S, Denmark. With permission.)

pressure of the outlet flow rate. The multistage centrifugal blower is generally operated by direct drive at 3,550 rpm (60 Hz) with a relatively low-pressure rise (< 90 kPa). The turbo compressor is driven via a gearbox at 6,000 to 40,000 rpm (typically some 20,000 rpm) with a pressure rise up to 160 kPa in a single-stage configuration.

Dynamic blowers and compressors are generally designed for larger airflow rates than PD-blowers and compressors. Multistage centrifugal blowers range from 500 to 75,000 m_N^3/h, single-stage turbo compressors from 3,000 to 120,000 m_N^3/h. Turbo compressors are also available in "compact" or "mini" configuration with airflow rates from 1,000 to 9,000 m_N^3/h.

The cutaway-view of a turbo compressor (Figure 4.19, compact type) displays the gearbox (right), the (single-stage) impeller (center left), and the open vanes of the discharge diffuser around the impeller, leading the air into the outlet channel. Figure 4.20 shows the covered impeller (center) and the variable prerotation system,

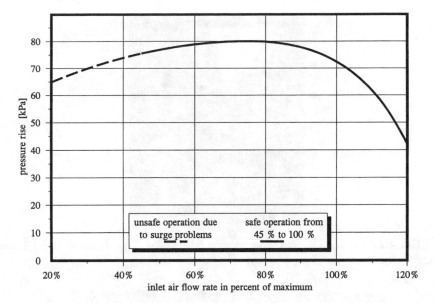

FIGURE 4.21 Typical operation characteristic of a turbo compressor indicating the range of safe operation.

which controls the inlet guide vanes for optimum flow rate control (left), and details of the adjustable discharge diffuser system in minimum position (right).

Although the airflow rate can be controlled by the rotational speed, higher efficiencies are obtained by operating the compressor at constant rpm. The flow rate is managed by the prerotation system and by opening or closing the outlet diffuser vanes. The operation characteristic, similar to that of a centrifugal pump, is very flat (Figure 4.21), indicating that the centrifugal compressor is sensitive to greater pressure changes. At low airflow rates, surging occurs. Below a certain minimum flow, the surge limit, the compressor performance is unstable and oscillates from zero to full capacity, resulting in vibrations and overheating. To prevent surging, turbo compressors are operated within a range of 45 to 100 percent of maximum capacity.

Multistage centrifugal blowers are manufactured with two to seven impellers and inlet airflow rates ranging from 500 to 75,000 m_N^3/h. The process of air compression is evident from the cutaway-view of a two-stage centrifugal blower in Figure 4.22 and the cross section of a six-stage centrifugal blower in Figure 4.23. The operation characteristic (compare Figure 4.21) depends on the type of impeller applied (radial or backward curved impellers or a combination of both types). The surge limit of multistage centrifugal blowers depends on the method of airflow control (next section). With conventional inlet throttling, it is 45 percent of the maximum capacity. When combined with inlet vanes, it may be reduced to about 30 percent.

4.3.2.3 Airflow Control of Blowers and Compressors

The rate of air delivery of blowers and compressors has to be controlled over a very wide range to match, rather exactly, the demands of the biological treatment

FIGURE 4.22 Cutaway view of a two-stage centrifugal blower. (From Hoffman Air & Filtration Systems, Syracuse, NY. With permission.)

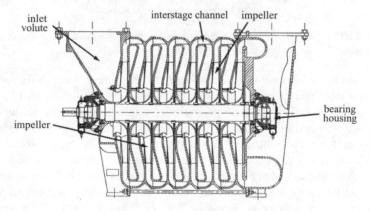

FIGURE 4.23 Cross section of a six-stage centrifugal blower. (From Hibon Inc., Dorval, Quebec, Canada. With permission.)

systems. Low supply can cause deficient treatment results; oversupply will create high oxygen concentrations in the reactors that can limit or even interrupt denitrification in the reactors to follow. Additionally, excess supply will waste energy. Frequently, a control range down to 20 percent of the maximum airflow rate is considered sufficient. When optimizing denitrification, however, a control range from 10 to 100 percent seems much more promising to constantly obtain low effluent nitrate concentrations. In this context, a comprehensive automated control system would comprise the on-line monitoring system, the control strategy, and the final control elements to carry out the required control action, viz. the adjustment of the airflow rate. The control strategy, based on conventional or advanced control theory, would be implemented into a programmable digital controller system. The following discussion is limited to the control of the airflow rate of blowers and compressors.

TABLE 4.6
Methods for Control of the Airflow Rate of Blowers and Compressors

	Applicable to		Cost Considerations	
Method of Control	PD-Blower Screw Compressor	Centrifugal Blower Turbo Compressor	First Cost	Energy Saving
Blow-off or by-pass	yes	no*	low	none
Inlet throttling	no*	yes	low	none
Inlet guide vanes	no	yes	medium	medium
Discharge diffuser	no	yes	medium	high
Variable speed driver	yes	no*	high	medium
On/off parallel units	yes	yes	low	none

* Theoretically applicable, but not useful in practice or even damaging.

The prime benefit of such a control system is the dynamic and long-term compliance with effluent requirements of the wastewater treatment plant, especially under conditions where the basis of plant performance control, with respect to the effluent nitrogen (nitrate nitrogen) parameter, is not (long-term) average effluent data but rather short-term samples (e.g., 2-h-composite). Next to this advantage, economic benefits can be achieved, mainly in terms of saving energy and its cost. When quantifying these benefits, a difficult question arises on how the capital cost for control of airflow rate is to be allocated, whether through (a) plant performance control, (b) aeration and its control, or (c) both. Frequently, however, this cost is allocated to the aeration system only, and the final decision with respect to "aeration" is taken on the aeration system cost including the total cost for control of the airflow rates.

Control of air delivery can be exerted in different ways, depending on the methods of achievement, which again may depend on the type of blower or compressor used.

Blow-off or by-pass: The excessive air is blown off into the atmosphere via a blow-off valve. This creates noise and will warm up the direct environment. A part of the discharged air is fed into the blower inlet again via a by-pass valve. The noise is limited to the direct surroundings. Continuous bypassing will increase the blower temperature, requiring cooling of the by-passed air.

Inlet throttling: Inlet throttling is a simple and effective means to reduce the airflow rate of a centrifugal blower by reducing the inlet pressure and increasing the required pressure rise (compare Figure 4.21). This method, however, is generally not applied with turbo compressor (see below). Since the capacity of PD-blowers is almost independent of pressure rise (Figure 4.16), inlet throttling is neither effective nor useful.

Variable inlet guide vanes: Turbo compressors are frequently equipped with variable inlet guide vanes (Figure 4.20). The flow rate control is exerted by turning the guide vanes to change the flow direction of the inlet air. Throttling losses are effectively reduced. This control method is also applied with centrifugal blowers.

Adjustable discharge diffusers: The adjustable diffuser vanes (Figures 4.19 and 4.20) control the airflow passage area ahead of the discharge without any hindrance of the air flow, i.e., without any additional friction loss and hence reduction of discharge pressure. Frequently, especially with turbo compressors, this method is combined with variable inlet guide vanes to minimize energy consumption.

Variable speed driver: A variable speed driver, e.g., a variable-frequency drive (frequencer), is the optimum capacity control for PD-blowers and screw compressors (compare Figure 4.16), allowing control of the airflow rate over a wide range. This advantage has to be paid for by the relatively high first cost of high capacity frequency converters and the energy loss of these devices amounting to two to five percent of the blower capacity.

Parallel operation of multiple units: Only in very small plants one blower or compressor plus one as a stand-by can be considered sufficient. In larger plants, more units are installed. With three PD-blowers or screw compressors the optimum capacity range of 1:10 can be easily reached, whereas three identical centrifugal units cover a range from 15 (one unit at 45 percent of maximum) to 100 percent. The combination of the control option, "parallel operation of multiple units" with any of the other alternatives allows one to continuously cover the entire required capacity range of control.

4.3.2.4 Power Demand of Blowers and Compressors

The power demand of a blower or compressor can be estimated by two different methods: (a) by using the equations developed for adiabatic gas compression or (b) by applying empirical equations derived from performance data of the manufacturers. The first approach has the advantage of physical exactness, but some coefficients (exponent for compression, various efficiency coefficients) have to be estimated. A precise assessment of the efficiencies (EPA, 1983) of blowers and compressors (50 to 80 percent), motors (95 percent), and gear box (95 percent) and of the resulting overall efficiency e_0 is difficult. The second method has the advantage of direct applicability, but the equation found for a certain type and size of blower or compressor may differ at other conditions. Both methods are discussed below, starting with the physical approach. Within this part, IS-units are used consequently. The standard airflow rate G_s is stated in m_N^3/s rather than in m_N^3/h as previously in the more applied part and again later when using the empirical equations.

For a positive displacement blower (Westphal, 1995), the power demand (WP in W) depends upon the airflow rate G_s (m_N^3/s), the differential pressure Δp (Pa, N/m^2) and the overall efficiency e_0.

$$WP = \frac{G_s \cdot \Delta p}{e_o} \qquad (4.25)$$

For a PD-blower delivering an airflow rate of G_s = 5,400 m^3/h (1.5 m^3/s) at a differential pressure of 45 kPa (45,000 N/m^2) with an estimated overall efficiency of 60 percent, the required wire power is as follows.

$$WP = \frac{1.5 \cdot 45,000}{0.6} = 112,500W = 112.5kW$$

The above pressure rise of 45 kPa represents a back pressure of $\Delta H = 4.59$ m WC (see Equation 4.1). From this ΔH, the specific energy $E_{\Delta H}$ required per m of ΔH for introducing 1 m_N^3 of air into an aeration tank of this (waste)water depth can be calculated.

$$E_{\Delta H} = \frac{WP}{(3,600 \cdot G_s) \cdot \Delta H} = \frac{112,500}{(3,600 \cdot 1.5) \cdot 4.59} = 4.54 \frac{Wh}{m_N^3 \cdot m} \qquad (4.26)$$

By this approach, all friction losses in the piping system and the diffusers are neglected (see Section 4.3.3.2).

From Equation 4.25 it follows, that the power demand of a PD-blower is directly proportional to the airflow rate and to the pressure rise. Consequently, the specific energy (Equation 4.26) is constant and independent of both parameters and is influenced only by the overall efficiency e_0.

With the centrifugal blower, turbo compressor, and screw compressor, internal air compression takes place, the power demand of which is given by (Metcalf and Eddy, 1991; Westphal, 1995; see Equation 2.47):

$$WP = \frac{\rho_a \cdot G_s \cdot RT_a}{K \cdot e_o} \cdot \left[\left(\frac{P_a + \Delta p}{P_a} \right)^K - 1 \right] = \frac{P_a \cdot G_s}{K \cdot e_o} \cdot \left[\left(\frac{P_a + \Delta p}{P_a} \right)^K - 1 \right] \qquad (4.27)$$

with
K = $(\kappa - 1)/ \kappa = 0.2857$
κ = 1.4 (adiabatic exponent)
ρ_a air density (1.293 kg/m³)
T_a inlet gas temperature (K)
P_a inlet pressure (Pa)
R gas constant (286.88 J/kg·K).

Contrary to the PD-blower, the power demand depends upon the inlet pressure P_a, and/or air density, ρ_s, and temperature. It increases less than proportional to the pressure rise Δp. This characteristic is illustrated by the following example for a turbo compressor at normal inlet pressure (101,325 Pa), the remaining data as in the foregoing example:

$$WP = \frac{P_a \cdot G_s}{K \cdot e_o} \cdot \left[\left(\frac{P_a + \Delta p}{P_a} \right)^K - 1 \right] = \frac{101,325 \cdot 1.5}{0.2857 \cdot 0.6} \cdot \left[\left(\frac{101,325 + 45,000}{101,325} \right)^{0.2857} - 1 \right]$$

$$= 98,154 \ W = 98 \ kW$$

from which the specific energy $E_{\Delta H}$ for this depth is obtained.

$$E_{\Delta H} = \frac{WP}{(3,600 \cdot G_s) \cdot \Delta H} = \frac{98,154}{(3,600 \cdot 1.5) \cdot 4.59} = 3.96 \frac{Wh}{m_N^3 \cdot m}$$

At a twofold pressure rise ($\Delta p = 90$ kPa or $\Delta H = 9.17$ m WC), the power demand would rise to only 177 kW, 181 percent of the former value. The specific energy would drop to $E_{\Delta H} = 3.56$ Wh/($m_N^3 \cdot m$), 90 percent of the former value.

The empirical equations are derived from manufacturers' data on wire power WP (W) as a function of pressure rise ΔH (m WC). For three blowers or compressors, the following type of relation has been derived (Pöpel and Wagner, 1994) with high accuracy (r > 0.99):

$$WP = G_s \cdot E_{\Delta H} \cdot \Delta H^{\Psi} \qquad (4.28)$$

with

G_s standard airflow rate in m_N^3/h
$E_{\Delta H}$ specific energy in Wh/($m_N^3 \cdot m$) related to $\Delta H = 1$ m WC pressure rise
ΔH pressure rise in m WC [$\Delta H = \Delta p/(\rho \cdot g)$]
Ψ empirical exponent (-).

The obtained parameters $E_{\Delta H}$ and Ψ for blowers with a capacity of around 5,000 m_N^3/h are given in Table 4.7. The parameters may differ for other capacities.

The empirical equation for the PD-blower with the exponent $\Psi = 1.00$ confirms Equation 4.25 with respect to the linear influence of pressure (Δp and ΔH). By equating 4.25 and 4.28 for this blower type the overall efficiency, e_0, for the empirical approach can be estimated. Care has to be taken, however, to use the necessary units for Δp (Pa), ΔH (m WC), and G_s (m_N^3/s and m_N^3/h), respectively, as shown by starting with Equation 4.28.

$$WP = G_s\left(m^3/h\right) \cdot E_{\Delta H} \cdot \Delta H^{\Psi} = 3,600 \cdot G_s\left(m^3/s\right) \cdot E_{\Delta H} \cdot \left[\frac{\Delta p}{\rho \cdot g}\right]^{\Psi}$$

$$= \frac{G_s\left(m^3/s\right) \cdot \Delta p}{e_o}$$

With $\Psi = 1$, e_0 can then be calculated to yield 63.3 percent, typical for PD-blowers.

$$e_o = \frac{\rho \cdot g}{3,600 \cdot E_{\Delta H}} = \frac{1,000 \cdot 9.806}{3,600 \cdot 4.3} = 0.633 = 63.3\%$$

TABLE 4.7
Empirical Blower and Compressor Parameters
(after Pöpel and Wagner, 1994; and Pöpel et al., 1998)

Type of Blower or Compressor	Depth Range (m WC)	Specific Energy $E_{\Delta H}$ [Wh/m_N^3·m]	Exponent Ψ
Positive displacement blower	0–9	4.3	1.00
Turbo compressor (single stage)	0–15	4.5	0.83
Screw compressor	0–>30	5.1	0.83

A similar check is performed for the empirical result of the turbo compressor by generating data with Equation 4.27 and analyzing the results by curvilinear regression with a power function (Equation 4.28 with Ψ as exponent). $E_{\Delta H} = 4.50$ and $\Psi = 0.830$ is obtained ($r = 0.9992$), when an overall efficiency of $e_0 = 70.7$ percent is applied, which is a characteristic value for this type of compressor.

In practice, the specific energy $E_{\Delta H}$ (Wh/m^3·m) is frequently calculated (Equation 4.26) by relating the required wire power to the airflow rate and the water depth H or diffuser submergence H_S, rather than to the pressure rise $\Delta H = H_S + \Delta H_\ell$. These approaches are illustrated by repeating Equation 4.26 and adding the other ways of calculation.

$$e_o = \frac{\rho \cdot g}{3,600 \cdot E_{\Delta H}} = \frac{1,000 \cdot 9.806}{3,600 \cdot 4.3} = 0.633 = 63.3\% \qquad (4.26)$$

$$E_H = \frac{WP}{(3,600 \cdot G_s) \cdot H} = E_{\Delta H} \cdot \frac{\Delta H}{H} \qquad H : \text{water depth} \qquad (4.26a)$$

$$E_{Hs} = \frac{WP}{(3,600 \cdot G_s) \cdot H_S} = E_{\Delta H} \cdot \frac{\Delta H}{H} \qquad H_S : \text{diff. submergence} \qquad (4.26b)$$

Since $\Delta H > H > H_S$, it follows that the three specific energies vary by $E_{\Delta H} < E_H < E_{Hs}$. In shallow tanks, the differences may be considerable, whereas they become negligible for deeper tanks.

The above check of the empirical equations shows the exactness and usefulness of this approach, which is applied in the following sections. On the other hand, the method allows the determination of the overall efficiency e_0 with high accuracy.

As an illustration of the foregoing calculations, the specific energy $E_{\Delta H}$ [Wh/(m_N^3·m)] is plotted versus the pressure rise ΔH (m WC) for different types of blowers and compressors (capacity about 5,000 m_N^3/h) in Figure 4.24. It is obvious that the PD-blower requires less power and energy at extremely low pressure rises (<1.5 to 2.5 m WC or <15 to 25 kPa). At higher pressure rises, there is a definite advantage of the two other types.

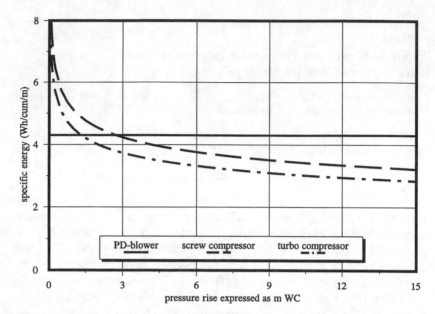

FIGURE 4.24 Specific energy (Wh per cum or air per m WC pressure rise) as a function of pressure rise and type of blower or compressor (example at 5,000 cum/h).

4.3.3 BLOWER AND COMPRESSOR SELECTION FOR OPTIMUM AERATION EFFICIENCY

4.3.3.1 Introduction

The primary goal of blower and compressor selection is to minimize the total cost for aeration, i.e., first cost and operating cost. The following discussion is limited to the main part of the operating cost, viz. the energy consumption by aeration. The energy consumption is caused by various features:

- type of blower or compressor and its overall efficiency
- airflow rate
- the required pressure rise affected by water or tank depth (diffuser submergence H_S) and by the flow resistance within the total air supply system
- airflow rate control system (throttling — by-pass — more advanced control systems)

All above factors have to be taken into account when selecting the optimum blower or compressor with respect to required power and energy consumption as is illustrated in the following section.

Since the oxygen transfer rate (Section 4.22) and the power demand (Section 4.3.2.4) can be modeled, the aeration efficiency AE (transfer rate over power demand: kg O_2/kWh) also becomes amenable to modeling. This will allow a closer comparison of different blowers and compressors with respect to their energy performance. Since OTR and power demand depend on the required pressure rise, an optimum depth with maximum aeration efficiency can be expected.

4.3.3.2 Factors Affecting Power Demand and Energy Consumption

Type of blower or compressor: The theoretical wire power depends on the type of air compaction and differs for positive displacement blowers, on the one hand, and centrifugal blowers and compressors and screw compressors, on the other hand. This relationship is quantified by Equations 4.25 (PD-blower) and 4.27 (other blowers and compressors). Second, the overall efficiencies of the discussed blowers and compressors together with their drive (and control devices) differ and affect the wire power correspondingly.

This relationship is illustrated by determining the ratio of wire power required for a turbo compressor (Equation 4.27; index TC with an overall efficiency of $e_{0,TC}$) to that of a PD-blower (Equation 4.25; index PD with $e_{0,PD}$).

$$\frac{WP_{TC}}{WP_{PD}} = \frac{e_{o,PD}}{e_{o,TC}} \cdot \frac{1}{K \cdot \Delta p / P_a} \cdot \left[\left(1 + \frac{\Delta p}{P_a} \right)^K - 1 \right] \qquad (4.29)$$

With increasing relative pressure rise $\Delta p / P_a$ (pressure rise Δp over inlet pressure P_a), the ratio of the wire powers decreases below one and even more so when the sample overall efficiencies (Section 4.3.2.4) are taken into account. This decrease is shown in Figure 4.25 illustrating the WP-ratio as a function of pressure rise at normal inlet pressure P_a at same and at sample overall efficiencies. Near the maximum pressure rise for PD-blowers (90 kPa), the power ratios amount to 78 percent and 70 percent.

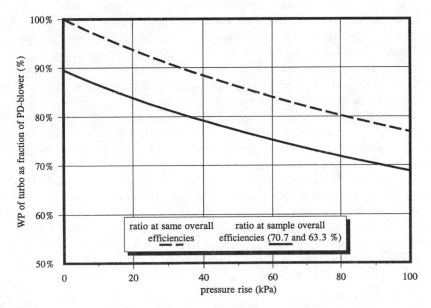

FIGURE 4.25 Wire power of a turbo compressor expressed as fraction of the wire power of a PD-blower as a function of pressure rise.

TABLE 4.8
Range of Airflow Rates (m_N^3/h) of Different Types of Blowers and Compressors

Type of Blower or Compressor	Minimum	Maximum
Positive displacement blower (roots type)	< 100	100,000
Screw compressor	300	60,000
Multistage centrifugal blower	500	90,000
Turbo compressor: conventional	3,000	150,000
Compact or mini TC	1,000	9,000

Airflow rate: Airflow rate would disfavor a PD-blower solely from the viewpoint of power requirement and energy consumption. Next to first cost, it has to be kept in mind, however, that PD-blowers are available also at very low airflow rates. Centrifugal blowers and compressors are designed for relatively large rates and are not economical at low airflow rates, except for the "compact" or "mini" types (see Table 4.8). When discussing this influence, the required maximum airflow rate is not of importance. Rather, this rate is divided by the designed number of units (two to four) for reasons of operational safety and control by parallel operation of multiple units (Section 4.3.2.3).

The range of typical airflow rates of different types of blowers and compressors is summarized in Table 4.8 and can serve for selecting blowers and compressors with respect to optimum airflow rate.

From Table 4.8, it is clear that PD-blowers and screw compressors are optimum for lower airflow rates; centrifugal blowers, screw and small turbo compressors are optimum for medium rates; and, all types are optimum for large airflow rates.

Required pressure rise: Next to the airflow rate, blowers and compressors are selected according to their maximum pressure rise Δp (kPa). Blowers can be applied for pressure rises up to 90 kPa, with single-stage turbo compressors up to 160 kPa and screw compressors for even higher pressure rises.

The required pressure rise is mainly caused by the depth of submergence of the diffusers below the wastewater level H_S (m WC) and the friction or head loss H_ℓ (m WC) in the entire air distribution system. This includes the loss across the diffuser membrane that may increase in time due to fouling effects.

The depth of submergence H_S is fixed by the system design, the friction loss H_ℓ and can be calculated using the Darcy–Weisbach equation (Metcalf and Eddy, 1991). Generally, the total friction loss lies within the range of 7 to 12 kPa (0.7 to 1.2 m WC). For preliminary calculations, a fixed value of $\Delta p_\ell = 8$ kPa (1.2 psi) or of $\Delta H_\ell = 0.8$ m (2.6 ft) WC can be applied. Hence, the required pressure rise is formulated in units of pressure, Pa, or m WC.

$$\Delta p = \frac{\Delta H}{g \cdot \rho} = \frac{H_S}{g \cdot \rho} + \Delta p_\ell \qquad (4.30a)$$

$$\Delta H = \Delta p \cdot g \cdot \rho = H_s + \Delta H_\ell = H_s + \Delta p_\ell \cdot g \cdot \rho \qquad (4.30\text{b})$$

Airflow rate and control system: The power requirement of blowers and compressors is theoretically (Equations 4.25 and 4.27) proportional to the airflow rate G_s. To what extent this "theoretical optimum" is reached under operational conditions depends mainly on the method of airflow rate control and its influence on power demand or energy consumption. This relationship has been discussed previously in Section 4.3.2.3 and summarized in Table 4.6. Blow-off, by-passing, and inlet throttling lead to no energy savings, meaning that at low and high airflow rates basically the same power is required. The other extreme is the combination of inlet guide vanes and discharge diffuser applied with centrifugal blowers and compressors, which comes close to the "theoretical optimum" at the expense of corresponding first cost.

Variable speed drivers (frequencers) may add up to two to five percent of the power requirement and energy consumption of positive displacement blowers, but within this range, a satisfactory energy management is possible at a broad spectrum of airflow rates.

4.3.3.3 Modeling of the Aeration Efficiency in Deep Tanks

The standard aeration efficiency SAE (kg/kWh of oxygen) as defined in Chapter 2 is

$$SAE = \frac{\text{mass rate of oxygen transfer at standard conditions (kg/h)}}{\text{wire power (kW)}}$$

The mass of oxygen transferred can be expressed as SOTR (Equation 4.3) or in combination with Equation 4.5 as $0.3 \cdot G_s \cdot SOTE$ (Equation 2.51). The wire power (W — not kW!) can be modeled with Equation 4.28. The required pressure difference, ΔH, is the sum of the depth of diffuser submergence, H_S, plus a fixed value for the total friction loss of $\Delta H_\ell = 0.8$ m WC ($\Delta p_\ell = 8$ kPa).

$$SAE = \frac{1000 \cdot SOTR}{G_s \cdot E_{\Delta H} \cdot \left(H_s + \Delta H_\ell\right)^\Psi} = \frac{300 \cdot SOTE}{E_{\Delta H} \cdot \left(H_s + \Delta H_\ell\right)^\Psi} \qquad (4.31)$$

The effect of water depth (depth of diffuser submergence) is brought about by two opposing tendencies:

- The SOTE increases with depth, but not linearly. The quantified effects, as depicted in Figure 4.6 (top), cause a lowering of this increase at greater depth due to oxygen depletion in the bubble air caused by oxygen uptake. Accordingly, this drop is more intense for aeration systems with a higher efficiency (greater SOTE$_{so}$).

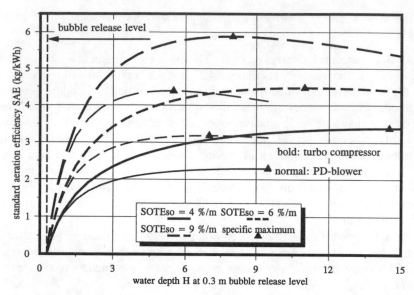

FIGURE 4.26 Standard aeration efficiency (kg/kWh) as a function of water depth (at 0.30 m bubble release level), type of compressor or blower, and basic specific oxygen transfer efficiency $SOTE_{so}$ (%/m).

- Multistage blowers and turbo and screw compressors require less power per m at greater tank depths and operate more efficient under conditions of higher pressure rise.
- As a consequence, the first aspect would call for relatively shallow tanks, the second for greater depths of submergence. Moreover, for shallow tanks, e.g., $H_S = 3.20$ m, the assumed fixed head loss of $\Delta H_\ell = 0.8$ m WC would account for $0.8/(3.2+0.8) = 20$ percent "energy loss". In a deep tank of $H_S = 9.20$ m, this effect would only cause $0.8/(9.2+0.8) = 8$ percent, which would again favor deeper tanks.

In summarizing the above effects, a maximum aeration efficiency can be expected at a certain depth. This result can be found by modeling the standard oxygen transfer efficiency (see Figure 4.6 — top), the numerator of Equation 4.31, and dividing the corresponding data by the wire power, calculated by the empirical Equation 4.28. The results for a PD-blower and a turbo compressor are depicted in Figure 4.26 for three basic specific oxygen transfer efficiencies $SOTE_{so}$ (%/m) as a function of water depth H, assuming a bubble release level of 0.30 m.

For all combinations, the standard aeration efficiency SAE (kg/kWh) increases very sharply at depths up to 3.00 m, especially with very efficient aeration systems (high $SOTE_{so}$) and with turbo compressors. Depending on $SOTE_{so}$ and the type of blower or compressor, the maximum aeration efficiencies, indicated as "specific maximum" in Figure 4.26, differ considerably. The PD-blower varies from 2.3 to 4.4 kg/kWh while the turbo compressor varies from 3.4 to 5.9 kg/kWh. They are also reached at different depths, the PD-blower from 5.50 m to 9.50 m and the turbo compressor from 8.00 to 14.50 m.

It is evident that very shallow tanks (less than 3.00 m water depth) and extremely deep tanks cause low aeration efficiencies. Since the maximal values are not very sharp, deviations from the "optimum depth" will not cause a very significant increase of power requirement and energy consumption.

4.3.3.4   Model Verification

The modeling of the SOTE, part of the discussed SAE model, has been verified earlier (Section 4.2.3). Also, the WP-part (empirical Equation 4.28) of this model has been validated (Section 4.3.2.4), at least when assuming specific overall efficiencies e_0 (63.3 percent for the PD-blower and 70.7 percent for the turbo compressor), on the basis of the well-known compression Equations 4.25 and 4.27. Hence, the basic structure of the SAE-model is confirmed, but numerical deviations from the presented data can be caused by overall efficiencies differing from the above values.

Nevertheless, the results of the full-scale experimental verification (see Section 4.2.3.2) are presented as follows (Pöpel et al., 1998). The SOTE data are taken from the foregoing discussion, and the power data are obtained in two ways.

The air pressure P_{el} (Pa) was measured in the lateral pipes directly ahead of the diffuser elements. Together with the standardized airflow rate G_s (m_N^3/h), the actual delivered power can be calculated.

$$DP = \frac{G_s}{3,600} \cdot P_{el} \qquad (4.32)$$

The wire power is calculated using Equation 4.28 with the coefficients (Table 4.7) for the applied screw compressor. The average results, averaged over the three diffuser densities, the five water depths, and the three airflow rates, are summarized in Table 4.9. All SAE values are very small, caused by the poor quality of the (self-made) membranes with poor and uneven air distribution. Hence, not the obtained data but rather the derived tendencies of the results with respect to the influence of diffuser density, water depth, and airflow rate can be applied in practice, i.e., serve as experimental verification.

From Table 4.9, it is evident that both aeration efficiencies (DP and WP) increase considerably with increasing diffuser density. With respect to water depth, both efficiencies exhibit a maximum at a certain depth: 5.35 kg/kWh at $H = 7.50$ m with respect to DP, 3.79 kg/kWh at $H = 10.00$ m. As usual, the highest aeration efficiencies are obtained at the lowest airflow rate. Altogether, the SAE values are relatively low due to the small transfer efficiencies of $SOTE_{sa} < 5.3$ %/m and of $SOTE_{so} < 6.6$ %/m (see Table 4.4).

A more detailed analysis of the data is accomplished by quantifying the above influences of diffuser density, water depth, and airflow rate on the standard aeration efficiency by linear regression. Following the tendencies of Table 4.9 and the regression (Equation 4.23), the influence of the diffuser density is quantified by its natural logarithm ($ln\ DD$) and the airflow rate by a linear relationship. For expressing the depth influence (H_S), the function must be curved in such a way that a maximum can

TABLE 4.9
Average SAE Values (kg/kWh O_2) Based on Delivered (DP) and Wire Power (WP; screw compressor) Obtained By the Full-Scale Experimental Verification

Diffuser Density (%)			Water Depth H (m)			Airflow Rate G_s (m_N^3/h)		
value	DP	WP	value	DP	WP	value	DP	WP
4.5	4.50	2.98	2.5	4.88	2.37	35.5	5.26	3.52
9.5	5.23	3.46	5.0	5.43	3.35	71.0	5.17	3.47
17.9	5.64	3.96	7.5	5.35	3.69	142.0	4.94	3.33
			10.0	5.13	3.79			
			12.5	4.77	3.72			

be found within the examined depth range. This figure is obtained by applying not only the depth of diffuser submergence H_S but additionally its square, H_S^2, as follows.

$$SAE = k_o + k_1 \cdot H_S + k_2 \cdot H_S^2 + k_3 \cdot \ln DD + k_4 \cdot G_s \qquad (4.33)$$

This approach allows one to calculate the depth $H_{S,max}$ of maximum SAE by differentiation with respect to H_S.

$$H_{S,max} = -\frac{k_1}{2 \cdot k_2} \qquad (4.34)$$

The regression calculations are performed for the SAE based on delivered power and on wire power, determined by the empirical approach for a PD-blower, a screw and a turbo compressor. Table 4.10 contains the regression coefficients k_0 to k_4 for all four configurations and the corresponding correlation coefficients. All coefficients are statistically highly significant. Table 4.10 also contains the depth of submergence $H_{S,max}$, calculated by Equation 4.34, at which the standard aeration efficiency is maximum.

As an illustration of the full-scale experimental results, the aeration efficiency is plotted as a function of the depth of diffuser submergence H_S for the applied screw compressor in Figure 4.27 following Equation 4.33. The three lines represent the results for the three diffuser densities DD. For each line, the respective data average of the airflow rate G_s is used in Equation 4.33. For comparison, the average of the data (averaged over the airflow rate G_s) is also plotted for the three diffuser densities and the depths of diffuser submergence. Figure 4.27 shows the good fit of the SAE-model to the data like the above correlation coefficients.

The model shows that the maximum aeration efficiency is reached at fairly great depths. Already from Figure 4.26, it was clear that depths below $H = 3.00$ m lead to distinctly lower aeration efficiencies. The (flat) maximum is reached between $H = 5.50$ to $H = 14.50$ m, depending on the transfer efficiency of the system ($SOTE_{so}$)

TABLE 4.10

Regression and Correlation Coefficients for the Standard Aeration Efficiency

Power and Type of Blower or Compressor	k_0	k_1 (H_S)	k_2 (H_S^2)	k_3 (DD)	k_4 (G_s)	r	$H_{S,max}$
Delivered power DP	6.926	0.212	−0.0174	0.814	−0.00283	0.921	6.09
WP: PD-blower	3.164	0.283	−0.0173	0.461	−0.00150	0.946	8.18
Turbo compressor	3.571	0.507	−0.0263	0.623	−0.00196	0.974	9.64
Screw compressor	3.150	0.447	−0.0232	0.550	−0.00173	0.974	9.63

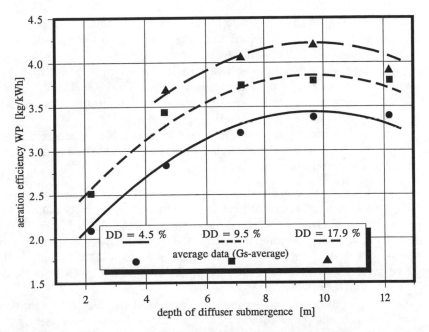

FIGURE 4.27 Aeration efficiency (wire power) as a function of diffuser submergence and density for a screw compressor (evaluation of full-scale experimental data).

and the type of blower or compressor applied. The full-scale verification validates this effect:

- The maximum SAE of the averaged depth data (Table 4.9) is at the data point $H = 7.50$ m on the basis of delivered power and at $H = 10.00$ m for wire power (screw compressor).
- The detailed regression analyses of the data together with the calculation of $H_{S,max}$ shows a maximum SAE at $H = 6.1$ m (DP) and $H = 8.2$ m (PD-blower) and $H = 9.6$ m (screw and turbo compressor) for wire power.
- Although the maximum is very flat, i.e., not sensitive with respect to the water depth H, it drops considerably below depths of 3 to 4 m (see Figure 4.27).

4.3.3.5 Model Application

The standard aeration efficiency SAE has been modeled by dividing the standard oxygen transfer efficiency SOTE by the modeled wire power demand (Equation 4.31). The SOTE has been modeled (Section 4.2.2) and applied (Section 4.2.4) to an extensive example, earlier. Application of the SAE-model means, therefore, combining the SOTE model with the most accurate estimate of the required wire power demand. This information can be obtained in a number of ways:

- precise estimate of the required pressure rise, estimating this information from the depth of diffuser submergence plus the calculated total friction loss Δp_ℓ or ΔH_ℓ, or taking a fixed value (e.g., 8 kPa or 0.80 m WC) for the losses;
- precise information on the wire power demand as a function of pressure rise and airflow rate obtained from the manufacturers of blowers and compressors, if publicly available;
- estimation of the wire power demand with Equations 4.25 (PD-blower) and 4.27 (centrifugal blower, turbo and screw compressor) applying an overall efficiency e_0 to be provided by the manufacturer or obtained by appropriate estimate (for example see Section 4.3.2.4);
- applying the derived empirical Equation 4.28 and the blower or compressor parameters of Table 4.7. Appropriate information could also be requested from the manufacturer in this case.

The model application is illustrated by proceeding from the example in Section 4.3.3.5 (oxygen transfer) and combining this information with wire power estimates using the empirical Equation 4.28 and the coefficients of Table 4.7. Applying the basic Equations 4.25 (PD-blower) and 4.27 (centrifugal blower, screw and turbo compressor) with the sample overall efficiencies of Section 4.3.2.3 would lead to the same results. The friction losses in the piping system plus diffuser are fixed at $\Delta p_\ell = 8$ kPa or $\Delta H_\ell = 0.80$ m WC.

The results of the model application are summarized in Table 4.11 for the depth of $H = 4.60$ m in the upper part and for $H = 9.20$ m in the lower section. The pressure rise in Equation 4.28 amounts to $\Delta H = 4.60 - 0.30 + 0.80 = 5.10$ m and to $\Delta H = 9.20 - 0.30 + 0.80 = 9.70$ m, respectively.

The first line of each section states the airflow rate, and line 2 states the SOTR. Lines 3 to 5 contain the calculated (Equation 4.28) wire power for a PD-blower, a screw compressor, and a turbo compressor, respectively. In lines 6 to 8, the SAE values are stated, obtained by applying Equation 4.31. Lines 9 to 11 contain the specific energy $E_{\Delta H}$ [Wh/(m$_N^3$·m)] required to introduce 1 m$_N^3$ into the (waste)water at a pressure rise of $\Delta H = 1.0$ m, calculated with Equation 4.26. Although the pressure rise (9.70 m) exceeds (almost) the capacity of a PD-blower, the pertinent data are nevertheless computed for comparison.

From the results, it is evident, that the wire power requirement decreases in the order of PD-blower, screw compressor, and turbo compressor. Therefore, the SAE increases in the same sequence. As usual, the highest SAE is obtained at the lowest airflow rate.

TABLE 4.11
Results of the SAE Model Application

Line	Parameter		Unit	G_s-1	G_s-2	G_s-3
	Conditions at H = 4.60 m water depth					
1	Airflow rate G_s		m_N^3/h	550	1,500	3,000
2	SOTR		kg/h	50	125	230
3	WP:	PD-blower	kW	12.1	32.9	65.8
4		Screw compressor	kW	10.8	29.6	59.2
5		Turbo compressor	kW	9.6	26.1	52.2
6	SAE:	PD-blower	kg/kWh	4.15	3.80	3.50
7		Screw compressor	kg/kWh	4.61	4.23	3.89
8		Turbo compressor	kg/kWh	5.23	4.79	4.41
9	$E_{\Delta H}$:	PD-blower	$Wh/(m_N^3 \cdot m)$		4.30	
10		Screw compressor	$Wh/(m_N^3 \cdot m)$		3.87	
11		Turbo compressor	$Wh/(m_N^3 \cdot m)$		3.41	
	Conditions at H = 9.20 m water depth					
1	Airflow rate G_s		m_N^3/h	614	1,653	3,262
2	SOTR		kg/h	100	250	460
3	WP:	PD-blower	kW	25.6	68.9	136.1
3		Screw compressor	kW	20.6	55.6	109.7
5		Turbo compressor	kW	18.2	49.0	96.8
6	SAE:	PD-blower	kg/kWh	3.90	3.63	3.38
7		Screw compressor	kg/kWh	4.84	4.50	4.19
8		Turbo compressor	kg/kWh	5.49	5.10	4.75
9	$E_{\Delta H}$:	PD-blower	$Wh/(m_N^3 \cdot m)$		4.30	
10		Screw compressor	$Wh/(m_N^3 \cdot m)$		3.47	
11		Turbo compressor	$Wh/(m_N^3 \cdot m)$		3.06	

In accordance with the findings of Section 4.3.3.4, especially with Figure 4.26, the SAE-values of the PD-blower are higher for the shallow tank: in the deep tank between 94 and 97 percent of the shallow tank SAE is reached. For the screw and turbo compressor, the situation is reversed: the deep tank SAE is between 105 and 108 percent of the shallow tank data. In each case, the lower percentage refers to the low airflow rate and the highest percentage to the maximum G_s.

The specific energies ($E_{\Delta H}$) are higher at low depth, primarily due to the great share of the total friction losses (ΔH_ℓ). They decrease in the sequence of PD-blower, screw compressor, and turbo compressor. The results are the same for each airflow rate G_s, since a constant friction loss has been assumed. In practice, however, ΔH_ℓ increases with increasing G_s, and the specific energy, therefore, will also increase (slightly).

Altogether, the model application shows that the aeration efficiency model is a valuable tool for optimizing wire power requirement and aeration efficiency, especially as a function of tank depth and type of blower or compressor.

4.4 NOMENCLATURE

A_{at}	m^2	bottom area of aeration tank
D	m^2/s	coefficient of molecular diffusion (of oxygen in (waste)water)
d_B	m	bubble diameter
DD	–, %	diffuser density: sum of gassing area of all diffusers over total aeration tank bottom area A_{at}
DP	W	delivered power
e_o	–, %	overall efficiency of a blower or compressor
$E_{\Delta H}$	$Wh/(m_N^3 \cdot m)$	specific energy required for blowing 1 m_N^3 of air into an aeration tank of any depth related to a back pressure rise of $\Delta H = 1.00$ m WC
G_s	m_N^3/h	airflow rate at standard conditions of temperature and pressure
H	m	(waste)water depth in aeration tank
h	m	(waste)water height above bubble release level
H_S	m	depth of submergence of diffusers below wastewater level
$H_{S,max}$	m	depth of diffuser submergence leading to a maximum SAE
o	–	index: parameter at very shallow tank depth ($H_S \rightarrow 0$)
P_a	Pa	atmospheric pressure
P_{el}	Pa	air pressure within air supply system directly ahead of the diffuser elements
P_t	Pa	total pressure (atmospheric plus water pressure)
R	$J/(kg \cdot K)$	universal gas constant (286.88 J/kg·K)
SOTE	–, %	standard oxygen transfer efficiency
$SOTE_s$	m^{-1}, %/m	specific standard oxygen transfer efficiency
$SOTE_{sa}$	m^{-1}, %/m	average specific standard oxygen transfer efficiency = SOTE/H
$SOTE_{so}$	m^{-1}, %/m	specific standard oxygen transfer efficiency at very shallow tank depth ($H_S \rightarrow 0$)
V	m^3	(waste)water volume of aeration tank
V_B	m^3	bubble volume
v_B	m/s	rising velocity of a bubble with respect to water
WP	W	wire power
ΔH	m WC	pressure rise of a blower or compressor expressed as water column: $\Delta p = \rho \cdot g \cdot \Delta H$ or $\Delta H = \Delta p/(\rho \cdot g)$
ΔH_ℓ	m WC	total friction loss within air supply system
ΔP	Pa	pressure difference caused by (waste)water pressure
Δp	Pa	pressure rise by blower or compressor
Δp_ℓ	Pa	total friction loss within air supply system
Ψ	–	empirical exponent
π	–	relative pressure
ρ	kg/m^3	(waste)water density
ρ_a	kg/m^3	air density (1.293 kg/m^3)

4.5 BIBLIOGRAPHY

ASCE (1991). *A Standard for the Measurement of Oxygen Transfer in Clean Water*, New York.

ATV (1979). *Arbeitsanleitung für die Bestimmung der Sauerstoffzufuhr von Belüftungssystemen in Reinwasser*, Korrespondenz Abwasser 26, 416–423.

ATV (1996). *Messung der Sauerstoffzufuhr von Belüftungseinrichtungen in Belebungsanlagen in Reinwasser und belebtem Schlamm, Merkblatt ATV-M 209*, Gesellschaft zur Förderung der Abwassertechnik e.V., Hennef, Germany.

ATV (1997). *ATV-Handbuch: Biologische und weitergehende Abwasserreinigung*, 4th edition, Ernst & Sohn, Berlin.

ATV-Arbeitsbericht (1996). *Hinweise zu tiefen Belebungsbecken*, Korrespondenz Abwasser 43, 1083–1086.

Diesterweg, G., Fuhr, H., and Reher, F. (1978). *Die Bayer-Turmbiologie, Zeitschrift Industrieabwässer*, 5, 7–17.

EPA (1983). *Development of Standard Procedures for Evaluating Oxygen Transfer Devices*, EPA-600/2-83-102.

EPA (1989). *Design Manual — Fine Pore Aeration Systems*, EPA/625/1-89/023, Cincinnati, OH.

Higbie, R. (1935). *The Rate of Absorption of a Pure Gas into a Still Liquid During Short Periods of Exposure*, American Institute of Chemical Engineers, 365–389.

Leistner, G., Müller, G., Sell, G., and Bauer, A. (1979). *Der Bio-Hochreaktor — eine biologische Abwasserreinigungsanlage in Hochbauweise*, Chem. Ing.-Techn. 51, 4, 288–294.

Metcalf and Eddy, Inc. (1991). *Wastewater Engineering, 3rd edition*, McGraw-Hill Publishing Co., New York.

Mortarjemi, M. and Jameson, G.J. (1978). "Mass Transfer from Small Bubbles — The Optimum Bubble Size for Aeration." *Chemical Engineering Sciences*, 33, 1415–1423.

Pasveer, A. (1955). *Oxygenation of Water with Air Bubbles, Sewage and Industrial Wastes* 27, 1130–1146.

Pöpel, H. J. and Wagner, M. (1989). *Sauerstoffeintrag und Sauerstoffertrag moderner Belüftungssysteme*, Teil 1: Druckluftbelüftung, Korrespondenz Abwasser 36, 453–457.

Pöpel, H. J. and Wagner, M. (1994). "Modeling of Oxygen Transfer in Deep Diffused-Aeration Tanks and Comparison with Full-Scale Plant Data." *Water Science and Technology*, 30, 71–80.

Pöpel, H. J., Wagner, M., and Weidmann, F. (1997). *O_2-deep, Computer Program for the Calculation of the Influence of Diffuser Depth of Submergence on SOTE and SAE Using Different Types of Blowers*, Darmstadt.

Pöpel, H. J., Wagner, M., and Weidmann, F. (1998). *Sauerstoffeintrag- und -ertrag in tiefen Belebungsbecken*, gwf-Wasser·Abwasser 139, 189–197.

Wagner, M. (1998). "Documentation on deep diffused aeration tanks in Europe." *Unpublished internal report*.

Westphal, G. (1995). *Leistungseintrag in Belebungsbecken — eine grundlegende Darstellung (Energy transfer into aeration tanks — a description of the underlying principles)*, Korrespondenz Abwasser 42, 353–1358.

WPCF (1988). *Aeration — Manual of Practice FD-13*, Manuals and Reports on Engineering Practice, 63.

5 Surface and Mechanical Aeration

5.1 INTRODUCTION

Mechanical aeration is defined in this text as the transfer of oxygen to water by mechanical devices so as to cause entrainment of atmospheric oxygen into the bulk liquid by surface agitation and mixing. In addition, equipment that causes dispersion or aspiration of compressed air, high purity oxygen, or atmospheric air by the shearing and pumping of a rotating turbine or propeller will also be included. One may classify mechanical aeration devices based on the physical configuration of the equipment and its operation. Classifications that will be used in this text include low speed surface aerators, motor speed (high speed), axial surface aerators, horizontal rotors, submerged, sparged turbine aerators and aspirating aerators. Detailed descriptions, applications, and performance ranges for these devices will be provided below.

It appears that mechanical aeration in wastewater was introduced to overcome problems with diffuser clogging in activated sludge systems. The concept was introduced in Europe in the late 1910s, predominantly in the UK, and spread to the U.S. slowly. By 1929, mechanical aeration plants outnumbered diffused aeration plants in the UK by two to one. In the U.S., a survey by Roe (1938) indicated that about 100 activated sludge plants employed mechanical aeration, 200 were using diffused aeration, and approximately 20 had combined aeration systems.

Porous tile diffuser clogging in Sheffield, England spurred the development of an Archimedian screw-type aerator in 1916. In 1920, Sheffield built a full-scale facility using submerged horizontal paddle wheels in narrow channels [1.2 to 1.8 m] (4 to 6 ft) that were about 1.2 m (4 ft) deep, called the Haworth System. Located midway between the channel ends that interconnected each aeration tank, the shaft rotated at 15 to 16 rpm producing a longitudinal velocity of (0.53 m/s) 1.75 ft/sec. The movement of wastewater along the channel created a wave action that allowed transport of oxygen from atmospheric air to the water. The power consumed was reported to be 0.114 kwh/m³ (576 hp-h/million gallons). The use of pumps to replace the paddles in moving wastewater along the channels did not provide sufficient oxygen transfer and were supplemented by submerged paddles to satisfy oxygen demand. Triangular paddles, which replaced the rectangular paddles in 1948, improved performance by 40 to 50 percent when the shaft was operated at twice the original rotational speed.

The Hartley aeration system was similar to that used at Sheffield but employed propellers fixed to inclined shafts. These units were located at the U-shaped ends of the shallow interconnected channels. A series of diagonal baffles were located at intervals along the channels. They were set at an angle in the direction of flow to reduce the velocity, prevent suspended solids separation, and create new liquid surfaces to come in contact with the atmosphere. These systems were used at Birmingham and Stoke-on-Trent in the UK. Neither the Haworth nor the Hartley system found service in the U.S.

Other horizontal shaft systems were also being developed in these early years. In 1929, pilot studies at Des Plaines, IL were described in which an aeration device employing a steel latticework was attached to a horizontal shaft to form a paddle-wheel. The paddlewheel, with a diameter of 66 to 76 cm (26 to 30 in), was suspended along the entire length of the aeration tank and was partially submerged so that when rotated, it would agitate the liquid surface. A vertical baffle, running along the entire basin length 46 cm (18 in) from the wall and located below the paddlewheel, terminated at the surface with a narrow trough located at right angles to the basin wall. The shaft was rotated in the range of 36 to 60 rpm by an electric motor. This rotation toward the wall caused mixed liquor to rise upward between the baffle and wall and fall downward in the main basin. The wave-like motion at the surface created new liquid surfaces to contact the air. Mixed liquor flowed in a spiral roll configuration down the aeration tank. This system was known as the Link-Belt aerator. Link-Belt aerators were installed in several U.S. plants in the 1930s but were not in production by the late 1940s.

Another horizontal rotor device often referred to as a brush aerator was developed in the U.S. and Europe in the 1930s. Called brush aerators because of the use of street cleaning brushes during early development, these devices were usually fastened to one longitudinal wall of the aeration tank and partially submerged below the liquid surface. Rotating at speeds ranging from 43 to 84 rpm, the brushes created a wave-like motion across the liquid surface and induced a spiral roll to the wastewater as it flowed down the aeration tank. Kessner employed brushes as well as a combination of brushes and submerged paddles in Holland as early as 1928. Similar in design and function as the brush aerators described above, Kessner employed the submerged paddles mounted on a horizontal shaft that rotated at 3 to 7 rpm. These paddles supplemented the brush and provided a reinforced spiral roll to the mixed liquor. The newer Kessner brushes employed acute triangles cut from stainless steel sheet in place of the brush. The aeration tank bottom was either rounded, or the sidewalls sloped near the bottom to enhance circulation.

An interesting modification of the horizontal paddle aeration system resulted in the combination of paddlewheels and diffused air developed in Germany and reported by Imhoff in 1926. The submerged paddles, made of steel angles and mounted on horizontal shafts running longitudinally along the aeration basin, were rotated counter to the upward flow of bubbles. Diffused air was provided longitudinally along the wall or center-line of the tank. More recent applications of this principle may be found in Chapter 3.

In addition to the horizontal rotor concept, vertical draft tube aerators were also being developed at this time. In the early 1920s the Simplex system was marketed in the UK. The Simplex system in its earliest version employed at Bury, England was a vertical draft tube device placed in a relatively deep hopper-bottom tank. A vertical steel draft tube with open bottom located about 15 cm (6 in) from the floor was suspended at the tank center. At the top of the tube was a cone with steel vanes. The cone was rotated at about 60 rpm drawing mixed liquor up through the draft

TABLE 5.1
Characteristics of Vertical Draft Tube Aerators in 1950

Manufacturer	Characteristic Construction	Number of Aerator Sizes	Variable Control
American Well Works	Down-draft by propeller at bottom of tube, aspirator orifice plate at top, radial inlet troughs		Time switch; adjustable orifice plate openings
Chicago Pump Company	Up-draft, propeller driven flow discharge against diffuser cone at top	10 propeller sizes	Time switch
Infilco, Inc.	Up-draft, induced by horizontal, radially vaned impeller at top		Time switch; adjustable impeller height
Vogt Mfg. Company	Down-draft produced by impeller in tube, radial inlet troughs		
Walker Process Equipment, Inc.	Down-draft by propeller at bottom of tube, aspirator orifice plate at top, radial troughs		Time switch; adjustable orifice plate openings
Yeomans Brothers Company	Up-draft induced by spiral vaned revolving cone at top	4 sizes of aerators	Time switch; optional variable speed

From Committee on Sewage and Industrial Waste Practice (1952). *Air Diffusion in Sewage Works- MOP 5,* Federation of Sewage and Industrial Waste Associations, Champaign, IL.

tube. The wastewater was then sprayed outward over the surface of the tank. Each draft tube was driven by its own motor through a speed reducer or by a line shaft with individual clutches. A number of vertical draft tube systems resembling the Simplex aerator became popular in the U.S. in the 1930s and 1940s. They were the predominant mechanical aerators in the U.S. by the 1950s. Their general character-istics are tabulated in Table 5.1.

The performance of these early mechanical aeration systems was reported as wire power required per unit mass of BOD_5 removed (kWh/lb BOD_5). Results of test conducted in the U.S. by Roe (1938) are given in Table 5.2. Note that a rough estimate of the AE in lb O_2 transferred/kWh can be calculated from these power values by assuming that the ultimate BOD is 1.4x BOD_5 and that no nitrification is taking place. These estimated values are presented in Table 5.2. Note that the values are not at standard conditions but are estimated in wastewater at field temperature and basin DO (not given).

In the 1950s and 1960s, many low-speed aerators were sold in the U.S. and they apparently performed satisfactorily. However, there was no generally accepted method of evaluating the units, and the main testing efforts were aimed at process performance. A major flaw soon became apparent relative to aerator maintenance and reliability, the gear reducers. Often gear reducers failed within a short period of time after initial start-up. Some lasted for a year or two, but many failed after only a few weeks or months.

TABLE 5.2
Power Consumption by Early Mechanical Aeration Plants

City	Make of Device	Wastewater Flow (MGD)	BOD$_5$ Reduction (mg/l)	Wire Power Consumed (kWh/lbBOD$_5$ removed)	Estimated AE based on wire power (lb O$_2$/kWh)
Buhl, MN	Yeomans	0.36	182*	0.205	1.58
Geneva, IL	Yeomans	0.7	116	0.210	1.54
Batavia, IL	American	0.611	178	0.220	1.47
Mitchell, SD	Yeomans	0.5	188	0.275	1.18
Woodstock, IL	Yeomans	0.733	100*	0.312	1.04
Chelsea, MI	American	0.15	156*	0.323	1.00
Waverly, IA	American	0.42	177*	0.407	0.80
Libertyville, IL	American	0.106	223	0.452	0.72
Christopher, IL	Yeomans	0.38	73*	0.480	0.68
Elmhurst, IL	Chicago	2.00	85	0.495	0.65
Princeton, IL	Yeomans	0.488	67*	0.498	0.65
Collingswood, NJ	Link-Belt	0.5	294	0.547	0.59
Flora, IL	Chicago	0.16	205	0.589	0.55
Rochelle, IL	American	0.326	206	0.61	0.53
Harlem, NY	—	0.500	95	0.627	0.52
Dane Co. Asylum, WI	—	0.042	275	0.651	0.50
Clintonville, WI	American	0.305	97	0.674	0.48
Holland/Kessner**	Kessner brush	Domestic	Domestic	0.13–0.19	2.49–1.71
Holland/Kessner**	Kessner brush	Industrial	Industrial	0.25–0.52	1.30–0.62
Muskegon, MI[†]	Combined#	1.37	131	0.348	0.93
Mansfield, OH[†]	Combined#	3.5	94	0.425	0.76
Escabana, MI[†]	Combined#	0.75	135	0.447	0.72
Phoenix, AZ[†]	Combined#	11.25	128	0.447	0.72
Jackson, MI[†]	Combined#	7.5	87	0.785	0.41

Note: MGD × 0.44 = m^3/s, kWh/lb × 2.205 = kWh/kg, lb/kWh × 0.453 = kg/kWh.

[*] Estimated; [**] After Kessner (in *Air Diffusion in Sewage Works*, 1952).
[†] From C. E. Keefer, (1940); [#] Combined-diffused air and mechanical aeration, [‡] estimated assuming BODu = 1.4 BOD$_5$ and no nitrification; nonstandard conditions; based on wire power.

From Committee on Sewage and Industrial Waste Practice (1952). *Air Diffusion in Sewage Works- MOP 5,* Federation of Sewage and Industrial Waste Associations, Champaign, IL.

From a performance perspective, the 1950s vintage impellers were almost all simple radial flow devices. A number of impeller designs were imported from Europe and adopted by U.S. suppliers. Innovative blade designs were developed as well by U.S. manufacturers. At that time and into the 1960s, no reliable test procedure was available to assess the value of these designs. The effects of impeller speed, basin geometry, and other important dependent variables were either unknown or poorly understood. As a result, it is likely that most systems were under designed. On the other hand, virtually all worked to the satisfaction of the operators with the exception of mechanical problems.

By the mid 1970s, the mechanical problems had been recognized and at least to a certain extent, addressed by the major aerator suppliers. The dynamics of the market were changing at that time, with the old-line equipment suppliers being squeezed by newer entrants. Since the 1960s Lightnin, a major manufacturer of mixers, made a big push in the low-speed aerator market using a very inexpensive impeller (a four-blade pitched blade turbine). Shortly thereafter, Philadelphia Gear's Mixing Division entered the market using specially designed reducers and new impellers that were less prone to cause failures. From a mechanical perspective, these new suppliers represented the best level of quality ever seen in the business at a cost that the older manufacturers found hard to match. At least as important, these mixing companies were very familiar with the best approach to blending liquids and suspending solids, and by the mid 1980s, the leading low-speed aerator manufacturers in the U.S. were Lightnin and Philadelphia Mixers. That situation still exists as we enter the twenty-first century since no new low-speed aerator suppliers have come into the market in the last 30 years.

Today, the low-speed surface aerator remains a very popular device in certain niches. High-purity oxygen suppliers have found that good low-speed aerators do the best job for their process, and Eimco continues to be successful in their Carrousel™ ditch process using the low-speed vertical shaft machines. In addition, many low-speed units are performing well in activated sludge systems.

For lagoon applications and situations where capital cost is a major factor, several manufacturers began to offer motor speed or high-speed aerators in the 1970s. Primarily of a floating configuration, the development aimed at lagoons and small-extended aeration facilities. All used marine propellers as the impeller of the nonsnagging type. In the early days of development, these devices were plagued by mechanical difficulties largely due to motor bearing failures as well as poor manufacturing quality control. The hydraulic forces were the main cause of bearing failure, and it took a while for manufacturers to find effective designs to ensure long-term service. New styles of motor speed devices are currently being designed and marketed. Because of their popularity, innovation continues to improve performance and reliability.

At the same time the low-speed aerator was being improved in the 1960s, the horizontal rotor became popular in oxidation ditch applications in the U.S. and Europe. A number of different rotor designs have been used, ranging from brushes to the more complex discs. Their efficiency is consistent with the radial flow style of low-speed aerator impellers, and similar concerns regarding mechanical integrity (gear reducer and bearings) have been addressed and largely overcome.

Also designed for lagoon applications, aspirating devices became popular in the U.S. in the 1970s. A number of different configurations have been used including a floating device that uses a marine propeller mounted at a shallow angle to the horizontal and a submersible pump unit using a vertical draft tube. Fashioned in a way that allows air to be aspirated through its hollow shaft, these devices are effective mixers, adding some oxygen in the process. These units have also experienced a series of historical mechanical difficulties, mainly associated with shaft-supported bearings located below the water surface. A number of approaches have been used in an effort to resolve these problems. The problems remain, and although very inexpensive, they do not provide top performance or trouble free operation.

Finally, in this brief historical overview, are the submerged, sparged turbine aerators that have been used for decades in a number of forms. Industrial mixing requirements often have called for the introduction of a gas into a liquid. The major mixing companies in the U.S. (Lightnin, Philadelphia Mixers, and Chemineer) were all familiar with the concept. In the 1960s and 1970s, several companies tried to improve the surface aerator performance by designing aerators that would disperse compressed air using what is essentially a mechanical mixer. Two general types were developed at that time: the radial and draft tube (radial) and an open-style axial flow type (down-pumping impeller above the sparger). These units were plagued with mechanical problems and did not perform as well as anticipated. As a result, they have fallen out of favor in today's market. The draft tube turbine aerator is similar in concept in that it uses a down-pumping impeller positioned above an air-release device. The impeller and sparger are located within a draft tube that assists flow direction and shearing action. These devices, used in deep basins (7.6 to 9.75 m) (25 to 32 ft), have experienced some early mechanical failures that have recently been overcome. A radial flow submerged turbine aerator uses a radial flow impeller positioned above a sparger. Offered in the early 1960s and still used today in aerobic digestion applications, its mechanical reliability is high.

This chapter will elaborate on mechanical aeration systems, their characteristics, applications, performance, design, and operation.

5.2 LOW-SPEED SURFACE AERATORS

5.2.1 DESCRIPTION

Low-speed mechanical aerators have an impeller positioned at the water surface and pull liquid directly upward in a vertical direction from beneath them. The liquid is then accelerated by the impeller vanes and discharged in essentially a horizontal direction at the impeller rim. The high-speed (supercritical) liquid plume at discharge, in contrast to the slow moving liquid in the tank (subcritical), results in a transition from supercritical to subcritical flow producing a hydraulic jump. The large interfacial area that is generated results in oxygen transfer. The gas phase may be considered continuous, and the liquid phase discontinuous. The reservoir of oxygen is infinite. Therefore, oxygen transfer is limited only by the rate at which the impeller can expose new liquid interfaces to the atmosphere. A relatively large quantity of liquid must be pumped in this process for two reasons: to maintain a high driving force of oxygen in the entraining liquid and to distribute the oxygen enriched liquid throughout the basin. Low speed aerators have extremely high pumping capacities.

The low-speed aerator typically uses impellers configured to pump liquid in a radial manner, so it is generally thought of as a radial flow device. There are, however, a number of impeller configurations (Figures 5.1 to 5.3). Some impellers are flat discs with rectangular or slightly curved vanes attached to the periphery of

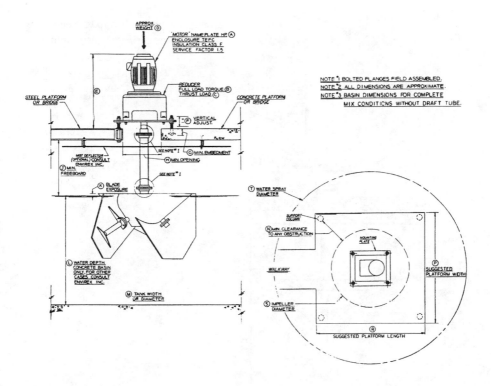

A	B	C	D	E	F	G	H	J	K	L		M		N	P	R	S	T
										MIN FT	MAX FT	MIN FT	MAX FT					
HP	FT LBS	LBS	LBS	IN	IN	IN	IN	FT	IN	FT	FT	FT	FT	FT	FT	FT	IN	FT
5	390	100	1,560	48	3	7	13	3.5	3	5	10	20	40	2	7	6.5	40	9
7.5	580	150	1,650	48	3	7	13	3.5	3	5	12	20	40	2	7	6.5	44	10
10	525	200	1,675	48	3	7	13	3.5	4	5	14	20	40	3	7	6.5	40	16
15	788	300	1,900	48	3	7	13	3.5	4	6	16	25	55	3.5	7	6.5	43	18
20	1,050	400	2,050	59	3	7	15	4.0	5	6	16	30	60	3.5	7	6.5	47	22
25	1,313	500	2,150	59	3	7	15	4.0	5	8	18	35	65	4.0	7	6.5	50	23
30	1,575	600	2,700	59	3	7	15	4.0	5	8	18	35	70	4.0	7	7	53	25
40	2,500	800	2,900	63	3	9	15	4.5	6.5	8	18	35	75	4.5	7.5	8	62	27
50	3,125	1,000	3,900	63	3	9	15	4.5	6.5	10	20	35	80	4.5	7.5	8	64	28
60	3,750	1,200	4,900	63	2	9	18	5.0	8	10	20	40	80	5.0	7.5	8	66	32
75	4,690	1,500	5,100	77	2	12	18	5.0	8	12	22	40	90	5.5	8	8	65	34
100	7,725	2,000	7,350	80	2	12	18	5.0	10	12	24	55	110	5.0	8	9	80	34

FIGURE 5.1 Low-speed surface aerator (courtesy of US Filter–Envirex, Waukesha, WI).

the disc lower surface. Others use inverted conical bodies with vertical blades originating at the center that may be located at top, bottom, or both sides of the cone. New designs include variations of pitched blade turbines, curved blade discs and reverse curvature discs. Most, if not all, surface aerators are hydraulically dependent on liquid level over the impeller (submergence). A small change in liquid level will generally cause a significant change in the head requirements of the impeller. This affects both power input and oxygen transfer. Many impeller configurations will have their own characteristic submergence-aeration efficiency-pumping rate curves. In some instances, a small change in submergence may result in as much as ±50 percent in power variation, whereas with the less sensitive impellers the variation may only be ±10 percent.

A

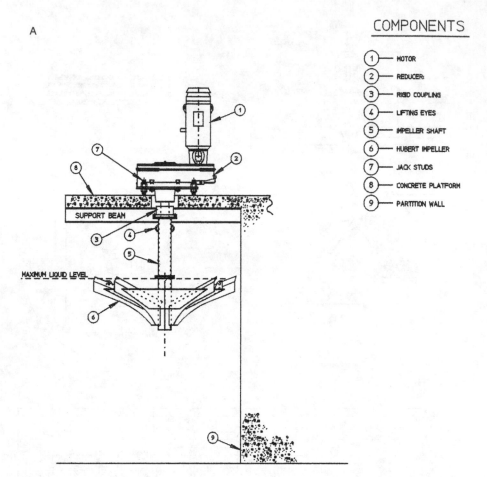

COMPONENTS

1 — MOTOR
2 — REDUCER:
3 — RIGID COUPLING
4 — LIFTING EYES
5 — IMPELLER SHAFT
6 — HUBERT IMPELLER
7 — JACK STUDS
8 — CONCRETE PLATFORM
9 — PARTITION WALL

SUPPORT BEAM

MAXIMUM LIQUID LEVEL

FIGURE 5.2 Low-speed surface aerators. [A) Courtesy of Baker Hughes, Houston, TX; B) courtesy of Philadelphia Mixers Corp., Palmyra, PA.]

Low-speed aerators typically operate at speeds in the range of 20 to 100 rpm. Thus, a gearbox is employed to reduce impeller speed below that of the motor. As described above, the early designs suffered from gear reducer failures. The problem was found to be associated with the reducers that were specified by the manufacturers. They had purchased gear reducers from the large U.S. gear manufacturers and had requested normal industrial reducers. The design of such machines was simply inadequate to handle the large hydraulic loads imposed by aerator duty, so the weakest link would fail. Usually, that was the bearings supporting the impeller shaft, but occasionally, the gears themselves would crater. The result was expensive, time-consuming, and, generally, a universal problem.

Different aerator suppliers dealt with the problem in different ways. Yeomans, for example, added a large bearing at the impeller (and, therefore, right near the water) to take the large loads. All suppliers increased the size of the reducers by increasing the service factor. (The service factor is defined as the calculated power

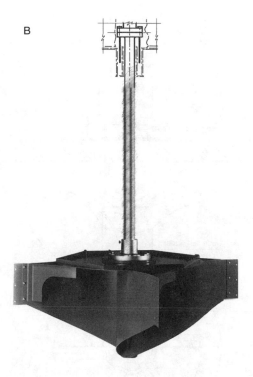

FIGURE 5.2 (continued)

transmission rating of the reducer divided by the actual amount of power used.) By using large reducers with service factors of 2.5 to 3.0, the manufacturers were able to reduce failures to a manageable level. At that point, failures began to occur at the impeller shaft, and so, the shafts were beefed up again reducing failure rates.

More progressive ways of reducing failures were adopted by some suppliers. For example, Lightnin introduced a new reducer design that was developed with Falk for heavy-duty mixer applications—the "hollow quill." That design protects the gears and bearings from the effects of hydraulic forces. A different approach was adopted by Infilco, who joined forces with Philadelphia Gear. They conducted field stress tests to quantify the magnitude of the hydraulic forces and tailored the right reducer to the application.

Low-speed surface aerators are typically bridge mounted because of their size and weight, but they can be float mounted where necessary. The shaft and impeller are suspended from the drive unit above. Platform or bridge designs must account for torque and vibration and should be designed for at least four times the maximum anticipated moment (torque and impeller side load). Some aerators will be equipped with submerged draft tubes to provide better flow distribution within the basin. They are typically used in deep basins (greater than 4.6 m [15 ft]) where the aerator alone may not provide sufficient dispersion of oxygen throughout the basin. The draft tube may also serve as a surge control device preventing wave generation in the tank and

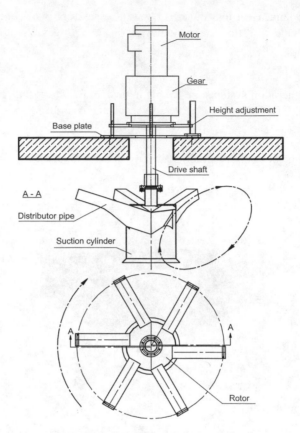

FIGURE 5.3 Low-speed surface aerator (courtesy of Geiger, Karlsruhe, Germany).

eliminating the pulsing loads on the gear-motor assembly. As an alternative to the draft tube, an auxiliary submerged impeller may be installed on the extended impeller shaft. The submerged impeller will increase the amount of liquid pumped from the bottom of the basin thereby increasing oxygen dispersion. The location and config-uration of the turbine will depend on basin geometry and the use of multiple units. Typically, radial flow impellers are used, but axial flow devices are also employed in practice. It should be noted that the additional impeller will result in greater power draw. The unsupported shaft will create high side loads that will create greater stress on the gearbox and must be considered in the design. Unsupported shaft lengths up to about 9 m (30 ft) have been used, but above that, supported shafts and bottom bearings are recommended.

Surface aeration devices create mists that can lead to freezing problems in the northern parts of the world. Furthermore, mists may generate odor problems and have been of concern in air-borne disease transmission. Mist shrouds are mounted above the impeller to restrict the flight of sprays and to reduce the accumulation of ice on the underside of the platform. A drive-ring hood may also be employed for ice control. Splashing effects can also be minimized with proper geometric design of the aeration tank. Heat loss induced by surface aerators is of concern in winter months and should be estimated in the design of the biological treatment system.

5.2.2 APPLICATIONS

The low-speed aeration systems are simple in design, easy to install, relatively easy to maintain with no submerged parts, and require little operational control. Units are available with motor power ranging from several kilowatts to over 150 kW. The low-speed surface aerators for the Carrousel ™ (oxidation ditch) process range from 4 to 150 kW. The very high pumping capacity of low-speed surface aerators allows them to provide excellent mixing and solids suspension in large volumes. It is important to note, however, that using only a surface impeller without a draft tube limits effective mixing depths. The units are flexible in turndown capacity, providing capability for 30 to 50 percent turndown with liquid level sensitive impellers. Typically, though, turndown in transfer rate and power consumption cannot be done independently of pumping capacity and mixing. Thus, oxygen uptake rates can limit the system design when dealing with high strength wastes in a high-rate system. Under variable flow and organic load conditions, the oxygen transfer rate for these units is controlled by the use of variable or dual-speed motors, variable frequency drives, or liquid-level sensitive impellers in conjunction with adjustable weirs.

Low-speed aerators were initially used in completely mixed aeration tanks of conventional and high-rate systems for design flows under about 0.6 m/s (13 mgd). Later, they were used in low-rate extended aeration facilities. Today, they are found in a number of different activated sludge configurations over a wide design flow capacity including tanks-in-series, oxidation ditches, and high-purity oxygen processes.

5.2.3 PERFORMANCE RANGE

The performance of low-speed surface aerators depends on a number of variables including impeller submergence, power input per unit basin volume, aerator pumpage, basin geometry, number and spacing of units, use of baffles, draft tubes and auxiliary impellers, temperature, and wastewater characteristics. Because of the complex hydraulic-pneumatic phenomena involved, it is not realistic to scale-up performance data from small shop tests or models. In general, small units have higher oxygen transfer rates per unit power than very large units. However, the volume of liquid under aeration for any given aerator, has an influence on the oxygen transfer rate, i.e., the smaller the liquid volume per unit of aerator pumpage (power consumption), the higher the transfer rate. Wastewater will affect the oxygen transfer rate as measured by alpha. Values of alpha depend on aerator type, power, basin configuration, and submergence as well as wastewater. Typical values of alpha are reported to range from 0.3 to 1.1 (Boyle et al., 1989; Stenstrom and Gilbert, 1981; WPCF, 1988) and are not very reliable. Few well-designed field studies have been performed with mechanical aeration equipment owing to sampling and measurement difficulties with these systems (See Chapter 7). A discussion of alpha will be found later in this chapter.

Today, performance of low-speed surface aerators are normally reported as standard aeration efficiencies (SAE) expressed as mass oxygen transferred per unit wire power per time (kg/kWh) under standard conditions of temperature, pressure, and DO concentration. Note that power may be measured as drawn wire power or as delivered shaft power (e.g., motor wire-power × motor efficiency × drive efficiency = delivered shaft power). In this chapter, power will be reported as wire power (hp or kW). In the U.S. and Europe, the standard test conditions are clean water (alpha = 1.0), T= 20°C, barometric pressure = 101.3 kP$_a$ (1.0 atm) and 0.0 dissolved oxygen (see Table 2.2).

The old radial flow impellers were found to perform in the range of 1.6 to 1.9 kg/kWh (2.6 to 3.1 lb/hp-h) in clean water at standard conditions. Today, good suppliers can now deliver performance in the range of 1.9 to 2.2 kg/kWh (3.1 to 3.7 lb/hp-h) under typical basin configuration/power situations.

5.3 HIGH-SPEED OR MOTOR SPEED AERATORS

5.3.1 DESCRIPTION

These axial-flow, vertical axis aerators usually have a propeller-type impeller driven by a motor without a gearbox, a shroud in which the impeller is located, and a flow-directing casing. Liquid is drawn upward through the volute. The design of the casing determines the direction of the liquid jets that discharge from the unit. The flow may be horizontal from the aerator, upward and away from the aerator, or downward and away from the unit. These liquid jets partially break into droplets, then entrain and disperse atmospheric air into bubbles on impingement into the bulk liquid of the tank. A large interfacial area is created that promotes oxygen transfer. The impellers typically used are smaller than those used for low-speed surface machines and have lower pumping capacity for a given motor size. Flow patterns are similar to the low-speed units, but bulk liquid rotation within the tank is virtually absent.

The high-speed aerators were initially designed using marine impellers of the nonsnagging type (Figure 5.4). In the 1980s, new styles of high-speed impellers were developed. One, using an Archimedes screw-style impeller, was developed in Europe and trademarked "screwpeller" (Figure 5.5). Another uses a high efficiency "scooped" impeller (Figure 5.6). These impellers provide higher water pumpage rates than the marine propellers and produce reduced hydraulic loads to the unit because of their smoother operation.

Because there is no gear reducer, the impeller rotates at the same speed as the motor. Speeds range from as high as 1800 rpm for the smaller units to about 900 rpm for the large aerators. Motor sizes range from 0.75 to 112 kW (1 to 150 hp). Because of the elimination of the gearbox, the high-speed aerator is lighter than the low-speed unit. Since they are lighter and have a limited shaft length, they are better suited for float mounting and are seldom fixed mounted. Floats are typically poly-urethane foam covered with a stainless steel jacket. In order to improve the effectiveness of the small impellers, a draft tube may be employed to extend the depth of influence. On the other hand, in shallow lagoon applications, an anti-erosion plate

FIGURE 5.4 High-speed surface aerator (courtesy of Aqua-Aerobics Systems, Inc., Rockford, IL).

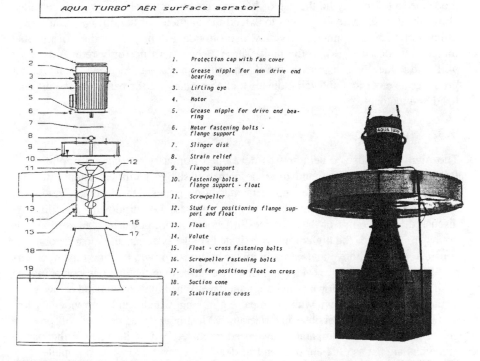

AQUA TURBO° AER surface aerator

1.   Protection cap with fan cover
2.   Grease nipple for non drive end bearing
3.   Lifting eye
4.   Motor
5.   Grease nipple for drive end bearing
6.   Motor fastening bolts - flange support
7.   Slinger disk
8.   Strain relief
9.   Flange support
10. Fastening bolts flange support - float
11. Screwpeller
12. Stud for positioning flange support and float
13. Float
14. Volute
15. Float - cross fastening bolts
16. Screwpeller fastening bolts
17. Stud for positiong float on cross
18. Suction cone
19. Stabilisation cross

FIGURE 5.5 High-speed surface aerator (courtesy of Aquaturbo Systems, Inc., Springdale, AR).

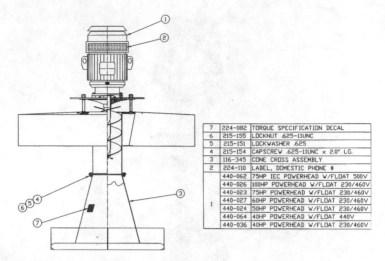

7	224-082	TORQUE SPECIFICATION DECAL
6	215-155	LOCKNUT .625-11UNC
5	215-151	LOCKWASHER .625
4	215-154	CAPSCREW .625-11UNC × 2.0" LG.
3	116-345	CONE CROSS ASSEMBLY
2	224-110	LABEL, DOMESTIC PHONE #
1	440-062	75HP IEC POWERHEAD W/FLOAT 500V
	440-026	100HP POWERHEAD W/FLOAT 230/460V
	440-023	75HP POWERHEAD W/FLOAT 230/460V
	440-027	60HP POWERHEAD W/FLOAT 230/460V
	440-024	50HP POWERHEAD W/FLOAT 230/460V
	440-064	40HP POWERHEAD W/FLOAT 440V
	440-036	40HP POWERHEAD W/FLOAT 230/460V

FIGURE 5.6 High-speed surface aerator (courtesy of Aeration Industries International, Inc., Minneapolis, MN).

may be attached to the bottom of the intake cone resulting in inflow from the sides of the cone rather than from below.

Like their low-speed counterparts, mists are formed from the discharge liquid jets. One solution may be the use of low-trajectory jets. A plate or ring is installed above the diffuser assembly so as to extend the diameter of the diffuser resulting in a flat spray. Some manufacturers may also provide a dome above the diffuser that directs flow downward into the tank. These devices will not only reduce mist but will also reduce heat loss from the discharging sprays. At least one manufacturer produces an electrical anti-icing device that eliminates ice formation on the diffuser head and motor.

5.3.2 APPLICATIONS

The high-speed surface aerator was developed primarily for lagoon applications. Presently, they are also found in some activated sludge facilities. Their low cost, portability, and flexibility are important marketing issues. On the other hand, they suffer from poor mixing characteristics and possess no turndown capability. As discussed later, the use of these devices in lagoons is promoted insofar as mixing is not as critical as for the high biomass activated sludge systems, and most lagoons are considered as facultative and partially mixed systems. In fact, floating mixer/aerators may be selected to improve overall lagoon performance by providing low-power mixing in situations where turndown is an important issue. When oxygen demand is low, some aerators may be shut down without impairing mixing, which can be provided by low power consuming submersible mixer/aerators. Most high-speed devices will produce greater cooling than a comparable low-speed machine. As described below, the marine propellers are not as efficient oxygen transfer devices as the low-speed impellers.

The high-speed aerator is also used in aerobic digesters. Unfortunately, mixing requirements often control design, and high-speed devices will often produce an over-aerated condition. The best equipment for this application includes jet aerators, submerged turbines, and combination aerator/mixers.

5.3.3 PERFORMANCE RANGE

The performance of high-speed aerators depends on basin configuration, unit spacing, power input per unit volume, pumpage rate, use of baffles, draft tubes or anti-erosion plates, impeller type, temperature, and wastewater characteristics among others. As with low-speed devices, performance scale-up from small test tanks is not advisable. The smaller the tank volume per unit pumping rate, the higher the transfer rate. Wastewater affects oxygen transfer in a manner similar to low-rate systems. Values of alpha are reported to fall in the same range as the low speed devices but the database is unreliable.

The high-speed axial propeller units typically produce standard aeration efficiencies in the range of 1.1 to 1.4 kg/kWh (1.8 to 2.3 lb/hp-h). The newer impeller designs claim values about 10 percent higher than the propellers.

5.4 HORIZONTAL ROTORS

5.4.1 DESCRIPTION

Horizontal rotor aerators were introduced early in the 20th century. Initially, they were used in rectangular tanks and placed along a longitudinal sidewall (see Section 5.1). More recently, the horizontal rotors are found in oxidation ditch applications. The earlier Kessner brushes had a horizontal cylinder rotor with bristles submerged in the wastewater at approximately the one-half diameter. Now, most devices use angle steel, other steel flat or curvilinear blades, plastic bars or blades, or plastic discs instead of the earlier bristles.

In the ditch configurations, the rotor spans the width of the channel and rotates so as to discharge a water jet or spray upstream and downward, while imparting a velocity to the liquid as the rotor blades rise out of the water. Oxygen is transferred at the air-water interfaces of the water droplets, or jets, as they are thrown outward from the blades or disc surfaces. Simultaneously, the liquid is propelled by the rotor, thereby mixing the basin and imparting a velocity to the bulk liquid along the basin length downstream. The velocity imparted by the rotor ranges from 0.3 to 1.0 m/s (1 to 3 ft/s) depending on rotor size and speed. Typical rotor lengths range from 3 to 9 m (10 to 30 ft) and are normally used in channels with liquid depths up to about 4 m (13 ft). The rotor is driven by a motor equipped with a gear reducer that provides a rotor speed ranging from 40 to over 80 rpm. A V-belt drive transfers power from the motor to the gear reducer. Speed changes may be provided by sheave changes at the V-belt or by staged bevel/spur gear reducers. The end of the rotor is independently supported by special heavy duty bearing systems that compensate for linear expansion and misalignments.

As described above, the rotor may be equipped with a number of different blade configurations. The steel bladed rotors are typically 69 to 107 cm (27 to 42 in) in diameter (Figure 5.7) and may be submerged 4 to 30 cm (1.6 to 12 in) depending

FIGURE 5.7 Horizontal rotor aerator (courtesy of Lakeside Equipment Corp., Bartlett, IL).

upon rotor diameter and power requirement. The disc aerators are wafer-thin circular plates (typically about 1.5 m [5 ft] in diameter) and submerged in the water for approximately one-eighth to three eighths of their diameter (Figure 5.8). Recesses or nodules located along the disc surface are used in some devices to provide additional lift of the entrained water into the air increasing oxygen transfer and mixing.

The power required to drive the rotor may be controlled by several processes. Standard aeration efficiency (SAE) is also controlled by these methods. These methods include rotor speed (RPM) and submergence of rotor blades, for all devices. For the disc units, power and SAE are also affected by the number of discs on the shaft and the reversal of disc rotation when nodules are employed on the disc surface. Daily variation in oxygen demand is most often met by changing wastewater depth (submergence) by variable weir adjustments. Baffles are often located downstream of the rotors to direct flow downward and to produce greater liquid turbulence. This process normally results in higher SAEs for a given power level.

FIGURE 5.8 Horizontal rotor aerator (courtesy of US Filter–Envirex, Waukesha, WI).

The operation of horizontal rotors is accompanied by liquid splash and mist. In cold climates, this effect may cause significant operational problems with ice build-up. Splash plates are often provided to protect the drive mechanism. Plastic and fiberglass covers and heated hoods are also available.

5.4.2 APPLICATIONS

Today, horizontal rotor aerators are used almost entirely to aerate oxidation ditches, although some units have been used in lagoon applications. The rotors are available in a range of blade configurations and lengths up to 9 m (30 ft). Motor drives are available in the range of 2 to 90 kW providing speeds of 40 to over 70 rpm. Problems associated with gear reducer and bearing reliability have been addressed and largely overcome today. The unidirectional discharge of the rotor is ideal for inducing circulation in a channel type system. Multiple units are often used along the ditch length to promote proper oxygenation and circulation. The disc-type rotors also offer the flexibility of adding or removing discs from the shaft to tune the rotor system to the aeration requirements of the process. Rotor aeration systems are capable of performing over a range of power inputs by several control strategies described above, thereby providing significant turndown capacity. Maintenance requirements are low and operational reliability is reported to be excellent.

5.4.3 PERFORMANCE RANGE

The efficiency of horizontal rotors depends on blade (disc) submergence, rotor speed, blade numbers and configuration, liquid temperature, and wastewater characteristics. The rotors perform in the same range as the radial flow style of low speed aerators. The range of SAE for horizontal rotors is 1.5 to 2.1 kg/kWh (2.5 to 3.5 lb/hp-h). Lesser submergence will decrease the oxygen transfer rate and power, but the SAE will remain approximately the same. One manufacturer claims that the range of oxygen transfer rate (SOTR) is in excess of six to one when both rotor speed and submergence are changed. Yet, SAE values remain relatively constant. As with other mechanical aeration devices, the value of alpha for rotor systems is not well documented.

5.5 SUBMERGED TURBINE AERATORS

5.5.1 DESCRIPTION

Submerged turbine aerators have been used for decades in a number of forms. The submerged turbine normally consists of an open-bladed turbine mounted on a vertical shaft driven by a gear motor assembly, with an air sparger located under the turbine. Both radial flow and axial flow configurations have been employed.

The open-style axial flow turbines use down-pumping impellers. They were designed primarily for basin depths ranging from 4.5 to 6.0 m (15 to 20 ft). Major mechanical difficulties have been encountered with this device caused primarily by the extremely high unbalanced hydraulic forces. The fluid forces acting on the long, overhung shaft and the critical speed considerations both dictate a low operating speed, and thus, a speed reducer. The rotational speed of the impeller is typically in the range of 50 to 100 rpm. Poor oxygen transfer was also obtained with these systems. As a result, these devices have fallen out of favor and are rarely seen today in wastewater treatment applications.

In an effort to improve performance, the down-pumping axial impeller (flat blade or airfoil) was placed within a draft tube along with the sparge ring (Figure 5.9). This change appears to assist flow direction and allows for high shearing action. These units are typically used in basins 7.6 to 9.8 m (25 to 32 ft) deep and achieve higher transfer efficiencies. The sparge ring is typically located at a mid-depth of 3.0 to 4.6 m (10 to 15 ft) depth, which allows for deep tank aeration at conventional depth blower pressures. Shaft lengths are smaller than the open-style units and the unit may be operated at higher speeds (130–180 rpm) and lower torque. As a result, smaller, less costly drive assemblies are required. Initially plagued by mechanical difficulties owing to the severe hydraulic forces, these problems have been overcome today.

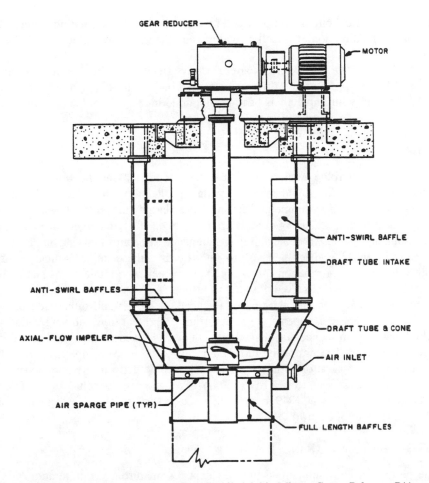

FIGURE 5.9 Draft tube aerator (courtesy of Philadelphia Mixers Corp., Palmyra, PA).

The radial flow submerged turbine uses a radial (horizontal) flow impeller positioned above the sparge ring. Several impellers are typically placed on the same shaft above the lower impeller. This was the typical industrial configuration and has been offered since the 1960s. It is less efficient than the axial units and finds limited application as an aeration device.

For all of these devices, oxygen transfer is affected by the high turbulence field provided by the impeller at the air bubble column discharging from the sparger. The high-energy field breaks up the bubbles and disperses them into the bulk liquid. Therefore, the mechanism of transfer is different than that of the surface aerators described above in that the fluid becomes the continuous phase, and the gas is the discontinuous phase. Oxygen supply is controlled by the rate of airflow to the system. Transfer rate depends on both airflow and the oxygen stripping efficiency of the impeller. Power is the sum of both the shaft input power to the turbine and the power to deliver the gas. Flow patterns are determined by three major components that include the vertical circulation provided by the impeller(s), the rotating water mass moving in the direction of impeller rotation, and the geometric effects of the basin and baffles.

Typically, all of these devices are driven by a standard direct connected gear-motor drive. Both motor and drive are mounted on a beam that spans the aeration tank. The shaft with impellers is bearing supported in the gear-reducing drive head. Shaft alignment is assured by a steady bearing.

5.5.2 APPLICATIONS

The submerged turbine aerators are best suited to deep tank applications. The draft tube turbines are also found in some oxidation ditches such as the barrier ditch process. The radial flow submerged turbine is used largely in aerobic digesters where independent mixing and oxygen transfer are desired. The submerged turbines offer high pumping capacity and the ability to independently control mixing and aeration by adjusting turbine speed and airflow rate. Further, these units eliminate spray-related ice and mist formation caused by the surface aeration units. This factor also minimizes heat loss observed for the surface units. The disadvantages of these devices include higher capital costs and the need for blower and submerged piping. As discussed above, the open style axial turbine and radial flow submerged turbine exhibit lower performance than that found with other surface aeration and submerged aeration systems. The draft tube turbine appears to be more competitive from the point of view of aeration efficiency. Available submerged turbine aerators match common motor sizes up to 112 kW (150 hp). Special designs include motors up to and exceeding 260 kW (349 hp). Airflow rates vary from 0.2 to greater than 8.0 m^3/min (8 to 300 scfm).

5.5.3 PERFORMANCE RANGE

The performance of submerged turbines depends on turbine configuration, basin geometry, airflow rate, turbine speed, temperature, and wastewater characteristics. The open-style axial submerged turbine has been reported to provide values of SAE in the range of 1.0 to 1.6 kg/kWh (1.75 to 2.75 lb/hp-h), whereas the radial flow turbines provide SAEs in the range of 1.1 to 1.5 kg/kWh (1.8 to 2.5 lb/hp-h). The improved draft tube turbine has been shown to provide substantially higher SAEs of 1.6 to 2.4 kg/kWh (2.7 to 4.0 lb/hp-h). However, in barrier ditch applications, these draft tube turbines produced low SAE values ranging from 0.8 to 1.2 kg/kWh (1.4 to 2.0 lb/hp-h) (Boyle et al., 1989).

5.6 ASPIRATING AERATORS

5.6.1 DESCRIPTION

Aspirating devices draw atmospheric air into a mixing chamber where wastewater is contacted with the air. The air-water mixture is subsequently discharged into the aeration basin.

A

FIGURE 5.10A Selected aspirator aerators (courtesy of Aeromix Systems, Inc., Minneapolis, MN).

At least two types of configurations are employed. The first uses a tube mounted at an angle in the water with a motor and air intake above the water surface. A propeller located below the water surface within the tube draws liquid down through the tube creating a low-pressure zone at the hub of the propeller. This low pressure draws air through the air inlet to the propeller hub where it intermixes with the water. Turbulence breaks up the air bubbles and the resultant air-water mixture discharges into the basin mixing the contents and dispersing the oxygen (Figure 5.10). These units may be mounted on booms or floats and can be mounted at various angles depending on basin geometry and aeration and mixing requirements. The degree of mixing, direction of flow, and speed of aspiration can be controlled.

Another aspirating device uses a submersible pump supplemented with a vertical air intake tube open to the atmosphere. The pumping of the liquid creates a low-pressure region at the impeller thereby drawing air down the shaft. Air and water are combined and discharge through diffuser channels into the aeration basin (Figure 5.11). Turbulence and flow created by the impeller break up the air bubbles and mix the basin. These units may be mounted on the basin floor, placed on removable guide rails, or fixed to a floating support.

5.6.2 APPLICATION

These devices are good, low-cost mixers but are not efficient aeration devices. They may be supplemented with small blowers to force more air into the unit, improving oxygen transfer rate but not efficiency. They are well suited for lagoon systems where supplemental mixing may be desirable for achieving more operational flexibility. The submersible pumping action may provide directional flow to move wastewater and/or

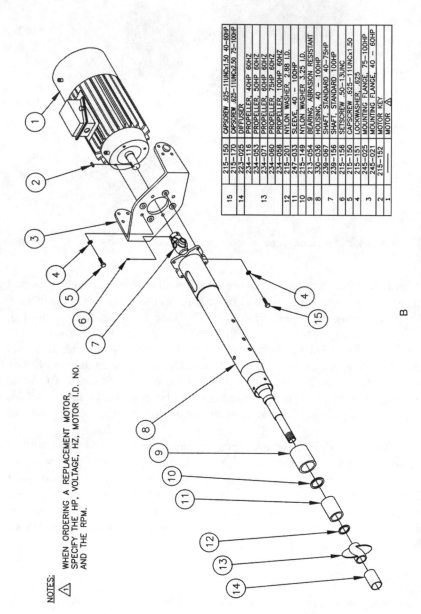

15	215–150	CAPSCREW .625–11UNCx1.50 40–60HP
	215–170	CAPSCREW .625–11UNCx2.50 75–100HP
14	223–025	DIFFUSER
13	234–116	PROPELLER, 40HP 60HZ
	234–053	PROPELLER, 50HP 60HZ
	234–071	PROPELLER, 60HP 60HZ
	234–060	PROPELLER, 75HP 60HZ
	234–058	PROPELLER, 100HP 60HZ
12	215–201	NYLON WASHER, 2.88 I.D.
11	247–033	SLEEVE, 40 — 100HP
10	215–149	NYLON WASHER 3.25 I.D.
9	213–054	BEARING, ABRASION RESISTANT
8	330–036	HOUSING, 40 — 100HP
7	239–067	SHAFT, STANDARD 40–75HP
	239–156	SHAFT, STANDARD 100HP
6	215–156	SETSCREW 50–13UNC
5	215–150	CAPSCREW .625–11UNCx1.50
4	215–151	LOCKWASHER, .625
3	245–025	MOUNTING FLANGE, 75–100HP
	245–021	MOUNTING FLANGE, 40 — 60HP
2	215–152	MOTOR KEY
1	———	MOTOR ⚠

NOTES:
⚠ WHEN ORDERING A REPLACEMENT MOTOR, SPECIFY THE HP, VOLTAGE, HZ, MOTOR I.D. NO. AND THE RPM.

B

FIGURE 5.10B (Courtesy of Aeration Industries International, Inc., Minneapolis, MN.)

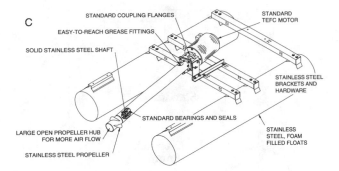

FIGURE 5.10C (Courtesy of Aeromix Systems, Inc., Minneapolis, MN.)

sludge to aerated zones. Furthermore, the units may be used during low oxygen demand periods to supply mixing and to allow cycling of other lagoon aerators (on and off) to meet system oxygen demand without sacrificing mixing. They also find applications aerobic digesters, post and preaeration systems, flow equalization tanks, for mixing stratified lakes, for ice control in harbors and as temporary supplemental aeration in wastewater treatment plants. The tube angle mounted units are available in sizes ranging from 0.75 to 75 kW (1 to 100 hp). The submersible units can be found in sizes ranging from 1.5 to 75 kW (2 to 100 hp).

5.6.3 PERFORMANCE RANGE

The aspirating aerator demonstrates low aerator efficiency (SAE), ranging from 0.4 to 0.9 kg/kWh (0.6 to 1.5 lb/hp-h). As described above, they are good low-cost mixers that are most effective in supplementing other aeration equipment and providing additional mixing energy to the system.

5.7 FACTORS AFFECTING PERFORMANCE

5.7.1 GENERAL

Two important relationships for agitated basins with a single-phase liquid are those for the dimensionless power number, P_0, and Reynolds Number, R_e,

$$P_0 = P/\rho N^3 D^5 \tag{5.1}$$

$$R_e = D^2 N \rho / \mu \tag{5.2}$$

where,

P = the mixer power input
D = the impeller diameter
N = the impeller speed
ρ = liquid density
μ = absolute viscosity

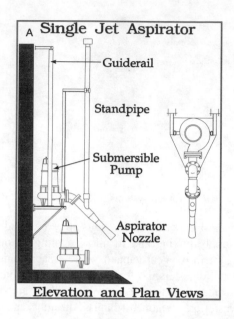

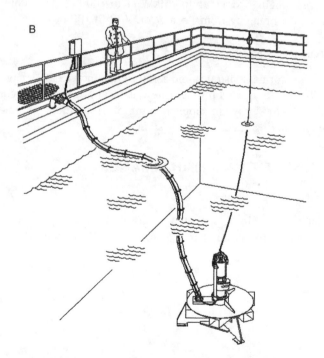

FIGURE 5.11 Selected aspirator aerators. [A) Courtesy of JetTech, Edwardsville, KS; B) and C) courtesy of Nopon Oy, Helsinki, Finland; D) courtesy of Aeration Industries International, Inc., Minneapolis, MN.]

C

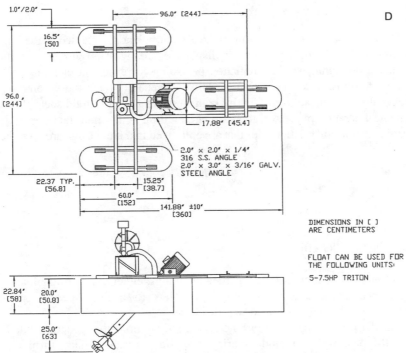

D

DIMENSIONS IN [ ]
ARE CENTIMETERS

FLOAT CAN BE USED FOR
THE FOLLOWING UNITS:

5-7.5HP TRITON

FIGURE 5.11 (continued)

In low viscosity fluids such as wastewater, turbulent flow at $R_e \geq 10^4$ is the logical operating condition. For baffled tanks at turbulent flow, the values for P_0 in ungassed liquids range from 0.1 to 10. For axial flow impellers (marine propellers, pitched-blade turbines) the value of P_0 ranges from 0.3 to 1.0 in baffled tanks, whereas the radial flow turbines produce values in the range of 2.5 to 5. This illustrates that radial flow machines will impart greater power to an ungassed fluid as compared with axial flow devices for the same operating conditions.

In mixer design, both the pumping capacity and impeller head characteristics are important and can be expressed by the following dimensionless groups.

$$N_q = Q/ND^3 \tag{5.3}$$

$$N_h = gH/N^2D^2 \tag{5.4}$$

where,

N_q = the pumping coefficient or pumping number
N_h = the head coefficient
Q = the pumping rate or circulation rate
H = the impeller head generated
g = gravitational acceleration

The head coefficient relates to the shear or turbulence generated at the impeller with high values being desirable for dispersive mixing. The pumping coefficient indicates a predominance of circulation providing for excellent blending and suspension of solids. Mixer operations are of two general types: those controlled by impeller flow and those controlled by both impeller flow and fluid shear (head). In mechanical aeration processes, the second condition is the important one.

Two other useful relationships that are employed in mixer design are given below.

$$P = \eta Q \rho g H \tag{5.5}$$

$$P_0 = \eta N_q N_h \tag{5.6}$$

η = the impeller efficiency.

In general, impellers can be regarded as pumps, either axial or radial, and their performance can be considered accordingly. Thus, for a given speed, as pumping rate increases, the pumping head will decrease.

Impeller types range from turbine-type configurations that generate radial flow to axial-types, which include marine impellers, pitched-blade turbines, and Archimedes screw-type impellers. With axial flow impellers, there is an option for choosing flow direction, either upward or downward. Radial flow turbines are primarily high shear devices used where a recirculation pattern is advantageous (solid/liquid mixing). Axial flow impellers are effective pumping devices. In aeration processes, the development of the fluidfoil (hydrofoil or high efficiency impeller) has been offered to minimize high localized shear in the impeller zone while providing a high pumping coefficient relative to the head coefficient. The pitch (the advance of fluid along the axis for one revolution) of an axial type impeller has been considered of importance in mixer design. As pitch increases the value of P_0 and N_q increases, indicating that greater power and pumping capacity is produced. For mechanical aerators, it appears that a pitch to impeller diameter ratio of 1.0 is desirable (a square pitch is defined as a pitch equal to the impeller diameter).

In reviewing the relationships above, it becomes evident that a given large diameter impeller running at low speed at a given power level produces a high fluid flow and a low level of fluid shear. In contrast, a smaller impeller running at high speed at the same power level produces a low quantity of fluid flow at a relatively high level of fluid shear. One may also deduce that for a given type of impeller, the pumping capacity per unit input power will be higher for the lower speeds. Although this discussion has centered on single-phase fluid mixing, the general concepts remain the same for gassed, two-phase fluids.

An additional performance parameter of importance is solids suspension brought about by basin mixing. Mixing continues to be poorly understood. As described below, engineers often rely on empirically measured parameters to provide some insight into solids suspension. Mixing requires that flow streams be developed within the basin. The greater the flow, the faster the mixing. Since the discharge rate per unit power input is widely variable for different devices, the volumetric power dissipation value (power/unit volume) may be misleading. Flow generation capacity per unit of basin volume or turnover rate is more pertinent but still does not define the quality of mixing. A small diameter, high velocity stream and a large diameter, low velocity stream may have the same mass flow rate, but they will produce different mixing results. Similarly, having all of the mixing energy injected at one point in a basin will produce a different result than having the energy injected at multiple points. The direction, location, and number of streams developed will have a significant affect on the resulting fluid regime of the basin.

It appears to be far more sensible to specify mixing performance of aeration equipment rather than power dissipation. Velocity specifications are popular, but velocity in itself is not sufficient. For example, if the entire basin is rotating in one direction at a given velocity, there would be little mixing insofar as the particles will move along the streamlines. The development of random mixing requires variation in velocity and its direction (G), thereby producing a measured zero velocity at some points within a well-mixed basin. If mixing is intended only to diffuse oxygen to all points within the basin, the mixing requirement will be significantly lower than if particulate solids are to be suspended. Generally, the best and most practical specification of mixing would be the specification of uniformity of distribution (of solids or DO or any desired characteristic) within the basin.

That being said, today's practice often finds system design based on volumetric power dissipation. Rules of thumb based on observation have been developed to that end. In general, for activated sludge systems, the power required for oxygen transfer will exceed that required for mixing. In lagoon systems, where oxygen demands per unit volume may be low, power for mixing may control aerator sizing. Chapter 3 (Section 3.4.4) discusses mixing characteristics. Using recommended values of G for particulate suspension (40 to 80 sec^{-1}), power dissipation values ranging from 1.6 to 6.4 W/m^3 (8 to 32 hp/MG) are calculated. Depending on aerator spacing and depth, power dissipation values ranging from 4 to 12 W/m^3 (20 to 60 hp/MG) are sometimes used in practice as empirical guidelines. Since the root mean square velocity gradient, G, is of great importance in solids suspension, the creation of turbulence is paramount in design. As described below, the use of baffling becomes an important design feature for these systems not only to insure that impeller shafts are not subjected to large forces, but also to provide an effective mixing regime.

5.7.2 FACTORS AFFECTING LOW-SPEED SURFACE AERATOR PERFORMANCE

The most significant factors affecting the oxygen transfer performance of low-speed surface aerators include impeller type, impeller speed, impeller submergence, basin geometry, and auxiliary mixing and baffling.

Impeller Type — The characteristics of the impeller have an important impact on the performance of the aerator. As discussed above, low-speed devices employ impellers that are configured to pump liquid in a radial manner. The pumping and shear characteristics and therefore, the oxygenation capability, will vary with impeller properties. Manufacturers have invested substantial research effort into providing impellers that will produce high air/fluid surfaces and promote effective mixing and entrainment of air bubbles. Each device has its own distinctive characteristics that are designed to achieve high SOTR at a low power input.

Impeller Speed — As indicated by the power, pumping, and shear equations provided above for submerged mixers in a single-phase liquid, the impeller speed is an important operational variable. Although the constants in these equations will differ for surface aerators, the trends are the same. A given surface impeller will pump more water at a lower speed for the same power input as one operated at a higher speed. Thus, it should transfer more oxygen up to a point where impeller shear falls too low and the water spray no longer provides effective gas transfer surfaces. The significance of speed depends on the impeller type, and each aerator will exhibit its own characteristic relationship (Figure 5.12).

Impeller Submergence — Low-speed surface aerators are located at the water surface, and the impeller is submerged partially in the water. Clearly, the more the impeller is submerged, the greater will be the power required to drive the device and the greater the power imparted to the fluid. As might be expected, the rate of oxygen transfer (SOTR) will increase with submergence to a point. The input power also increases but not necessarily in the same manner. Therefore, there will normally be an optimum submergence to achieve the highest SAE (Figure 5.12). This characteristic of the aerator provides a useful tool for controlling surface aerator performance by using water surface elevation as a control parameter.

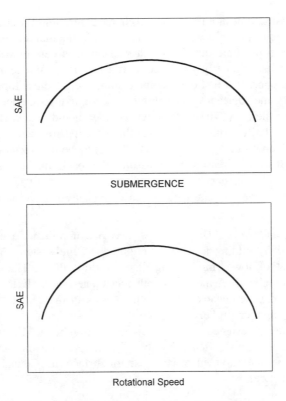

FIGURE 5.12 Effect of rotor speed and submergence on SAE for low-speed surface aerators.

Basin Geometry — The geometry of the aeration basin will significantly impact performance. Since the flow-fluid shear ratio at a given power level can be varied by varying the ratio of impeller size to tank size, it is of great importance to know the optimum combination of flow and shear for a given process. For a given impeller, there will be an optimum ratio that should be selected. Although some rules-of-thumb have been generated for an aerator in various geometric configurations, these relationships are strictly empirical and the factors affecting performance are not well understood. Engineers often use aerator power input per unit area or volume to attempt to quantify and scale-up performance. It is apparent from experimental observation that as power input per unit area (or volume) decreases, the aeration efficiency decreases for a given device placed in different sized basins. The reasons for this are presumed to be due to changes in turbulence and the impact of wall effects. This phenomenon has been reported in Oldshue (1956), von der Emde (1968), EPA (1983). Other geometric relationships such as basin depth and impeller diameter to basin diameter ratio have been related to aerator performance. It is important that the designer determines the effects of basin geometry on aerator performance of a given device and specifies specific geometric constraints in aerator specifications and testing. In general basin geometry impacts system turbulence and circulation.

Another important design feature of mechanical aeration devices deals with the incorporation of multiple units within a basin. Considering that the aerator is a pump, it is reasonable to presume that each unit will pump a certain volume of water, thereby defining its volume of influence within the basin. For optimal efficiency, each unit should operate at its full capacity. If units are too close together, each will compete for the same adjacent water volume resulting in interference that will result in a loss of unit efficiency. These interferences may include the creation of surface turbulence that will affect submergence and therefore, performance. If machines are too far apart, there will be a volume of water that may be unaffected, and therefore, poorly aerated. It may be reasonable to assume that each unit serves an equivalent cell and that the cell boundary bisects the distance between adjacent aerators. This *de facto* volume per aerator can then be reasonably used to estimate performance in each cell.

Ancillary Mixing, Baffling, and Draft Tubes — The zone of influence of a surface aerator can be extended by adding ancillary equipment to the system. Mixing devices generally consist of lower mixing impellers or draft tubes. These devices may improve circulation but often have little influence on aerator SAE. However, in some instances, there may be negative effects on SAE as in the case of supplementary down-pumping impellers in very deep tanks. Draft tubes may also have a negative impact if improperly designed so as to increase friction head. The use of baffles in circular or square tanks is often prescribed for mixers to avoid vortexing and to promote greater system turbulence. Their use for surface mechanical aeration is not often seen in practice.

5.7.3 FACTORS AFFECTING HIGH-SPEED AERATOR PERFORMANCE

The factors affecting the performance of high-speed aerators are generally the same as low-speed devices with the exception of impeller submergence.

Impeller Type — The type of impeller used in high-speed aerators is an axial flow type that pumps upward. Marine propellers and Archimedes-type screw impellers are most often used. Both appear to produce similar results related to oxygen transfer and circulation.

Impeller Speed — Impeller speed affects performance of high-speed machines in the same way that it affects low-speed devices. Generally, speed control is not considered as an operational variable for these devices.

Basin Geometry — Basin geometry appears to affect the high-speed aerators in the same way as it affects low-speed devices. Little work has been done, however, to develop good data on effects of geometry for high-speed aerators. Because a majority of these aerators are used in lagoon applications, there is little wall effect and most of the transfer appears to occur within the area of aerator spray. Most of these devices are not efficient mixers and, therefore, have limited effectiveness in deep basins. The rules governing multiple units are similar to those for low-speed units.

Auxiliary Mixing, Baffling, and Draft Tubes — Draft tubes are used to improve circulation in deep basin applications. Supplementary mixing by addition of submerged

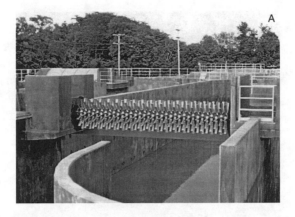

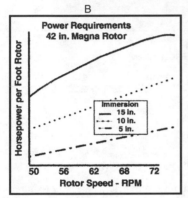

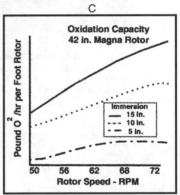

FIGURE 5.13 Effect of rotor speed on SAE for horizontal rotors (courtesy of Lakeside Equipment Corp., Bartlett, IL).

impellers is not practiced, but additional mixing units may be added to large basins to provide greater flexibility in aeration turndown control. These supplementary operations do not appear to appreciably affect oxygen transfer efficiency but may improve basin circulation and solids suspension.

5.7.4 FACTORS AFFECTING HORIZONTAL ROTOR PERFORMANCE

The performance of horizontal rotor aerators primarily depends on rotor speed and the submergence of the rotor blade. As rotor speed increases, the oxygen transfer rate (SOTR) increases up to some maximum speed characteristic of the device. Power also increases with speed although not at the same rate (Figure 5.13). As a result, the SAE for a given rotor will exhibit a curve with a unique rotor speed. Oxygen transfer rate also increases with rotor submergence as does input power. Again, there may be an optimum submergence for SAE. As described above for low-speed aerators, submergence may be a useful tool in oxygen transfer control. Figure 5.14 shows the impact of submergence on power draw and oxygen transfer for a horizontal rotor aerator operating at 90 rpm.

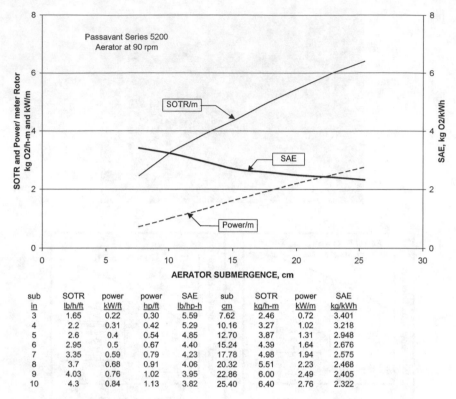

FIGURE 5.14 Effect of aerator submergence on oxygen transfer and power requirements for horizontal rotors (EPA, 1978).

Horizontal rotors are affected to some extent by basin geometry. Most applications of horizontal rotors today are in ditch configurations. Since they are used to transfer oxygen, suspend mixed liquor, and rotate the basin contents, the horizontal velocity profile that is generated by these devices is an important design factor. Maximum water depths are often prescribed. Baffles may be used to direct flow along the ditch. Both tiltable and fixed guide baffles are found in practice. Baffles do not typically result in improved SAE values. Since mixing is an important issue in ditch designs, the use of ancillary horizontal mixers in conjunction with the rotors may be considered.

5.7.5 FACTORS AFFECTING SUBMERGED TURBINE AERATOR PERFORMANCE

The submerged turbine aerator employs separate gas sparging as an incremental part of the system. Since air or high purity oxygen is introduced under the impeller, the fluid density is decreased, resulting in a decrease in impeller power. This may be expressed as

$$P_g = KP_m \tag{5.7}$$

where

P_g = the gassed mixer power

P_m = the ungassed mixer power

K = proportionality factor, which depends on gas flow/area or gas flow/impeller area

For these machines, the rate of oxygen transfer depends on both impeller (mixer) power and compressor power. Oldshue (1955) showed that for a given type of impeller, tank geometry, and wastewater

$$OTR \propto P_g^x P_c^y \tag{5.8}$$

where

P_c = compressor power

x and y = experimental exponents

To obtain adequate aeration capabilities, the power can either be put into the compressor or mixer. When $P_g/P_c \ll 1.0$, the high airflow rates result in large bubbles, impeller flooding, and low oxygen transfer efficiencies. When $P_g/P_c \gg 1.0$, too much impeller power is being expended in fluid mixing. The optimum power split will vary with tank geometry and impeller but is typically near 1.0. Variation in oxygen demand can be easily adjusted by varying airflow rate under the impeller.

To eliminate swirling and vortexing, baffles are normally required in square or round submerged turbine aeration tanks. In round tanks, baffles are provided at the quarter points; in square tanks two baffles are provided at opposite sides. In rectangular tanks greater than a length to width ratio of 1.5, no baffles are required.

5.8 PERFORMANCE OF MECHANICAL AERATION DEVICES

There is a paucity of good, reliable data on the performance of mechanical aerators. Testing of mechanical aerators in clean or process wastewater should be performed in accordance with an acceptable methodology. The testing of aerators is described in detail in Chapter 7. Until about the mid 1980s, test procedures were highly variable, and results of these tests were often of questionable value. Testing may be performed in the manufacturer's or consultant's test facility or in the full-scale facility. Scale-up of test data from shop tests to full-scale facilities is perilous and should be avoided for mechanical devices unless a reliable, substantial database is available. As described earlier, SAE values tend to decrease as the volumetric (or areal) power dissipation value (W/m³) decreases.

Tables 5.3 and 5.4 provide a selection of good, reliable data for a variety of mechanical aerators in given geometric configurations. These examples of clean water performance have been selected primarily based on a careful review of test procedures and methods of evaluation. They are not intended to be used for design purposes but to illustrate the typical ranges of performance observed for these devices.

TABLE 5.3
Typical Clean Water Performance of Selected Mechanical Aerators

Type	Description	Input Power kW(hp)	Power Dissipation W/m³(hp/MG)	Power Dissipation W/m²(hp/kft²)	n rpm	Subm cm(in)	(wire)SAE kg/kWh (lb/hp-h)	Notes	Reference
HS	Arch.screw/float	11.3 (15.3)	51 (257)	196 (24.4)	1450	—	1.51 (2.47)	Shop/d = 3.83 m	GSEE, 1997a
HS	Arch.screw/float	18.9 (25.3)	85 (431)	326 (40.4)	1460	—	1.54 (2.53)	Shop/d = 3.83 m	GSEE, 1997a
HS	Arch.screw/float	28.8 (38.6)	129 (657)	497 (61.7)	1465	—	1.46 (2.39)	Shop/d = 3.83 m	GSEE, 1997a
HS	Arch.screw/float	43.0 (57.6)	50 (253)	180 (22.9)	1175	—	1.15 (1.88)	Shop/d = 3.71 m	GSEE, 1997b
HS	Marine prop/float	48.1 (64.5)	55 (282)	210 (25.8)	1180	—	1.20 (1.97)	Shop/d = 3.71 m	GSEE, 1997b
HS	Arch.screw/float	51.1 (68.4)	59 (301)	220 (27.4)	1175	—	1.27 (2.09)	Shop/d = 3.71 m	GSEE, 1997b
HS	Marine prop	51.0 (68.4)	13 (68)	48 (6.0)	1200	—	1.17 (1.92)	Field/d = 3.58 m	Dausman, 1995
HR	Magna/40hp/4units 1.07 mØ, 7.6 m long	135 (181)	16 (83.4)	64 (8.0)	72	22 (8.5)	1.88 (3.10)	Field ditch/d = 4.5 m	GSEE, 1989
HR	Magna/40hp/4units 1.07 mØ, 7.6 m long	95 (128)	11 (59.0)	45 (5.7)	61	22 (8.5)	1.88 (3.10)	Field ditch/d = 4.5 m	GSEE, 1989
HR	Magna/40hp/3units 1.07 mØ, 6.1 m long	95 (128)	29 (147)	68 (8.4)	—	32 (12.7)	1.79 (2.96)	Field ditch/d = 2.28 m	GSEE, 1984
HR	Magna/40hp/3units 1.07 mØ, 6.1 m long	69 (92)	21 (106)	49 (6.1)	—	22 (8.5)	1.77 (2.93)	Field ditch/d = 2.28 m	GSEE, 1984
HR	Magna/1unit/1.07 mØ, 6.4 m long	24.4 (32.7)	19 (35.3)	57 (7.1)	72	23 (9.2)	1.72 (2.85)	Field circular ditch/d = 3.05 m	Lakeside, 1992

Type	Name							Notes	Reference
ST	Draft tube turbine/2spd	98 (132)*	58 (293)	443 (55)	—	—	1.8 (3.0)	Shop test/d = 7.6 m	Mixco, 1984
		66 (88)*	39 (195)	296 (36.7)	—	—	1.7 (2.8)	Gas flows = 775/520 scfm	
ST	Conical gas diffuser 75hp/1unit	88 (118)*	40 (228)	410 (51.3)	—	—	1.66 (2.72)	Field/d = 9.2 m gas flow = 800 scfm	Stenstrom, 1989
ST	Conical gas diffuser 60hp/1unit	53 (71)*	27 (137)	250 (30.8)	—	—	1.54 (2.52)	Field/d = 9.2 m gas flow = 480 scfm	Stenstrom, 1989
ST	Conical gas diffuser 40hp/1unit	45 (60)*	23 (116)	210 (26.1)	—	—	1.55 (2.55)	Field/d = 9.2 m gas flow = 420 scfm	Stenstrom, 1989
ST	Draft tube turbine/75hp 2 units	123 (165)*	24 (110)	76 (9.4)	—	—	1.05 (1.73)	Field/total barrier	Boyle et al., 1989
		60 (80)	11 (53)	37 (4.6)	—	—	1.13 (1.86)	ditch/d = 3.49 m	
		106 (142)	20 (94)	65 (8.1)	—	—	0.76 (1.25)		
		182 (244)	35 (162)	112 (14.0)	—	—	1.22 (2.01)		
ST	Draft tube turbine/50hp 1unit	45 (60)	16 (80)	69 (8.6)	—	—	0.95 (1.57)	Field/ total barrier	Boyle et al., 1989
		22 (30)	7 (40)	33 (4.3)	—	—	1.13 (1.86)	ditch/d = 4.0 m	
AS	Float/20hp	14 (19)	47.8 (242)	—	2900	65 (26)	0.40 (0.66)	Field/ditch/d = 2.47 m	Kayser, 1992
AS	Float/15hp	14.4 (19.3)	48.3 (245)	—	1455	80 (31)	0.83 (1.37)	Field/ditch/d = 2.47 m	Kayser, 1992
AS	Float/7.5hp	5.8 (7.8)	24.3 (123)	—	2930	50 (20)	0.56 (0.92)	Field/ditch/d = 2.47 m	Kayser, 1992
AS	Float/5.5hp	4.9 (6.6)	20.5 (104)	—	1430	50 (20)	0.92 (1.51)	Field/ditch/d = 2.47 m	Kayser, 1992

Type: HS — high-speed surface; HR — horizontal rotor; ST — submerged sparged turbine; AS — aspirating aerator; wSAE — standard wire efficiency.

* Total power, including turbine and blower wire power.

TABLE 5.4
Performance of Low-Speed Surface Aerators in Clean Water*

Description**	No./ Spacing (m)	Depth m (ft)	Input Power kW (hp)	Power Dissipation W/m³ (hp/MG)	Power Dissipation W/m² (hp/kft²)	Speed rpm	(wire) SAE kg/kWh (lb/hp-h)
75kW/2spd/35°∇	4/12.5	5.8 (18.9)	65 (87)	71 (359)	400 (51)	56/42	1.8 (3.0)
56kW/1spd/35°∇	4/11.0	8.2 (27.0)	48 (64)	50 (253)	415 (52)	56	1.9 (3.1)
75kW/1spd/CSO	9/13.4	5.2 (17.2)	65 (87)	66 (332)	345 (43)	45	1.8 (3.0)
75kW/1spd/35°∇	9/13.4	6.1 (20.0)	65 (87)	61 (306)	370 (46)	46	2.4 (3.9)
45kW/1spd/CURV	2/11.6	4.6 (15.0)	39 (52)	63 (319)	290 (36)	56	2.3 (3.8)
75kW/2spd/CSO	4/16.8	5.5 (18.1)	65 (87)	42 (213)	230 (29)	45/34	1.9 (3.2)
75kW/2spd/30°∇	4/13.1	5.8 (19.2)	56 (75)	66 (332)	385 (48)	47/35	2.0 (3.3)
30kW/2spd/CSO	2/8.5	4.7 (15.4)	26 (35)	63 (319)	300 (37)	56/42	2.1 (3.5)
112kW/2spd/CC	4/18.9	7.9 (25.8)	97 (130)	63 (319)	280 (35)	42/32	1.8 (3.0)
45kW/1spd/CURV	4/13.7	4.9 (16.0)	39 (52)	39 (200)	192 (24)	56	2.0 (3.3)
112kW/2spd/CC	4/15.8	6.0 (19.6)	112 (150)	74 (372)	440 (55)	42/28	2.1 (3.5)
45kW/2spd/CURV	4/11.6	4.7 (15.6)	39 (52)	61 (306)	290 (36)	56/42	2.2 (3.6)
37kW/1spd/25°∇	4/9.8	6.3 (20.8)	37 (52)	61 (306)	385 (48)	56	2.3 (3.8)
37kW/2spd/35°∇*	4/9.8	4.7 (15.6)	32 (44)	66 (332)	310 (39)	68/45	2.0 (3.3)
112kW/2spd/∇	3/10.7	8.0 (26.4)	60 (80)	82 (412)	660 (82)	37/28	2.1 (3.4)
75kW/2spd/CC	4/11.6	4.8 (15.8)	65 (87)	95 (480)	460 (57)	42/32	1.8 (2.9)
56Kw/1spd/30°∇	9/12.8	5.0 (16.5)	48 (65)	61 (306)	305 (38)	56	2.5 (4.1)
56kW/2spd/CSO	9/16.5	6.1 (20.0)	48 (65)	39 (200)	240 (30)	56/42	2.1 (3.4)
75kW/1spd/CURV	9/13.7	4.6 (15.2)	65 (87)	71 (360)	330 (41)	47	1.9 (3.1)
75kW/1spd/CURV	9/13.7	4.5 (14.8)	60 (80)	66 (332)	300 (37)	47	1.9 (3.1)
112kW/2spd/CURV	4/30	4.6 (15.0)	97 (130)	56 (279)	230 (29)	47/35	2.0 (3.3)
75kW/2spd/CURV	6/14.5	4.2 (13.8)	60 (80)	50 (253)	224 (28)	47/35	2.1 (3.4)
45kW/2spd/35°∇	4/12.8	4.5 (14.7)	39 (52)	53 (266)	240 (30)	56/42	1.9 (3.1)

* All units equipped with draft tube but * unit equipped with lower turbine; wSAE–clean water wire efficiency at maximum output power.
** Impeller configuration: x°∇–flat blade cone; CSO–curved or flat blade plate; CURV–curved blade on shaft; CC–curved blade in cone.

Modified from Stukenberg, 1984.

The strong dependence of mechanical aeration performance on system geometry cannot be overstated. The fundamental relationships for scale-up of mechanical aeration processes are given by the principles of hydraulic similarity. There are three types that are important: geometric similarity, kinematic similarity, and dynamic similarity. Geometric similarity is concerned with ratios of dimensions in sizes of the system, and the impeller must be completely evaluated in terms of fluid mechanics, power characteristics, and scale-up characteristics to predict hydraulic similarity. At present, there is not sufficient data on the role of the dimensionless groups including Reynolds Number, Froude Number, and Weber Number, to use dynamic similarity in scaling up mechanical aeration devices. Presently, scale-up of these systems to maintain a given process result relies on an index which describes the condition within the mixing basin as a function of tank size. To date, the most widely used indices are power per unit volume and power per unit surface area (power dissipation). Unfortunately, these indices are not very reliable and, as a result, the design is often compromised.

The reader is urged to use caution in interpreting oxygen transfer data even when acceptable testing methods are employed. The scale-up issue is not trivial. The best solution to this dilemma is to require compliance testing of the aerator under field conditions using acceptable standard testing procedures.

The impact of wastewater on oxygen transfer has long been of great concern to designers. Discussion of this effect has been elaborated in Chapter 3 for diffused air systems. The literature would suggest that, in general, mechanical aeration devices appear to be less affected by wastewater characteristics (they appear to exhibit higher values of alpha). Hwang and Stenstrom (1979) provide two intuitive explanations for this effect based on surfactant as the major causative agent. First, they point out that fine bubble aerators provide a maximum surface area normal to transfer. Thus, the surfactant is unlikely to significantly increase surface area but essentially has a major impact on transfer by reducing the film coefficient. The coarse bubble system and the mechanical aeration systems (that generate larger bubbles) might benefit from the increase in surface area caused by the surfactant. The second explanation relates to the rate of surfactant adsorption at the bubble surface. Since the transfer coefficient is greatest at the instant of bubble formation, and decreases with bubble age, those devices with high surface renewal rates (mechanical surface aeration devices) might be less affected by surfactants that require time to adsorb. Their experimental laboratory studies demonstrated that indeed, the fine bubble system exhibited lower values of alpha than either the high-speed aerator or the submerged turbine when exposed to a surfactant. They also showed that the value of alpha for the two mechanical aerators depended on power input, increasing at increased power levels.

Field studies conducted by Mueller (1983) at two sites in the northeast U.S. were performed to evaluate process water oxygen transfer testing procedures. At Haverstraw, NY, 22.4 kW (30 hp) low speed surface aerators were used to treat a mixture of industrial and municipal wastewater (1:1). Weir levels were used to control transfer rate. The average value of αSAE ranged from 1.15 to 1.24 kg/kWh (1.9 – 2.0 lb/hp-h) when the plant loading ranged from an F/M of 0.1 to 0.7. The higher loaded conditions appeared to appreciably lower the value of αSAE (1.4 kg/kWh at F/M = 0.1 and 0.9 kg/kWh at F/M = 0.7). The volumetric power dissipation at this facility was approximately 30 W/m³ (145 hp/MG). At Miller's Falls, MA, 15 kW (20 hp) low speed surface aerators were used to treat a wastewater that consisted of 60 percent paper mill waste and 40 percent residential waste. Two impellers were used per shaft, one at the surface and one that was submerged to provide supplementary mixing. The tests were conducted both as a batch operation (to control highly variable oxygen uptake rates) and continuous flow. Values of αSAE ranged from 1.25 to 1.32 kg/kWh (2.1 to 2.2 lb/hp-h) for both batch and continuous flow (F/M = 0.27) conditions. The volumetric power dissipation at Miller's falls ranged from 14 to 28 W/m³ (68 to 140 hp/MG). In both of these examples, values of αSAE were significantly lower than values for clean water (Table 5.4) for low speed surface devices. No clean water data was available for these specific devices, but using an average SAE of approximately 2.0 kg/kWh for low speed surface aerators (Table 5.4), the estimated value of alpha would range from 0.5 to 0.7.

Boyle et al. (1989) conducted extensive testing of draft tube turbines in total barrier oxidation ditches at Opelika, AL. At an F/M of 0.14, they found values of wire αSAE ranging from 0.7 to 0.8 kg/kWh (1.15 to 1.30 lb/hp-h). The estimated values of alpha for this facility based on clean water field tests ranged from 0.7 to 0.8. At South Hill, VA, the field tests revealed values of wire αSAE ranging from 0.80 to 0.97 kg/kWh (1.3 to 1.6 lb/hp-h) for a process loading (F/M) of 0.09. Alpha values at this site were calculated to range from 0.81 to 0.87.

Based on this scanty database, it would be difficult to predict the value of alpha for a given mechanical aeration device. A review of available data would suggest, however, that sparged turbines generating fine bubbles and, perhaps, aspirating aeration devices would produce lower alpha values than the surface aeration devices for a given system. Process loading and wastewater characteristics would play an important role in this evaluation. The engineer must be very cautious in selecting values of alpha for mechanical aeration devices. A comprehensive database similar to that found for diffused air devices is needed.

5.9 DESIGN

5.9.1 GENERAL

The elements of design for mechanical aeration systems follows that described for diffused aeration systems, up to the selection of diffusers, found in Section 3.5. The decision to select mechanical devices over diffused air systems is often based on client or engineer preference. If oxidation ditches are selected as the preferred process, the use of mechanical aeration equipment is most often dictated. Also, when aerated lagoons are considered, mechanical devices are often the system of choice although the decision as to aeration system is less apparent.

The selection of a specific type of mechanical aeration system will often be based on factors described in Sections 5.2 through 5.6 above. Factors used in the decision making process will include capital and operation/maintenance costs, system reliability, flexibility and operational control, aeration efficiency, and mixing capability. Environmental considerations including noise, aerosol production, susceptibility to cold weather operation, and stripping of volatile substances may also be considered in process selection.

In addition to performance related issues, one important consideration is mechanical integrity. It is important for the engineer, when specifying the equipment, to understand and react properly to mechanical issues. Among the questions to be considered are:

- motor details including efficiency and enclosure type. Premium efficiency motors are not mandated by law for special applications, and mechanical aerators often fall into that category. If premium efficiency is desired, it must be spelled out in the specifications.

- gear reducer details, most significantly, service factor and bearing lives must be specified and detailed. Reducer failures can be common and expensive, and repairs may take a long time.
- Details of any other significant mechanical components (shafts, bearings) should be described in the specifications.
- Warranty details should be carefully studied. A good supplier will give a warranty of at least two, and often three, years of service.

5.9.2 DESIGN EXAMPLE

A simplified design example has been developed using the design data presented in Section 3.5.2. For this example, low speed surface aerators will be employed. Instead of four parallel basins, two parallel basins will be used. Each basin will be 12 m (39 ft) wide, 48 m (158 ft) long, and 4.6 m (15 ft) deep as shown in Figure 5.15. It will be assumed that oxygen requirements along the basin will be uniform with only small variation from influent to effluent. To evaluate the hydraulic flow pattern for this configuration, the correlation developed by the Water Research Center in Stevenage, England (EPA, 1989) was used.

$$N = 7.4 \, LQ(1+r)/Wd$$

where
N = equivalent number of tanks-in-series
Q = wastewater flow, m³/s
L = tank length, m
R = return recycle ratio
W = tank width, m
d = tank depth, m

The value of N for this tank configuration was calculated to be 1.1 indicating that the assumption of a completely mixed basin was appropriate. Some gradient in oxygen demand may exist.

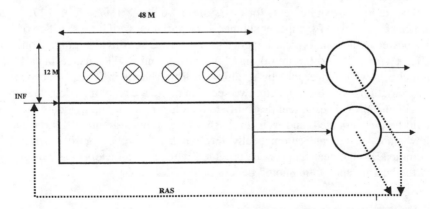

FIGURE 5.15 Activated sludge system for surface aerator design example.

TABLE 5.5
Design Parameters for One Basin — 20 Year Design[†]

Item	Minimum Month	Maximum Month Nitrifying	Peak Day Nonnitrifying
AOR = OTR_f (kg/d)	1620	5256	5516
Temp (°C)	10	25	25
DO (mg/L)	2.0	2.0	0.5
Alpha	0.75	0.80	0.70
OTR_f/SOTR	0.567	0.579	0.636
SOTR (kg/d)	2857	9078	8673
SOTR (kg/h-basin)[*]	59.5	189.1	180.7

[†] from design example, Section 3.5.2.
[*] two basins.

In reviewing the design problem in Section 3.5.2, it appears that the controlling OTR_f will be at either maximum month nitrifying or peak day nonnitrifying. Turndown will be dictated by minimum month. Table 5.5 is produced from data in Section 3.5.2, and appropriate parameters are added.

In this tabulation, note that alpha values for the slow speed surface aerators are assumed somewhat, but conservatively, higher than for the diffused air system. The calculations presume a completely mixed basin. Four slow speed units will be selected per basin providing a spacing between units of 12 m (39 ft) and 6 m (19.8 ft) from the walls. Maximum month controls and a typical SAE for these aerators would be 2.0 kg/kWh (3.3 lb/hp-h). Thus, the total power for aerators per basin would be 94.5 kW (127 hp) or approximately 24 kW (32 hp) per unit. Select four 30 kW units (40 hp) per basin. This number would produce a volumetric power dissipation of approximately 44 W/m³ (225 hp/MG), which should be adequate for mixing the basin. It is recommended that draft tubes be used for the aerators to insure good circulation and solids suspension. Baffles within the basin would not be necessary.

The turndown requirements for this system would be 189/59.5 or 3.17:1. This ratio can be achieved by impeller submergence variation with little sacrifice of SAE with power reduction. Typical impeller submergence sensitivities range from 0.08 to 0.2 kW/mm (2.7 to 6.8 hp/in), and both power and SOTR change linearly and parallel in a wide range, thereby resulting in little change in SAE. To account for OTR_f variations with basin length, two speed aerators are recommended. This number will also provide excellent flexibility to handle turndown requirements. Aerators would be pier mounted, and weir levels should be provided with electric drives to allow plants to operate automatically through a DO probe control device. It is recommended that gear reducers be provided with service factors of 2.0 or greater, and the bearing span ratio should be less than 4:1.

5.10 NOMENCLATURE

AE	lb O$_2$/kWh	aeration efficiency under process conditions
AOR	kg/d	actual oxygen requirements = OTR$_f$
D	m	impeller diameter
d	m	tank depth
F/M	lb BOD$_5$/d-lb MLSS	food to microorganism ratio
g	m/s^2	gravitational acceleration
G	s^{-1}	root mean square velocity gradient
H	m	impeller head generated
K		proportionality factor
L	m	tank length
n	rpm	impeller speed
N	rad/s	impeller speed
N$_q$		pumping number, dimensionless
N$_h$		head coefficient, dimensionless
OTR	kg/h, lb/h	oxygen transfer rate
OTR$_f$	kg/h, lb/h	oxygen transfer rate under process conditions
P$_g$	kW, hp	gassed mixer power
P$_m$	kW, hp	ungassed mixer power
P$_0$		power number, dimensionless
P	kW, hp	mixer power input
Q	m^3/s	circulation rate
Q	m^3/s	wastewater flow rate
R$_e$		Reynolds number, dimensionless
SAE	kg/kWh, lb/hp-h	standard aeration efficiency
r		recycle ratio for return activated sludge
SOTE	–, %	standard oxygen transfer efficiency
SOTR	kg/h, lb/h	standard oxygen transfer rate
W	m	tank width
α		wastewater correction factor for oxygen transfer coefficient
η		impeller efficiency
ρ	kg/m^3	liquid density
μ	N-s/m^2	absolute viscosity

5.11 BIBLIOGRAPHY

Boyle, W.C. et al. (1989). "Oxygen Transfer in Clean and Process Water for Draft Tube Turbine Aerators in Total Barrier Oxidation Ditches," *J. Water Pollution Cont. Fed.*, 61, 1449.

Committee on Sewage and Industrial Waste Practice (1952). *Air Diffusion in Sewage Works-MOP 5*, Federation of Sewage and Industrial Waste Associations, Champaign, IL.

EPA (1978). *A Comparison of Oxidation Ditch Plants to Competing Processes for Secondary and Tertiary Treatment of Municipal Wastewaters,* EPA 600/2-78-051, Municipal and Environmental Research Laboratories (MERL), Cincinnati, OH.

EPA, (1983). *Development of Standard Procedure for Evaluation of Oxygen Transfer Devices,* EPA 600/2-83-102, MERL, Cincinnati, OH.

EPA (1989). *Fine Pore Aeration System- Design Manual,* EPA 625/1-89-023, Center for Research Information, Cincinnati, OH.

GSEE (1984). *Oxygen Transfer Evaluation of Cage Rotor Installed at Jackson, OH WWTP,* November.

GSEE (1989). *Evaluation of Oxygen Transfer Capabilities of Lakeside Corp. Magna Rotors Installed at Woodland, CA WWTP,* February.

GSEE (1997a). *Oxygen Transfer Evaluation of Aqua Systems International N.V. Aerators, Northwest Water Company, Davy Hulme, England WWTP,* September.

GSEE (1997b). *Oxygen Transfer Capability Comparisons of Aeration Industries, Aqua Aerobics, and Aquaturbo Systems, Inc. Floating High Speed Aerators for International Paper Corp.,* June.

Hwang, H.J. and Stenstrom, M.K. (1979). *The Effects of Surfactants on Oxygen Transfer,* UCLA School of Engineering and Applied Sciences, Report 7928, UCLA, Los Angeles, CA.

Kayser, R. (1992). *Oxygen Transfer Test in Clean Water at Sewage Treatment Plant at Uxeim,* Grimm Data Team GmbH, Berkenthin, Germany.

Keefer, C.E. (1940). *Sewage Treatment Works,* McGraw-Hill Book Co., New York.

Lakeside Equipment Co. (1992). *Evaluation of Oxygen Transfer Capability of Magna Rotor Aerators, Berks Co. PA WWTP,* August.

Mixco (1984). *Draft Tube Turbine Aerator Oxygen Transfer Test Report for the Sanitary District of Elgin, IL, (Analysis by W. Boyle, May, 1986),* test, November.

Mueller, J.A. (1983). *Non-Steady State Field Testing of Surface and Diffused Aeration Equipment,* Department of Environmental Engineering and Science, Manhattan College, New York.

Oldshue, J.Y. (1955). "Theory and Design of Mixers for the Aeration of Wastewater." *10th Purdue Industrial Waste Conference,* Lafayette, IN, 391.

Oldshue, J.Y. (1956). "Aeration of Biological Systems Using Mixing Impellers" *Biological Treatment of Sewage and Industrial Waste,* 1, Edited by: Bro. J. McCabe and W.W. Eckenfelder Jr, Manhattan College, Reinhold Publishing Corp., New York.

Roe, F.C. (1938). "The Activated Sludge Process- A Case for Diffused Aeration." *Sewage Works Journal,* 10, 999.

Stenstrom, M.K. and Gilbert, G. (1981). "Effect of Alpha, Beta, and Theta on Specification, Design, and Operation of Aeration Systems." *Water Research (GB),* 15, 643.

Stenstrom, M.K. (1989). Estimation of Oxygen Transfer Capacity of Full-Scale Pure Oxygen Activated Sludge Plant." *Journal Water Pollution Cont. Federation,* 61, 208.

Stukenberg, J.R. (1984). "Physical Aspects of Surface Aeration Design." *Journal Water Pollution Cont. Fed.,* 56, 1014.

Von der Emde, W. (1968). "Aeration Developments in Europe." *Advances in Water Quality Improvement Symposium 1,* Ed: E. Gloyna and W.W. Eckenfelder, Jr., Univ. of Texas Press, Austin, TX.

Water Pollution Control Federation (1988). *Aeration- Manual of Practice FD-13,* Washington, DC.

6 High-Purity Oxygen Aeration

The use of pure oxygen instead of air significantly increases the oxygen mass transfer driving force for aeration. Figure 6.1 shows a schematic of the increased driving force available for oxygen transfer at 20°C. With a 100 percent oxygen gas phase, the saturation value is increased from 9.09 to 43.4 mg/L. This value provides a driving force for transfer almost five times greater for the pure oxygen system and allows for design of a somewhat higher DO in the aeration tanks. The objective of high-purity oxygen (HPO) systems is to provide higher gas phase oxygen concentrations than air systems, allowing faster treatment rates with higher mixed liquor suspended solids and smaller aeration tanks. Figure 6.2 and the following section trace the development of this system into a commercially viable aeration process for activated sludge systems.

6.1 HISTORY

6.1.1 INITIAL DEVELOPMENTS

Before 1940, oxygen generation costs were prohibitively high for use in wastewater treatment. Due to the possibility of a breakthrough in the manufacture of cheap oxygen, Pirnie (1948) suggested a method developed the following year by Okun (1949) at Harvard. Using an upflow fluidized bed reactor with preoxygenation of the wastewater, Okun obtained 90 percent removal at MLSS concentrations of 5000–8000 mg/L. He found no marked difference in microbial biomass. Later, from laboratory studies, Okun (1957) concluded that the only benefits were to eliminate anoxic conditions in aeration tanks. Use of the process was thought economically unfeasible due to low oxygen transfer efficiencies in aeration tanks (Okun and Lynn, 1956).

In 1953, Budd and Lambeth (1957), under the auspices of Dorr-Oliver, conducted studies on a 55–160 m³/d (8–30 gpm) bio-precipitation pilot plant at Baltimore's Black River treatment plant. Sludge settling characteristics were optimum at $\frac{F}{M} = 0.2 - 0.4 \frac{kgBOD_5}{day - kgMLSS}$. As a follow up to the Baltimore study, a more sophisticated 270 – 380 m³/d (41 – 58 gpm) pilot plant incorporating fluctuating inflows was constructed at Stamford, CT. A BOD removal efficiency of 90 percent was obtained at an upflow rate of 36.7 m³/m²/d (900 gpd/sf). The power requirements of this unit, including oxygen generation, were equal to those of a conventional activated sludge plant while tank area requirements were reduced by as much as 50 percent and volume by 30 percent.

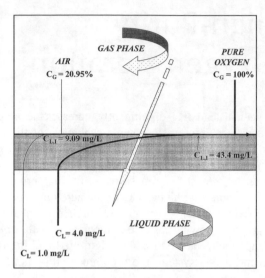

FIGURE 6.1 O_2 transfer schematic for air and high-purity oxygen.

Robbins (1961) utilized Okun's bioprecipitation process to treat Kraft mill sulfite wastes. A BOD removal of 90 percent was obtained on this semi-chemical waste at an $\frac{F}{M} = 3.3 \frac{kgBOD_5}{day - kgMLSS}$ in an eight-hour detention time. He suggested that the capital investment would be less than that for a conventional plant. Pfeffer and McKinney (1965) conducted oxygen enriched air laboratory studies on industrial wastewater. They concluded that with the high transfer rates, the size and capital investment of new plants could be reduced and for existing overloaded plants, improved efficiency could be obtained without new tankage installation.

6.1.2 COVERED TANK DEVELOPMENTS

To develop the HPO system into a commercially viable process, typically more than 90 percent of the oxygen must be transferred to the liquid phase due to the significant cost of oxygen generation. Departing from the previous pilot studies, which used preoxygenation of raw sewage, Union Carbide Corporation developed the UNOX® process using covered aeration tanks in series. Extensive full-scale (1–3 MGD) studies, (Albertsson et al., 1970; Stamberg, 1972) were conducted at Batavia, NY in 1969 to compare the performance of the UNOX® system to a parallel air system. Oxygen was injected into the first stage of covered aeration tanks, flowing sequentially from stage to stage in the headspace above the mixed liquor, until it was vented from the last stage. Gas from the headspace was recycled to the mixed liquor in each stage through hollow-shaft turbine aerators. In this closed system, the degree of venting in the last stage was controlled to attain the desired 90 percent oxygen utilization. This process provided the significant breakthrough in technology needed to justify commercial development.

In comparison to the single-stage air system at Batavia, only one-third the aeration time was required for the UNOX® system with a 30 percent reduction in

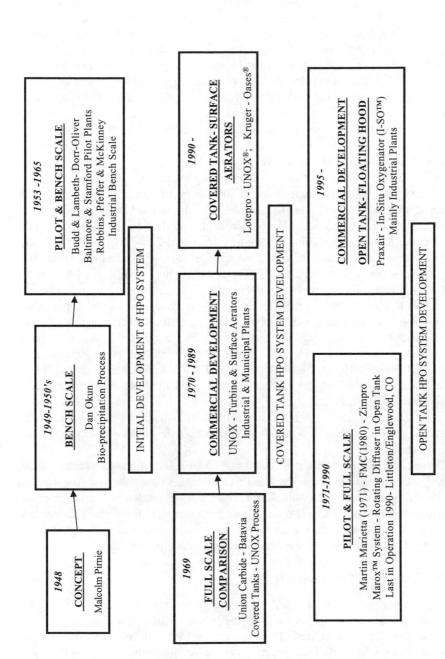

FIGURE 6.2 History of HPO system development.

waste activated sludge. An economic comparison from plants from 6 to 100 MGD indicated that UNOX® costs would be, respectively, 80 to 70 percent the costs for air systems, due mainly to reduced sludge disposal requirements. This reduction provided the incentive for construction of numerous full-scale industrial and municipal plants using turbine and later surface aerators (Figure 6.3). For large cities, the reduced land area requirements continue to make this process attractive. In 1976, (Brenner, 1977) indicates that 152 covered plants were operational, under construction or being designed, most in the U.S. Of the above, 16 plants were in Japan and one each in Canada, Mexico, England, Germany, Denmark, Switzerland, and Belgium. Approximately 25 percent of the plants were industrial treating 10 percent of the total flow. Most of the covered plants utilized surface aerators with only seven plants with submerged turbines. However, these latter were large municipal plants comprising 30 percent of the total flow treated in covered tank HPO systems.

In 1981, the Lotepro Corporation, a subsidiary of Linde AG, obtained the registered trademark UNOX® and became the provider of the UNOX® process. The turbine aeration mode has been dropped from the product line due to costs. The manufacturer's brochures (Gilligan, 1998) indicate that 220–300 UNOX® systems treating both municipal and industrial wastewater have been installed worldwide since its introduction in the late 1960s. The latest emphasis (Gilligan, 1999) is on the UNOX® Biological Nutrient Removal (BNR) design approach, as shown in Figure 6.4. This approach incorporates flexibility for front-end anaerobic phosphorous removal, selector zones, single- or two-step nitrification and denitrification, and an open reactor for the last stage to elevate the pH. Recently, BNR plants have been constructed or upgraded in Monterrey, Mexico; Morgantown, NC; Lancaster, PA; Mahoning County, OH; and Cedar Rapids, IA. The City of Hagerstown, MD plant will be upgraded to include an anoxic/anaerobic step for front-end denitrification and phosphorous removal and an open last stage for CO_2 stripping. The New Salem, MA plant has a first stage selector to control bulking which can be run in either in an anaerobic mode with a nitrogen blanket or in an oxic mode with oxygen in the gas head-space.

Kruger, Inc., in the 1990s, provided a closed tank staged reactor process called "Oases®", developed previously by Air Products and Chemicals, Inc. The Kruger website (Krugerworld.com, 1998), listed 39 Oases® processes in North America, five treating pulp and paper wastewater and the remainder municipal. Kruger has also replaced the original turbine aeration system in the Middlesex County Utilities Authority (MCUA) plants in New Jersey, using Philadelphia mixers.

6.1.3 OPEN TANK DEVELOPMENTS

In 1971, development of an open tank HPO system was underway in Denver, CO by Martin Marietta Company. Initially, a fixed fine bubble diffuser system creating minute (~50–200μ) bubbles was utilized. This system provided a large surface for oxygen transfer and a slow rise rate allowing high oxygen utilization without covering the tanks. Further development of the system by FMC utilized a rotating diffuser that formed a fine mist. A comparison at the Denver Metro plant (Fullerton and Pearlman, 1979),

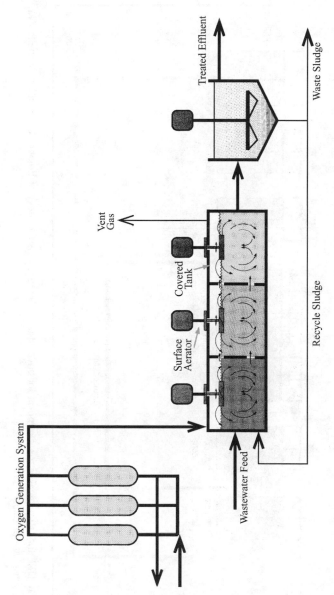

FIGURE 6.3 Covered tank HPO systems (courtesy of Lotepro Corporation, Valhalla, NY, a subsidiary of Linde AG).

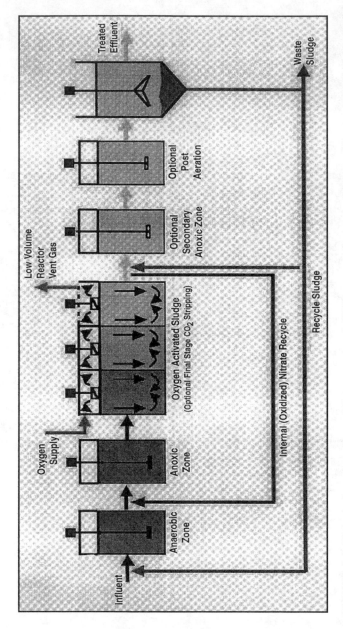

FIGURE 6.4 Biological nutrient removal for covered tank HPO systems (courtesy of Lotepro Corporation, Valhalla, NY, a subsidiary of Linde AG).

between open tank air and oxygen systems in 1979, indicated that the Marox™ oxygen system would require 39 percent less power. In 1976, there were five Marox plants operational (Brenner, 1977), one full-scale in Littleton/Englewood, CO. Demonstration plants were in Denver Metro #2, Minneapolis, MN, East Bay Municipal Utility District #2 in Oakland, CA, and a pharmaceutical wastewater plant in Osaka, Japan. In ~1980, the process was transferred to Zimpro, Inc. (Rakness, 1981). No plants presently utilize the Marox system, with Littleton/Englewood, CO removing it in 1990. The Littleton plant was expanded at that time using air instead of oxygen with fine pore ceramic diffusers to reduce operator involvement, maintenance on the cryogenic reciprocating compressors, and costs (Tallent, 1998).

The latest development in open tank technology, by Praxair, Inc., formerly the Linde Division of Union Carbide, utilizes a floating hood to capture the oxygen into a small headspace. It fits somewhere between the fully open tank of the Marox system and the fully closed tanks of the UNOX® and Oases® systems. Liquid circulation is created by the downward pumping action of a helical screw impeller. The first commercial installation went into operation at the Schuck Tannery in Novo Hamburgo, Brazil in December, 1992 (Bergman and Storms, 1994) where 95 percent oxygen utilization has been measured in field tests. The aeration unit is referred to as an In-Situ Oxygenator™ (I-SO™) and is shown in Figure 6.5. As of April 1998, there were 38 locations either operational or under contract in North and South America with nearly 200 I-SO™ units installed by April 2001 (Storms, 1998a, 2001). Two additional locations involving four units were undergoing tests in Spain. All applications to date have been at industrial sites except for one municipal plant in Brazil and two in the US, Cedar Rapids, IA and Merced, CA. Seven sites have been new activated sludge plants while the majority of the others add additional capacity to existing activated sludge or aerated lagoon systems. The I-SO™ unit has also been installed for sludge digestion, fish growing operations, and in activated sludge using ozone for color removal.

6.1.4 PUMPED LIQUID SYSTEMS

In pumped liquid systems, a portion of the wastewater is pumped to a high pressure, two to seven atmospheres, oxygen injected, and then returned to the main flow through dispersion pipes or eductors. In the sidestream pumping system developed by Praxair, Inc. in the 1960s (Storms, 1995), 90 percent of the oxygen was dissolved in the pipeline. Since the system required a relatively high power input, a variation was developed by SIAD, a Praxair affiliate, in the 1980s called the Mixflo™ System. In this system, Figure 6.6, mixed liquor is continually recirculated through a pipeline contactor at two to three atmospheres pressure. It is then reintroduced into the aeration tank with liquid-liquid ejectors or eductors. This method provides aeration tank mixing as well as 90 percent or greater oxygen transfer efficiency. It has been used in over 150 secondary treatment installations worldwide. For new plants, it is not as economical as the newly developed I-SO™ unit discussed above. It may find future application for remediation of hazardous waste sites where it was used successfully for in-situ slurry phase biotreatment at the French Limited superfund site in Crosby, TX (Bergman et al., 1992).

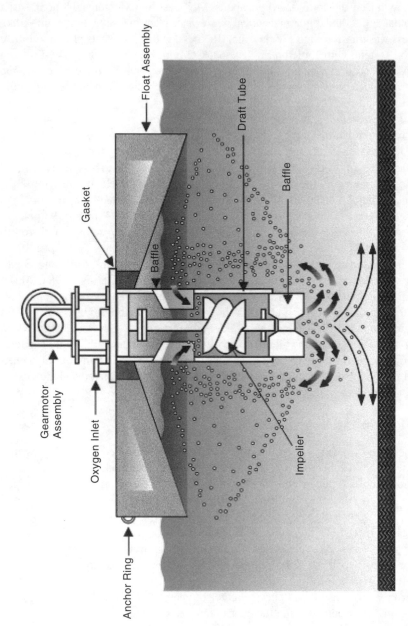

FIGURE 6.5 Praxair, Inc.'s patented (U.S. patent 6,135,430) I-SO™ oxygen dissolution system. (Used with permission.)

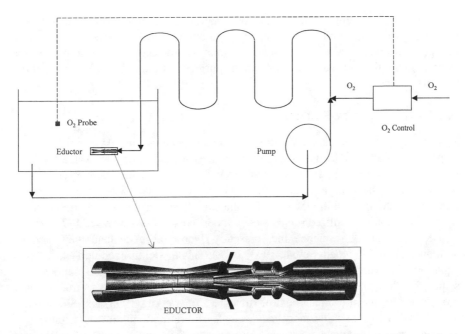

FIGURE 6.6 Praxair, Inc.'s proprietary Mixflo™ oxygen dissolution system. (Used with permission.)

6.2 COVERED TANK SYSTEMS

The covered HPO processes use a series of well-mixed reactors employing co-current gas-liquid contact. Feed wastewater, recycled sludge, and oxygen gas are introduced into the first stage where the highest reaction is exhibited. An average DO in the reactors is typically 4–6 mg/L.

The oxygen gas is fed at low pressure, 5–10 cm water, to the headspace in the first stage. With the older turbine aeration systems, recirculating gas compressors in each stage pumped the gas through a hollow shaft to a rotating sparger. The present practice of using surface aerators eliminates the need for gas recirculating compressors with associated piping and maintenance. The surface aerators often have a bottom impeller for mixing purposes. A design study was conducted by Pettit et al. (1997) at the East Bay Municipal Utilities District for a plant upgrade from a submerged turbine to a surface aeration system. It showed the installed power would be reduced from 3800 kW (5100 hp) for the original turbine system, a third of which was for the recirculation gas compressors, to 1790 kW (2400 hp) for surface acrators.

Openings in the interstage walls allow gas flow from stage to stage with venting from the last stage. The control of oxygen flow to the system is typically accomplished by a pressure controller and control valve on the oxygen feed line. The valve setting on the vent gas line is typically set to insure a high oxygen utilization, typically ~ 90 percent. The vent-gas phase composition will typically be about 50 percent oxygen with the remainder carbon dioxide and nitrogen. Due to the net transfer of gas to the liquid, the vent-gas flow rate will be a fraction (10–20 percent) of the oxygen gas feed rate.

Two safety systems are provided. Combination vacuum/pressure relief valves in the headspace of the first and last stages open automatically if excessively high or low pressures occur. A second system continuously monitors hydrocarbon concentrations in the first and last stages so an air purge can be initiated if concentrations become unacceptable.

6.2.1 GAS TRANSFER KINETICS

To properly design the aerators in the closed tank systems, the oxygen supply must be properly balanced with the oxygen demand in each reactor, similar to an air aeration system. The major difference between the two systems is that the oxygen partial pressure in the headspace of the HPO is not known as it is with an air system. It requires a mathematical model to predict this concentration. In its early development work, Union Carbide utilized such a model. Independent of this work, in 1973 a model was developed at Hydroscience, Inc. (presently Hydroqual) to evaluate the system for an industrial client (Mueller et al., 1973; 1978). This model utilized non–steady state equations that were rapidly solved numerically to obtain steady-state solutions. Subsequently, Clifft (1988; 1992) solved the non–steady state equations as true dynamic models and began to evaluate control strategies. Yuan et al. (1993) and Stenstrom et al. (1989) developed similar models to evaluate calibration requirements and to use in oxygen transfer compliance testing. More recently, Yin and Stenstrom (1996) have evaluated both feed forward and feed back control strategies. This section will present the basic principles involved in the models with the steady-state results.

Gas transfer occurs for at least four constituents when pure oxygen is introduced into an aeration tank as shown in Figure 6.7. Oxygen is transferred from the gas to the liquid phase. Nitrogen and inert gases such as argon, originally present in the liquid phase or produced in a prior denitrification reaction, are transferred to the gas phase. Carbon dioxide, produced by the biological reaction, is transferred to the gas phase. Since dry gas is introduced into the gas phase from the oxygen generation equipment, water vapor is transferred to the gas phase until it reaches the saturated vapor pressure.

Using the CSTR schematic in Figure 6.8, two mass balance equations are required for each parameter of concern, one for the liquid phase and one for the gas phase.

Liquid Phase:

$$V_{L,n}\frac{dC_{L,i,n}}{dt} = Q\left(C_{L,i,n-1} - C_{L,i,n}\right) + K_L a_{f,i,n} V_{L,n}\left(C^*_{\infty f,i,n} - C_{L,i,n}\right) + r_{v,i,n} V_{L,n} \quad (6.1)$$

Gas Phase:

$$V_{G,n}\frac{dC_{G,i,n}}{dt} = G_{n-1}C_{G,i,n-1} - G_n C_{G,i,n} - K_L a_{f,i,n} V_{L,n}\left(C^*_{\infty f,i,n} - C_{L,i,n}\right) \quad (6.2)$$

An additional equation defining the gas phase concentration as a function of partial pressure in the gas phase is as follows from Chapter 2.

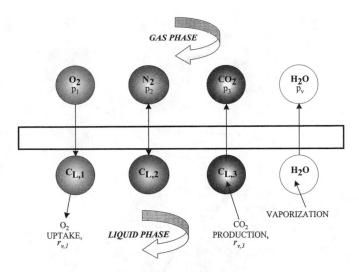

FIGURE 6.7 Gas transfer constituents in HPO system.

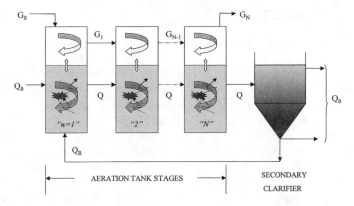

FIGURE 6.8 HPO completely mixed series tank reactors (CSTR) schematic.

$$C_{G,i,n} = \frac{M_i}{RT} p_{i,n} \tag{6.3}$$

The linkage between the two phases is provided by using the Henry's law relationship, corrected for field conditions as given in Chapter 2. Note that the pressure correction factor (Ω) is not included since the actual partial pressure as defined in Equation 6.3 is used to define the gas phase concentration.

Both Phases:

$$C^*_{\infty f,i,n} = \frac{\tau \beta \delta C_{G,i,n}}{H_{20,i}} \tag{6.4}$$

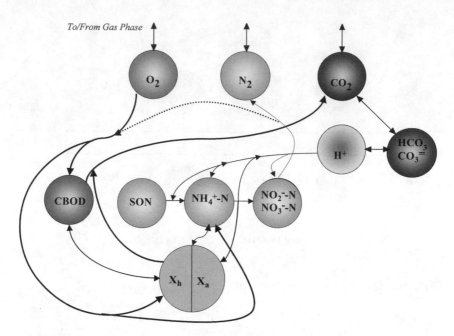

FIGURE 6.9 Biological and chemical reaction schematic in liquid phase.

Equations 6.3 and 6.4 provide the gas phase concentrations and the saturation values as a function of the headspace partial pressures in each stage. For each parameter of concern, the liquid phase concentration and gas phase partial pressure are unknown with Equations 6.1 and 6.2 available to solve them. However, an additional unknown always exists in the gas phase, i.e., the gas flow. A final equation, defining the total pressure in the gas phase, allows simultaneous solutions of the equation set for each stage.

$$\sum_{i=1}^{I} p_{i,n} = p_{T,n} - p_v \tag{6.5}$$

The equations using first order BOD removal kinetics and bacterial respiration for the reaction rate term, r_v, were originally solved numerically to a steady-state solution using CSMP. Later, when more complex nitrification and sulfite oxidation kinetics were utilized with bacterial growth, Famularo (1975) developed a solution technique using the steady state equations by stepping up the recycle stream in small increments.

A simplified schematic of the biological reactions occurring during carbon oxidation, CBOD removal, is shown in Figure 6.9. The buildup of CO_2 in the gas phase causes a significant reduction in the tank pH. The amount of CO_2 production is related to the oxygen consumption by the respiratory quotient, RQ, with the pH calculated from the first equilibrium constant for the CO_2 system.

$$pH = pK_1 + \log \frac{[HCO_3^-]}{[CO_2]} \tag{6.6}$$

If the reactor is designed for nitrification, significant reductions in bicarbonate alkalinity will occur while denitrification will produce alkalinity. A further discussion of the alkalinity effects and a more complex evaluation of the carbonate equilibrium system are given in Mueller et al. (1978; 1980).

In Equations 6.1 and 6.2, the mass transfer coefficient for oxygen can be calculated from measured field calibrations. It can also be determined from aeration equipment specifications as a function of the power level as given in Chapter 2.

$$\left.\begin{aligned} K_L a_{20} &= \frac{SAE}{C^*_{\infty 20}} \left(\frac{WP}{V} \right) \\ C^*_{\infty 20} &= \delta C^*_{s20} = \delta 9.09 \ mg/L \end{aligned}\right\} \tag{6.7}$$

Note that the above SAE value is based on the manufacturer's specifications using air and not high purity oxygen. The mass transfer coefficients for the other gases can be corrected for diffusivity. This has some impact when large volatile organics are being stripped from solution. For the smaller inorganic gases, O_2, CO_2, and N_2, the diffusivity difference is small and has often been ignored. In laboratory experiments, Speece and Humenick (1973) have shown that CO_2 has the same K_L value as O_2 and that N_2 is 89 percent that of O_2. The field transfer coefficients are then determined, similar to the air aeration systems.

$$K_L a_{f,i,n} = \alpha_n \theta^{t-20} K_L a_{20,i,n} \tag{6.8}$$

6.2.2 APPLICATIONS OF STEADY-STATE KINETICS

Figure 6.10 shows the gas phase parameters for the Batavia Phase III data (Albertsson et al., 1970). The measured data for oxygen and gas flow are given with the solid lines representing calculated values from a model employing the above mechanisms (Mueller et al., 1973). Gas flow significantly decreases from the influent to the vent from the last stage in this three-stage reactor system. Oxygen partial pressure decreases successively from stage to stage as CO_2 and N_2 increase. Figure 6.11 provides the liquid phase concentrations along with the pH. CO_2 increases in successive stages due to its high solubility, yielding effluent concentrations >250 mg/L with a resulting pH of 6.3. In the parallel air system at Batavia, the pH remained near the raw wastewater pH of 7.1 due to continual CO_2 stripping to the atmosphere. Lower RQ values result at higher organic loading rates, probably due to incomplete oxidation of the organics.

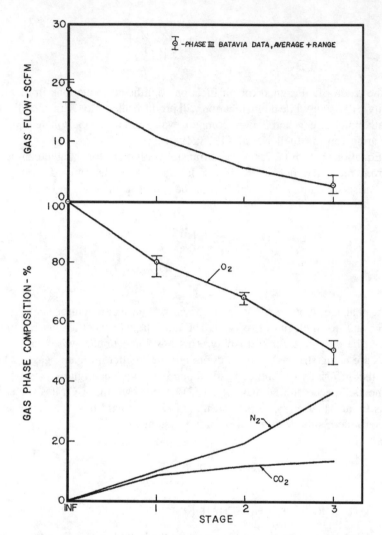

FIGURE 6.10 Gas phase parameters for Batavia, NY, HPO plant. (From Mueller, J. A., Mulligan, T. J., and DiToro, D. M. (1973). "Gas Transfer Kinetics of Pure Oxygen System." *J. Environ. Eng. Div.,* ASCE, 99(EE3), 269–282. With permission.)

The nitrogen behavior is interesting. Nitrogen in the influent is assumed saturated and in equilibrium with air. In the first stage of the aeration tank, N_2 is stripped out of solution into the gas phase causing a decrease in the liquid phase concentration. However, in the second and third stages, due to the continuing utilization of oxygen, the equilibrium shifts, and N_2 is transferred back to the liquid phase causing the dissolved concentration to rise.

In application of the above kinetics to an industrial wastewater with an alkalinity of 100 mg/L and a pH of 6.0, Mueller et al. (1973) show the impact of gas flow on the dissolved oxygen concentration and O_2 utilization. The volume of the first stage was designed at twice the size of the latter two stages to provide sufficient area and

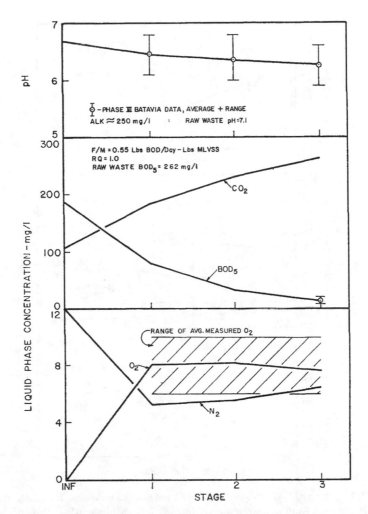

FIGURE 6.11 Liquid phase parameters for Batavia, NY, HPO plant. (From Mueller, J. A.,
Mulligan, T. J., and DiToro, D. M. (1973). "Gas Transfer Kinetics of Pure Oxygen
System." *J. Environ. Eng. Div.,* ASCE, 99(EE3), 269–282. With permission.)

volume for the surface aerators. Figure 6.12 shows that at 90 percent O_2 utilization,
the gas flow, G_{90}, would maintain about 2 mg/L DO in the three stages. Maintaining
a desired level of 4 mg/L as specified by the client would require a 25 percent increase
in the gas flow, resulting in an oxygen utilization efficiency of 70 percent. The aeration
tank pH in the above system would be about 5.5. At higher wastewater alkalinities and
initial pH of 8 or above, the DO would easily be maintained above 4 mg/L at the G_{90}
except at very high loading rates. This highlights the effect that changing wastewater
chemistry and organic loading rate have on system operation and ultimately economics.

To illustrate further applications of the above kinetics, Mueller et al. (1978)
applied them to the treatment of a wastewater from a chemical plant with the
following conditions:

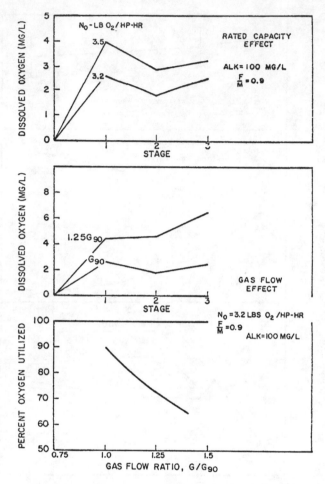

FIGURE 6.12 Effect of SAE and gas flow on DO and oxygen utilization for a three-stage HPO design for a Kraft mill wastewater. (Mueller, J. A., Mulligan, T. J., and DiToro, D. M. (1973). "Gas Transfer Kinetics of Pure Oxygen System." *J. Environ. Eng. Div.,* ASCE, 99(EE3), 269–282. With permission.)

Q = 15.4 MGD
BOD_5 = 1144 mg/L at 24 h peak load
BOD_5 = 608 mg/L at average load
Alkalinity = 500 mg/L
pH = 10.3
SAE = 3.0 lb O_2/hp-hr
BOD removal = 80 percent

Using a 2-1-1-tank configuration, the chemistry effects and power levels required were compared with air systems. Figure 6.13, using an RQ of 0.63, illustrates the high CO_2 concentration in the HPO system compared with the air system with resulting lower pH values.

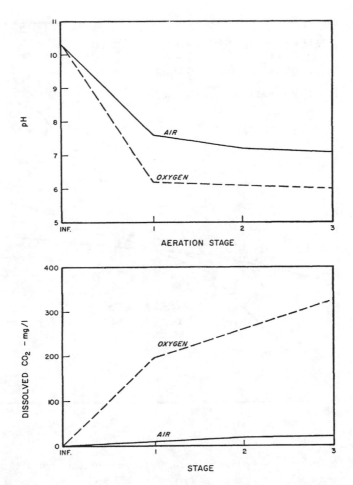

FIGURE 6.13 Comparison of pH and dissolved CO_2 for three stage air and HPO designs for a chemical wastewater (Mueller et al., 1978).

For the maximum load condition, the HPO system power requirements, including a generation power for the oxygen of 1.25 hp/scfm, as well as aerator power, are shown in Figure 6.14 to be less than for the air system. This effect is due to the high oxygen partial pressures of >60 percent existing in all stages of the oxygen system. At higher field oxygen transfer capabilities, differences between air and oxygen systems become less.

In the design of HPO systems, a trade-off can be made between oxygen gas flow and aerator power level in achieving optimum operation. The optimum aerator power should minimize total treatment power. Curve (a) in Figure 6.15 is a design curve for the peak load of 1144 mg/L BOD. Portions of the curve at low aeration power correspond to high O_2 gas flows, which require 1.25 (range from 0.88 – 1.29) hp/scfm of oxygen fed to the system. As aeration power is increased, less O_2 is required to maintain a DO of 2 mg/L, and the O_2 utilization increases (curve b). It

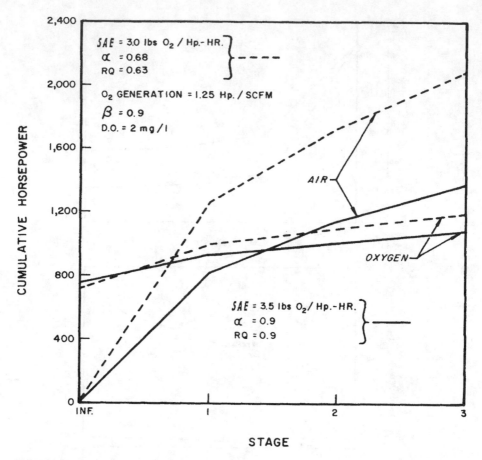

FIGURE 6.14 Comparison of power requirements for three-stage air and HPO designs for a chemical wastewater (Mueller et al., 1978).

is apparent that optimum operation corresponds to use of as little oxygen as possible. Figures 6.13 and 6.14 correspond to an aeration power level of 450 hp. The total power requirements for average conditions using the same aerator power for the maximum conditions are shown as curve (c). For 90 percent oxygen utilization, DO values of 12 to 17 mg/L result with total power levels from 820 to 900 hp. Slightly less oxygen could be fed with somewhat higher oxygen utilization to maintain DO levels around 4 mg/L.

Figure 6.16 shows a design curve to maintain 4 mg/L DO at the average loading. The aeration power of 250 hp and total power levels between 650 to 700 hp, adequate for the average condition, is unable to achieve the desired DO of 2 mg/L during peak loading periods. For this plant, if constant power aerators are employed, the permissible aerator power must be between 390 and 460 hp to handle peak loads. Clearly, the large difference between peak and average demands of this system suggests evaluating a dual speed or variable submergence aerator. Since the cost of dual speed aerators is more than double that of single speed units (Geselbracht et al.,

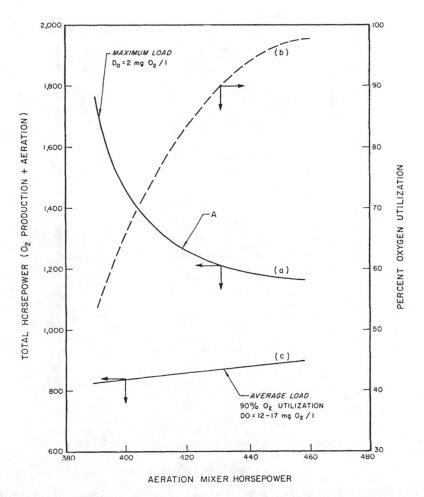

FIGURE 6.15 Average and peak load power consumption for a three-stage HPO design for a chemical wastewater (Mueller et al., 1978).

1997), the economics would favor designing for average demand and working at low utilization efficiencies for short term peak demands.

6.2.3 FULL-SCALE APPLICATIONS

Studies conducted at the Joint Water Pollution Control Plant (JWPCP) in Carson, CA, for a consent decree that mandated upgrading by 2002 have provided insightful results for various modes of operation (Pettit et al., 1997). A portion of the plant is a four-stage covered HPO System with surface aerators. Two problems were encountered with the operation of the process. Foaming and poor settling floc (bulking) occurred in the secondary clarifiers. Low pH in the effluent caused corrosion problems with the iron, steel, and concrete in the plant as well as the 12.9 km (8 mile) effluent tunnel and outfall system. The plant also operated at high DO concentrations from 10 to 15 mg/L.

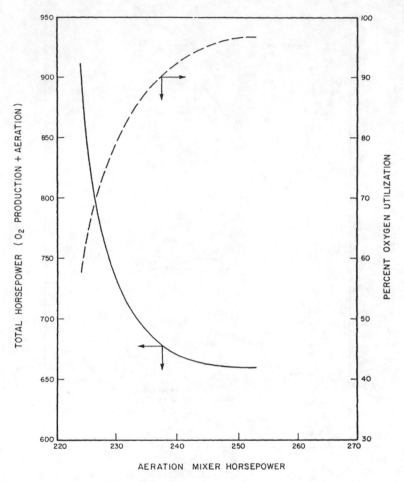

FIGURE 6.16 Total power consumption to achieve 4 mg/L DO at average load for a three-stage HPO design for a chemical wastewater (Mueller et al., 1978).

To obtain less power draw and to stop destroying gearboxes, the blade diameter in the first stages were cut shorter, and extensions on the blades were removed in the latter stages to reduce the turbine blade diameter. This provided lower K_La values with a significant decrease in power. When the selector was utilized as the first stage with no aeration, the extensions were returned on the latter stages to get adequate oxygen transfer in the total system. The selector process successfully controlled bulking and the CO_2 purge increased effluent pH.

Figure 6.17 shows the effect of the selector and the CO_2 purge on the headspace CO_2 concentrations. With all stages of the system using the full aeration capacity, the headspace CO_2 increased from stage to stage to discharge at almost 15% CO_2 by volume. The first stage selector was operated with 98 percent oxygen in the headspace but mixing only for two hours per day to prevent solids buildup on the tank bottom. The headspace CO_2 profile was similar to the full aeration system except for a lag in the first stage.

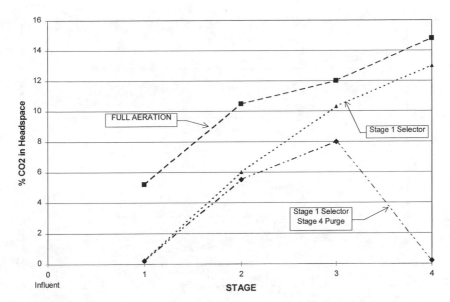

FIGURE 6.17 Effect of selector and CO_2 purge on headspace CO_2 concentrations for JWPCP four-stage HPO system (adapted from Pettit et al., 1997).

The CO_2 purge used air feed into the fourth stage venting the HPO gases from the third stage. Thus the CO_2 in the headspace dropped to <0.5 percent as it was released to the atmosphere by using a large fan to move air across the surface. This effect provided an increase in the effluent pH to 6.7 from the 6.2 typical of the process without the purge as shown in Figure 6.18. Note the higher pH values in stages one and two with the system having the stage four purge. This should be due to the recycle sludge having much lower dissolved CO_2 concentrations than the system without the purge.

The expansion of the JWPCP will incorporate a combination of selector and CO_2 stripping reactors. The anaerobic selector, as the first stage for foaming and bulking control, will be designed with bottom mixers, not aerators. An air purge will be used in the fourth stage for CO_2 stripping to minimize corrosion of concrete structures that cannot be dewatered for installation of protective coatings.

6.3 OPEN TANK SYSTEMS — FLOATING COVER

6.3.1 DESCRIPTION

The In-Situ Oxygenator (I-SO™) was developed to improve the dissolution technologies for HPO systems. It uses lower power than the pump-type oxygenation technologies. According to Bergman et al. (1992), it also eliminates the severe foaming problems that can occur with covered tank surface aeration systems, does not have a confined space, and is easy to install.

As shown previously in Figure 6.5, liquid is directed into a draft tube by an upper conical baffle. Vertical plates within the baffle form vortices, which entrain oxygen in

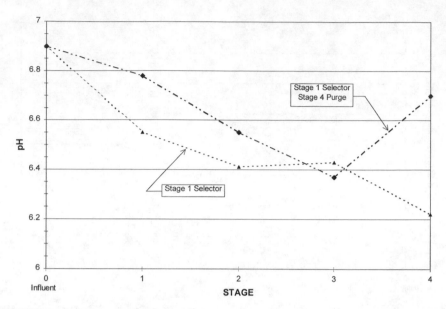

FIGURE 6.18 Effect of selector and CO_2 purge on pH for JWPCP four-stage HPO system (adapted from Pettit et al., 1997).

the headspace above the liquid. The gas/liquid dispersion is pumped downward through the draft tube by a rotating impeller. A second set of baffles at the draft tube discharge continues to direct the rotating gas/liquid dispersion downward, where it passes through the bulk liquid as a jet and slows as it expands and entrains more fluid. The system is designed so that the liquid momentum at the tank floor is lower than the gas bubble buoyant force. This design allows the majority of the gas bubbles to release before passing beyond the off gas hood circumference and rise to the surface where the hood captures them for recirculation. The bottom current, now devoid of most gas bubbles, flows along the tank floor suspending solids with settling velocities lower than the liquid velocity. Fresh oxygen is injected under the hood at a rate sufficient to maintain the desired oxygen dissolution rate. The liquid level within the hood automatically adjusts to the changes in pressure. At high liquid levels, more liquid flows over the inlet weir to the pump with less gas. When oxygen is injected and liquid level reduced, less liquid flows over the weir with more gas drawn into the impeller. This allows oxygen to be dissolved automatically at about the same rate that it is injected. Data obtained at the first commercial scale installation at the Shuck tannery in Brazil give an indication of the operating mode of the I-SO™. Off gas testing in a lagoon showed the gas flow escaping the hood to be about five percent of the injected gas flow, the majority of the gas flow escaping within 1 m of the hood. The off gas leaving the hood was observed to consist of small bubbles, ~1 mm in diameter, and large bubbles >5 mm in diameter. The smaller bubbles represent moderate losses while the larger bubbles gross losses. Figure 6.19 shows that with increasing power density of the pump, greater O_2 injection rates are allowable before large bubble formation. In design of the I-SO™ units, large bubbles must be eliminated to obtain acceptable O_2 utilization. Some small

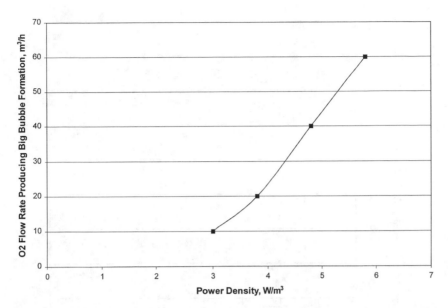

FIGURE 6.19 Effect of mixer power density on allowable O_2 flow rate before big bubble formation for 11.2 kW (15 hp) in-situ oxygenator (adapted from Bergman and Storms, 1994).

bubble loss will occur due to their greater travel distance, as shown in Figure 6.20, increasing with greater O_2 injection rates. Manual control of the oxygen injection rate is presently the norm with the operator adjusting a valve in response to observed aeration tank DO (Storms, 1998b). To prevent gross oxygen loss, an upper limit of valve opening is recommended to the operator.

In the above tests, no analyses of the off gases were reported. The transfer efficiency estimated only from the gas volumes was 95 percent. Actual transfer efficiencies were probably somewhat greater since off gas oxygen partial pressure had to be lower than the influent. Liquid level was varied to obtain the maximum aeration efficiencies at constant wire power densities. Considering only oxygen injection rates with no gross O_2 losses yielded a maximum aeration efficiency of about 4.3 kg/kWh (wire). This early data was obtained on a retrofitted system using an existing belt-driven motor driving the pump shaft at only 60 percent efficiency. The present design uses gear motor driven pumps having 91–93 percent efficiency.

From six locations where I-SO™ units were installed from 1992 to 1995, the average aeration efficiency has been 5.5 kg/kWh (wire) with an average OTE of 92 percent (Cheng and Storms, 1995). For three industrial locations, average power savings of 40 to 50 percent occurred when Mixflo™ units were replaced by I-SO™ units. Higher power reductions (66 to 80 percent) occurred when surface aerators and fine pore diffusers were replaced with I-SO™ units. However, generation power was not included. The above installations did not have any motor turndown capability so that they were operating at less than their full oxygen dissolution capacity (oxygen flow). Also α and β were not known for these plants so the above values are not comparable to SAE values.

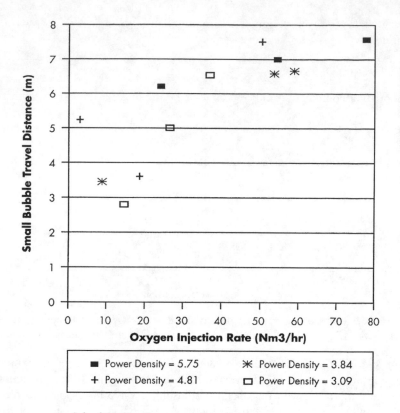

Schuck In-Situ Oxygenator Tests

FIGURE 6.20 Effect of O_2 injection rate on small bubble travel distance for 11.2 kW (15 hp) in-situ oxygenator. (From Bergman, T. J. J. and Storms, G. E. (1994). "Odor and VOC Emission Minimization by In-Situ Oxygenation." *Water Environment Federation Conference on Odor and Volatile Organic Compound Emission Control*, Jacksonville, FL.)

For comparison purposes with air aeration systems, the manufacturer presently uses an AE_{HPO} of 10.1 kg/kWh (delivered) or 16 lb/hp-h (delivered) in clean water at 20°C. The tank DO value is zero, and the oxygen partial pressure in the gas phase is taken as 99.5 percent purity obtained from liquid oxygen. These high values are due mainly to the higher O_2 partial pressures in HPO systems providing a higher O_2 transfer driving force than air systems. They are not true SAE values, which are based on air saturation and would be significantly lower than the above.

At this early stage of development, little published data is available on the system. The composition of the off-gases is not available nor is pH data in the aeration tanks. At these high oxygen utilization rates, CO_2 and N_2 must build up in the headspace under the hood similar to the closed tank system. Due to the low turbulence levels outside the hood diameter and minimal off-gassing, little CO_2 stripping should occur. Thus, the pH should decrease as much or greater than the covered tank HPO systems.

Foaming by detergents is minimal due to the down pumping action of the impeller and the low turbulence outside the hood diameter. This effect has interesting ramifications for *Nocardia* proliferation. If no foam is generated, *Nocardia* may not proliferate allowing operation at any sludge age level without chlorine addition of return sludge. However, design of the aeration tank should provide an overflowing weir outlet so any *Nocardia* growth will not accumulate on the surface of the system.

6.3.2 I-SO™ DESIGN EXAMPLE

A proprietary computer spreadsheet is used by the manufacturer for design of the I-SO™ systems. The manufacturer was requested to provide a design for the conditions shown in Table 6.1. These conditions are similar to those used earlier for the fine pore system except the hydraulic detention time was reduced to 2 h from the prior 6 h, reflecting the greater transfer capabilities of the HPO system. This would require about 3200 mg/L MLSS, triple that required in the air system. Note that an α value of 0.5 is used in the design for comparison to the fine pore system. Manufacturer's tests in municipal wastewater have shown α to be above 0.8 (Storms, 2001) thus making this design conservative.

From preliminary designs using four tanks, each with three zones similar to the air system design, it became obvious that mixing controlled the design with much greater power utilization than required for oxygen transfer. Therefore, three aeration tanks operating in parallel, each completely mixed were chosen for the final design. The tanks were also circular to eliminate dead spaces where sludge settling might occur.

The results of this design yielded a 40 hp (29.8 kW) unit in each of the three tanks as shown in Table 6.2. At the design power level, the diameter of influence or the mixing diameter is significantly greater than the tank diameter, which should provide complete suspension of the solids. The design capacity of the 29.8 kW Oxygenator unit is 20 percent higher than that needed for the peak load and about 50 percent higher than that needed for the average load. Thus, the generating unit would be operated at significant turndown from full capacity. Peak hourly loads would require minimum liquid oxygen due to this available capacity.

The aeration efficiencies, 3.5 kgO$_2$/kWh (wire) at peak to 2.8 kgO$_2$/kWh (wire) at average monthly conditions, are somewhat lower than those reported from field units, 4.3 to 5.5 kgO$_2$/kWh (wire). This may be due to the low α value of 0.5 used in the design example as mentioned previously. When the generation power in Table 6.3 is taken into account for the average load, the field aeration efficiency decreases to 1.24 kgO$_2$/kWh (wire).

Figure 6.21 shows the fraction of the area covered by the floating hood varied from 8.6 to 23 percent of the total tank surface area as a function of aeration tank depth. A clarifier design was also conducted to get a sense of the relative size of the two units. Using a range of realistic overflow rates in Figure 6.22, the clarifier surface area is significantly greater than the aeration tanks, typical of HPO systems. Figure 6.23 gives a schematic layout of the plant using 9.1 m (30 ft) deep aeration tanks with two clarifiers at 24.4 m^3/m^2/d (600 gpd/sf). Table 6.3 summarizes the monthly power requirements and total costs of the I-SO™ system including the generation costs using a single-bed vacuum pressure swing adsorption (VPSA) system and liquid oxygen (LOx) costs to handle load variability. All equipment would be leased, the lease costs estimated as 73 percent of the total monthly costs. These are not bid values and may be lower under competitive bidding. The unit costs per volume treated for the above cost estimates are $0.039/m^3 ($0.148/1000 gal).

TABLE 6.1
HPO Design Conditions

Q = 5.3 MGD = 0.232 m³/s

Tank Type	peak day	max mo	avg mo	avg no nit	min mo
			BOD_5, lb/d		
Influent	12800	7700	6600	6600	5500
			BOD_5, kg/d		
Influent	5805	3492	2993	2993	2494
			OTRf, lb/d		
CSTR	12160	11588	9685	5412	3575
PLUG FLOW SYSTEM					
zone 1	6187	5430	4613	2904	2109
2	4053	4147	3513	1804	1191
3	1920	2010	1559	704	275
			OTRf, kg/d		
CSTR	5515	5255	4392	2454	1621
PLUG FLOW SYSTEM					
zone 1	2806	2463	2092	1317	956
2	1838	1881	1593	818	540
3	871	912	707	319	125
			MLSS, X, mg/L*		
All Tanks	6252	3761	3223	3223	2686

# tanks	4	maximum
	2	minimum
HRT, hr	2	
Vol, m³	1671	
SWD, m	4.57	minimum
	≥ 9.14	maximum
ELEV =	1000	ft = 305 m
Pb =	14.21	psi = 97.95 kPa
OMEGA	0.97	
ALPHA	0.5	all zones
BETA	0.99	
SRT, day	4	
DO in tanks	4	mg/L

* Assuming net sludge wastage (ΔM) = 0.45 * BOD_5 load.
For SRT = $V*X/\Delta M$; X = $SRT*\Delta M/V$.

TABLE 6.2
I-SO™ Design

No. Aeration Tanks	**3**				
Vol./tank	557.1	m³	19672	ft³	0.147 MG
Depth	9.1	m	30	ft	
Diameter	8.810	m	28.9	ft	

No. I-SO™ Units	**3**		1/tank		
Impeller Size	0.610	m	24	in	
Motor Power	29.83	kW	40	hp	
Actual Power	21.85	kW	29.3	hp	
Hood Dia.	3.657	m	12	ft	
Hood As	10.51	m²	113.1	ft²	
Power Level	39.22	W/m³	199.1	hp/MG	
Mixing Dia.	19.45	m	63.8	ft	

Oxygen Transfer Capabilities					
I-SO™ Capacity	92.1	kgO₂/h	203	lbO₂/h	
Peak Req'd.	76.6	kgO₂/h	169	lbO₂/h	
Avg. Req'd.	61.0	kgO₂/h	135	lbO₂/h	
Peak AEf	3.51	kgO₂/kWh	5.76	lbO₂/hp-h	mixer power only
Avg AEf	2.79	kgO₂/kWh	4.59	lbO₂/hp-h	mixer power only
Avg AEf	1.24	kgO₂/kWh	2.13	lbO₂/hp-h	mixer and generation power

No. Secondary Clarifiers	**2**				
Overflow Rate	24.44	m³/m²/d	600	gpd/sf	
Diameter	22.9	m	75	ft	

TABLE 6.3
I-SO™ Design Power Requirements and Costs (April 1998)

Item	Number	Type	Unit Cost	Monthly Cost
I-SO™ Units, 40 hp, 24"	3	Lease/mo*	$1,500	$4,500
VPSA, single-bed (2% downtime)	1	Lease/mo*	$13,000	$13,000
LOx, 1000 cf (Supplemental + Backup)	70	Purchase/mo	$5	$350
Power for Generation, 1000 kWh (Avg O₂ Demand incl. turn down)	60	Purchase/mo	$50	$3,000
Power for Aeration, 1000 kWh	47.8	Purchase/mo	$50	$2,391
Total Power and Lease Costs				$23,241
Site Preparation, i = 8%, n = 20 yr		Construction	$75,000	$627
Total Monthly Cost				$23,868

* Conservative estimate, not actual bid value.

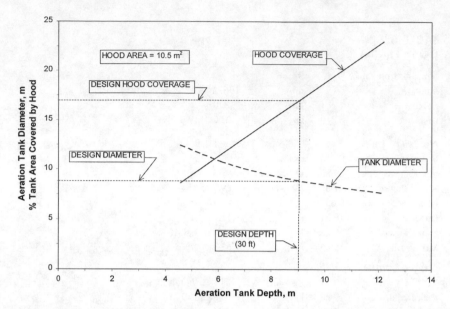

FIGURE 6.21 Effect of depth on aeration tank diameter for I-SO™ design example using three aeration tanks each with a 29.8 kW (40 hp) motor, 0.61 m (24 in) impeller and a 3.66 m (12 ft) off gas hood.

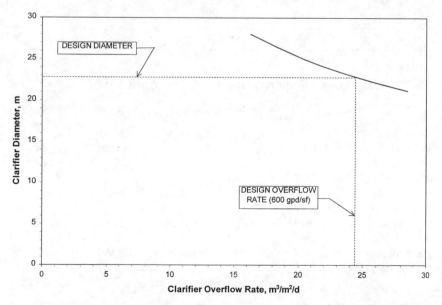

FIGURE 6.22 Effect of overflow rate on secondary clarifier diameter for I-SO™ design example using two clarifiers.

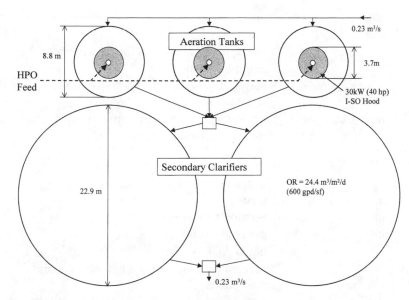

FIGURE 6.23 Schematic of in-situ oxygenator layout for 9.1 m deep aeration tanks.

6.4 NOMENCLATURE

AE_f	kg/kWh, lb/hp-h	aeration efficiency under process conditions
C_G	mg/L	bulk gas phase concentration
C_L	mg/L	bulk liquid phase concentration
C_{s20}^*	mg/L	surface saturation concentration at 20°C, 9.09 mg/L
$C_{\infty 20}^*$	mg/l	clean water oxygen saturation concentration at diffuser depth and 20°C
$C_{\infty f}^*$	mg/l	process water oxygen saturation concentration
F/M	lb BOD$_5$/d-lb MLSS	food to microorganism ratio
G	m³/h	gas flow rate
G_{90}	m³/h	gas flow rate that obtains 90% oxygen utilization
H_{20}	(mg/L)$_{gas}$/(mg/L)$_{liquid}$	Henry's constant at 20°C, 29.8 from Table 2.1
HRT	h	hydraulic detention time
$K_L a_f$	h⁻¹	oxygen transfer coefficient under process conditions
$K_L a_{20}$	h⁻¹	clean water oxygen transfer coefficient at 20°C
LO_x		liquid oxygen
M	g/mole	molecular weight
N_o	lb/hp-h	standard aeration efficiency = SAE
OR	m³/m²-d, gpd/sf	clarifier overflow rate
p	atm	partial pressure of constituent in gas phase
P_b	kPa, psia	barometric pressure

p_t	atm	total pressure
p_v	atm	vapor pressure
pK_1		first equilibrium constant for CO_2 system, 6.35 at 25°C
Q	m^3/h	liquid flow rate
R	m^3-atm/gmole-K	universal gas constant ($8.205*10^{-5}$ m^3-atm/gmole-K)
r_v	mg/L-h	reaction rate
RQ	mole CO_2/mole O_2	respiratory quotient, CO_2 production/O_2 utilization
SAE	kg/kWh, lb/hp-h	standard aeration efficiency
SWD	m	sidewater depth
T	°K	absolute temperature
t	°C	temperature in aeration basin
t	h	time
V	m^3	aeration tank volume
V_G	m^3	gas phase volume
V_L	m^3	liquid phase volume
VPSA		vacuum pressure swing adsorption system
WP	kW, hp	wire power
X	mg/L	mixed liquor suspended solids concentration, MLSS
α		wastewater correction factor for oxygen transfer coefficient
β		wastewater correction factor for oxygen saturation
δ		depth correction factor for oxygen saturation
ΔM	kg/d, lb/d	net sludge production rate
θ		temperature correction factor for oxygen transfer coefficient
τ		temperature correction factor for oxygen saturation
Ω		pressure correction factor for oxygen saturation

subscripts

i		constituent
n		reactor number

6.5 BIBLIOGRAPHY

Albertsson, J. G., McWhirter, J. R., Robinson, E. K., and Vahldieck, N. P. (1970). "Investigation of the Use of High Purity Oxygen in the Conventional Activated Sludge Process." *17050DNW*, Federal Water Quality Administration (FWQA), Washington, D.C.

Bergman, T. J. J., Greene, J. M., and Davis, T. R. (1992). "An In-Situ Slurry-Phase Bioremediation Case with Emphasis on Selection and Design of a Pure Oxygen Dissolution System." *In-Situ Treatment of Contaminated Soil and Water Symposium*, Cincinnati, OH.

Bergman, T. J. J., and Storms, G. E. (1994). "Odor and VOC Emission Minimization by In-Situ Oxygenation." *Water Environment Federation Conference on Odor and Volatile Organic Compound Emission Control*, Jacksonville, FL.

Brenner, R. C. (1977). "Status of Oxygen-Activated Sludge Wastewater Treatment." *EPA-625/4-77-003a*, USEPA, Cincinnati, OH.

Budd, W. E., and Lambeth, G. F. (1957). "High Purity Oxygen in Biological Treatment." *Sewage and Industrial Wastes*, 29(3), 237–253.

Cheng, A. T. Y., and Storms, G. E. (1995). "Oxygen Based Aeration Systems for Reducing Volatile Emissions and Increasing Wastewater Treatment Capacity." *P-8017*, Praxair, Inc., Tarrytown, NY.

Clifft, R. C. (1988). "Gas Transfer Kinetics in Oxygen Activated Sludge." *J of Environmental Engineering*, 114(2), 415–432.

Clifft, R. C. (1992). "Gas Phase Control for Oxygen-Activated Sludge." *J of Environmental Engineering*, 118(3), 390–401.

Famularo, J. (1975). "Purox User Manual — Pure Oxygen Plant Program." Hydroscience, Inc., Westwood, NJ.

Fullerton, D. G., and Pearlman, S. R. (1979). "Full Scale Demonstration of Open Tank Pure Oxygen Activated Sludge Treatment in Upgrading an Existing Basin at Metro Denver." *Water Pollution Control Federation 52nd Annual Conference*.

Geselbracht, J., Clark, J., Horenstein, B., and Benson, B. (1997). "Surface Aerator Performance in a Confined Headspace." *WEFTEC'97 — 70th Annual Conference of the Water Environment Federation*, Chicago, IL, 605–615.

Gilligan, T. (1998). "Lotepro Memo on UNOX Update." Personal communication.

Gilligan, T. P. (1999). "High Purity Oxygen Biological Nutrient Removal (BNR)." *Journal of the New England Water Environment Association*, 33(1), 1–16.

Krugerworld.com. (1998). "Oases® Pure Oxygen Activated Sludge System." Website Data.

Mueller, J. A., Famularo, J., and Mulligan, T. J. (1978). "Chap. 26. Application of Carbonate Equilibria to High Purity Oxygen and Anaerobic Filter Systems." *Chemistry of Wastewater Technology*, A. J. Rubin, ed., Ann Arbor Science, 465–491.

Mueller, J. A., Famularo, J., and Paquin, P. (1980). "Nitrification in Rotating Biological Contactors." *J. Water Pollution Control Federation*, 52(4), 688–710.

Mueller, J. A., Mulligan, T. J., and DiToro, D. M. (1973). "Gas Transfer Kinetics of Pure Oxygen System." *Journal of the Environmental Engineering Division, ASCE*, 99(EE3), 269–282.

Okun, D. A. (1949). "System of Bio-Precipitation of Organic Matter from Sewage." *Sewage Works Journal*, 21, 763–792.

Okun, D. A. (1957). "Discussion of High Purity Oxygen in Biological Sewage Treatment." *Sewage and Industrial Wastes*, 29(3), 253–257.

Okun, D. A., and Lynn, W. R. (1956). "Preliminary Investigation into the Effect of Oxygen Tension on Biological Sewage Treatment." *Biological Treatment of Sewage and Industrial Wastes*, Reinhold Pub. Co., New York.

Pettit, M., Gary, D., Morton, R., Friess, P., and Caballero, R. (1997). "Operation of a High-Purity Oxygen Activated Sludge Plant Employing an Anaerobic Selector and Carbon Dioxide Stripping." *WEFTEC'97 — 70th Annual Conference of the Water Environment Federation*, Chicago, IL, 595–604.

Pfeffer, J. T., and McKinney, R. E. (1965). "Oxygen Enriched Air for Biological Waste Treatment." *Water and Sewage Works*, 112(10), 381–384.

Pirnie, M. (1948) "Presentation at the 21st Annual Meeting." *Sewage Works Association*, Detroit, MI.

Rakness, K. R. (1981). "Feasibility Study of Open Tank Activated Sludge Wastewater Treatment." *600/S2-81-095*, EPA.

Robbins, M. H. J. (1961). "Use of Molecular Oxygen in Treating Semi-Chemical Pulp Mill Wastes." *16th Purdue Industrial Waste Conference*, Purdue University, Lafayette, IN, 304–310.

Speece, R. E., and Humenick, M. J. (1973). "Carbon Dioxide Stripping from Oxygen Activated Sludge." *J. Water Pollution Control Federation*, 45, 412–423.

Stamberg, J. B. (1972). "EPA Research and Development Activities with Oxygen Aeration." EPA, New York.

Stenstrom, M. K., Kido, W., Shanks, R. F., and Mulkerin, M. (1989). "Estimating Oxygen Transfer Capacity of a Full-Scale Pure Oxygen Activated Sludge Plant." *J. Water Pollut. Control Fed.*, 61(2), 208–220.

Storms, G. E. (1995). "Oxygen Dissolution Technologies for Biotreatment Applications." *P-7710A*, Praxair, Inc., Tarrytown, NY.

Storms, G. E. (1998a). "In-Situ Oxygenator (I-SO™) Installations." Praxair, personal communication, 6 April 1998.

Storms, G. E. (1998b). Telephone communication, 30 April 1998.

Storms, G. E. (2001). Email communication, 13 April 2001.

Tallent, J. (1998). "Discussion of Littleton, CO Wastewater Plant Upgrade," telephone communication, 29 April 1998.

Yin, M. T., and Stenstrom, M. K. (1996). "Fuzzy Logic Process Control of HPO-AS Process." *J. of Environmental Engineering,* ASCE, 122(6), 484–492.

Yuan, W. W., Okrent, D., and Stenstrom, M. K. (1993). "Model Calibration for the High-Purity Oxygen Activated Sludge Process — Algorithm Development and Evaluation." *Water Science & Technology*, 28(11–12), 163–171.

7 Testing and Measurement

7.1 INTRODUCTION

Historically, many methods have been used to test and specify aeration equipment. Over time varied methodologies have led to confusion and misrepresentation of equipment performance. Furthermore, equipment suppliers, consultants, and users often employ differing nomenclature when they report equipment capabilities.

Performance guarantees for oxygen transfer devices have long been the topic of lively discussion by engineers all over the world. It is important that the engineer/owner have some guarantee from the manufacturer ensuring efficient and effective performance of the proposed aeration equipment.

In the design of an aeration system, the engineer/owner must first select a process or processes that will meet discharge permit requirements. There is substantial latitude in process selection, but the choice is often made on the basis of engineer/owner experience, process and operational reliability, and capital and operating costs. Often, several alternatives may be initially selected, and evaluations are made to objectively select the best system. It is likely that the oxygen transfer system will play an important role in this selection process since it usually represents a significant portion of the total process power cost. From that point of view, it would be highly desirable for the engineer/owner to obtain guarantees on aeration performance under actual process conditions.

Typically, once a process is selected, the engineer may estimate actual oxygen requirements (ΛOR), which depends on wastewater characteristics, mean cell residence time (MCRT) or F/M, and requirements for nitrogen transformations among other process variables (see design example in Chapter 3). The AOR is subsequently used to estimate the field oxygen transfer rate (OTR_f). If an in-process oxygen transfer efficiency guarantee is available (usually expressed as mass/time power or percent efficiency), the engineer can estimate power requirements for each competitive system. Once the oxygen transfer system is selected, it is necessary to verify the guarantee by means of compliance testing.

For this scenario, the engineer must provide all process information that may impact aeration performance in order for the manufacturer to provide an in-process guarantee. The manufacturer can then apply their equipment to the prescribed process using their most favorable equipment, layout patterns, gas flow rates, and other physical considerations and based upon experience with their equipment, estimate alpha and beta for the prescribed wastewater and operating conditions. The manufacturer then may estimate a guaranteed oxygen transfer under process conditions.

In order for in-process guarantees to be successful, therefore, it is important that the following elements are accurately and clearly fulfilled:

- the engineer's specifications relative to the AOR, process, physical layout, operational parameters, and wastewater characteristics
- the manufacturer's knowledge of the factors that affect their aeration system performance including equipment, operation, and wastewater characteristics
- the verification method for the in-process guarantee, or compliance specification, which must include the test method to be used, the test protocol, and procedures and test methods for test evaluation

Typically, the first two elements are technically feasible although often misunderstood, but the third, field verification, is still in its infancy and creates the single biggest impasse to the successful application of in-process guarantees for oxygen transfer devices. As a result, most compliance specifications are written for clean water performance. Thus, the engineer/owner must make the decisions on aeration system performance under process conditions and estimate clean water performance requirements that will meet the required field conditions.

At present, there are standard methods in the U.S., Europe, and other countries that have been written for both clean water and in-process performance testing of aeration equipment. These methods are discussed below. Other testing methods are also required for aeration equipment. In recent years, there have been reported instances where installed fine pore diffuser systems did not meet specified requirements when tested in full scale. Since performance tests were conducted near the end of the construction period, failure to meet performance requirements resulted in delay of start-up. Recent work has produced guidelines for quality assurance of fine-pore diffusers at the construction site. To better understand and evaluate diffused air devices, methodologies have also been developed to characterize diffuser elements in new and used condition.

7.2 AERATION TANK MASS BALANCE

In deriving the equations for the analysis of the data collected from aeration systems, a mass balance of oxygen around a completely mixed aeration tank, Figure 7.1 is constructed.

$$Q_i C_i - Q_i C_L + K_L a_f \left(C_{\infty f}^* - C_L \right) V - RV = V \frac{\Delta C_L}{\Delta t} \tag{7.1}$$

Dividing by the aeration tank volume and taking the limit as $\Delta \to 0$, yields the differential equation.

$$\frac{dC_L}{dt} = \frac{C_i - C_L}{t_0} + K_L a_f \left(C_{\infty f}^* - C_L \right) - R \tag{7.2}$$

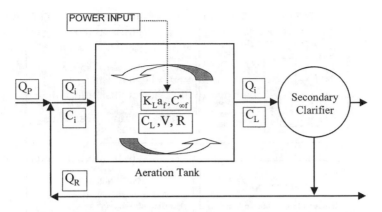

FIGURE 7.1 Mass balance on a completely mixed aeration tank.

This is more general than Equation 2.26 since it is not limited to a clean water batch system with the subscript "*f*" relating to field conditions. It includes the oxygen transport rate as well as the oxygen transfer rate and oxygen uptake rate (OUR), R. In Equation 7.2, t_0 is the detention time in the aeration tank based on the total influent flow, Q_i, to the aeration tank, including the primary flow, Q_P, and the return activated sludge flow, Q_R.

$$t_0 = \frac{V}{Q_i}; \qquad Q_i = Q_P + Q_R$$

7.3 CLEAN WATER PERFORMANCE TESTING

Consensus procedures for the evaluation of aeration equipment in clean water are now in place in the U.S. and Europe and have been adopted by a large number of engineering firms and manufacturers worldwide. The ASCE *Standard-Measurement of Oxygen Transfer in Clean Water* (ASCE, 1991) was first published in 1985 and was reedited and adopted in principle in Europe as a European Standard in 2000 (CEN/TC, 2000). The method covers the measurement of the oxygen transfer rate (OTR) as a mass of oxygen per unit time dissolved in a volume of water by an oxygen transfer system operating under given gas and power conditions. The method is applicable to laboratory-scale oxygenation devices with small volumes of water as well as the full-scale system with water volumes found in activated sludge treatment processes. The process is valid for a variety of mixing conditions and process configurations. The ASCE method also includes measurement of gas rates and power.

A schematic of the clean water testing technique is given in Figure 7.2. The test is conducted using clean (tap) water under batch (nonflowing) conditions. The non-steady-state method is based on dissolved oxygen (DO) removal from the test water volume by the addition of sodium sulfite in the presence of cobalt catalyst. These steps are followed by transfer measurements of reoxygenation to near saturation concentrations. Test water volume DO inventory is monitored during the reoxygenation period by measuring DO concentrations at several points selected to best

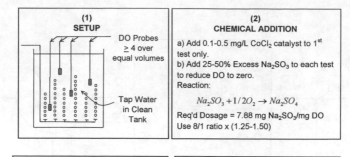

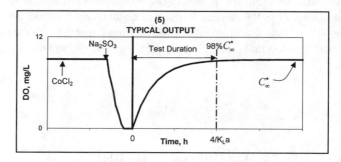

FIGURE 7.2 Clean water test schematic.

represent the tank contents. These DO concentrations are measured *in situ* or on samples pumped from the tank. The method specifies minimum sample number, distribution, and range of DO measurements at each sample point.

Equation 2.26 describes these conditions. Letting $D = C_\infty^* - C_L$ and $dD = -dC_L$ provides the following.

$$\int_{D_0}^{D} \frac{dD}{D} = -K_L a \int_{0}^{t} dt$$

$$\left. \begin{array}{c} \ln \dfrac{D}{D_0} = -K_L a t \\[2mm] D = D_0 e^{-K_L a t} \end{array} \right\}$$ (7.3)

Analysis of data using the above equation is referred to as the "log deficit" technique and is one of the oldest methods used in the field. Due to difficulties in interpreting results from the above approach when exact values of oxygen saturation are not known, the ASCE Committee on oxygen transfer has recommended using Equation 7.3 in terms of concentration.

$$C_L = C_\infty^* - \left(C_\infty^* - C_0\right)e^{-K_L at} \qquad (7.4)$$

Data obtained at each sample point are then analyzed using a nonlinear regression analysis of Equation 7.4 to estimate three parameters including the apparent volumetric mass-transfer coefficient ($K_L a$), the equilibrium spatial average DO saturation concentration (C_∞^*), and the initial DO concentration (C_0). The nonlinear regression, NLR, computer program developed by the ASCE committee to fit the DO - time profile measured at each sampling point during reoxygenation also provides statistics on the best-fit parameters and the residuals to the model equation. For a viable test, no trend in residuals should occur. Typically, the coefficient of variation on $K_L a$ will be < 5 percent and the standard deviation on C_∞^* < 0.1 mg/L.

Figure 7.3 shows the use of both "log deficit" and NLR techniques on a typical set of clean water field data. The NLR fit is excellent with very low residuals. Note that if any lingering effects of sulfide addition exist in the system, a lag in the exponential increase will occur giving an initial "S" shape to the curve. This initial data must be truncated during data analysis since only the exponential portion of the curve is analyzed by Equation 7.4. The log deficit results depend on the choice of the saturation value. When the C_∞^* value is too high, the semi-log plot tails upwards as the deficit approaches zero. The reverse is true when C_∞^* is too low. Errors in $K_L a$, between 13 and 23 percent, occurred for this data set for the <1 percent change in saturation value. However, when the log deficit is performed on the measured DO data using only values up to 80 percent of saturation, as recommended by Boyle et al. (1974), then an error of only 2 to 4 percent in $K_L a$ occurs. This result is shown in Figure 7.4.

From the above results, it is recommended that the NLR technique always be used in final data analysis. For rapid on-site estimates, the log deficit technique should provide $K_L a$ values within 5 percent of the NLR value when data up to ~ 80 percent of saturation is analyzed.

For results presentation, the $K_L a$ and C_∞^* values for each individual sampling location, i, are adjusted to standard conditions as indicated in Chapter 2.

$$K_L a_{20i} = K_L a_i \theta^{20-t}$$

$$C_{\infty 20i}^* = \frac{C_{\infty i}^*}{t\Omega}$$

The tank SOTR is then calculated by using the estimates of $K_L a$ and C_∞^* adjusted to standard conditions at each sample point.

$$\left.\begin{aligned} SOTR_i &= K_L a_{20i} C_{\infty 20i}^* V \\ SOTR &= \frac{1}{n}\sum_{i=1}^{n} SOTR_i \end{aligned}\right\} \qquad (7.5)$$

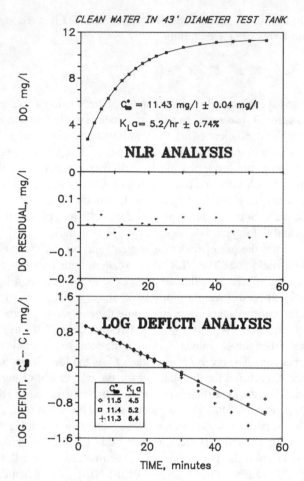

FIGURE 7.3 Clean water data analysis techniques.

In the above equations, V is the total tank volume and n is the total number of measurement locations. SOTR represents the average mass of oxygen transferred per unit time for the total tank at zero DO concentration, water temperature of 20°C, and barometric pressure of 101.3 kPa (1.0 atm), under specified gas flow rate and power conditions. The test is conducted in clean water (alpha presumed to be 1.0) as specified in the standard. Results may also be presented as a standard oxygen transfer efficiency (SOTE), obtained by dividing SOTR by the mass flow of oxygen in the gas stream (Equation 2.50), or as standard aeration efficiency (SAE), by dividing the SOTR by the power input (Equation 2.45). Although there is no way to verify method accuracy, it is precise within ± 5 percent (Baillod et al., 1986).

The foundation and key elements of the oxygen transfer measurement test are the definition of terms used during aeration testing, subsequent data analysis, and final result reporting. A consistent nomenclature has been established with more logical and understandable terminology than the numerous and varied symbols used historically.

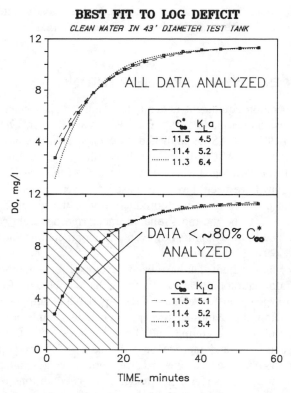

BEST FIT TO LOG DEFICIT

CLEAN WATER IN 43' DIAMETER TEST TANK

FIGURE 7.4 Effect of data truncation on log deficit analysis.

The clean water compliance test may be performed in the full-scale system or in the manufacturer's shop test facility. If performed at the shop test facility, it is important to ensure that the test results will properly simulate the field scale system. Scale-up would include geometric similarity (e.g., water depth, length to width, and width to depth ratios), gas flow rates per unit and volume, power input per unit volume, density of diffuser placement, and distance between aeration units, to name a few considerations. Potential interferences resulting from wall effects and any extraneous piping or other materials in the tank should be minimized. Where necessary (e.g., long, narrow diffused aeration tanks), testing of tank sections may be required where there is little circulation of water between adjacent sections. Sealed partitions are used to ensure that oxygen does not interchange between units.

Although most projects require a shop or field test to verify diffuser performance, SOTR can also be measured in the laboratory to aid in characterizing diffusers both new and used. These tests are not intended to be a substitute for shop or field-testing or for predicting field OTR. They are most often used to determine relative differences in performance between diffusers or to assess effectiveness of cleaning methods. A typical laboratory setup will include a small column, 61 to 91 cm (2 to 3 ft) in diameter and 2 to 3 m (7 to 10 ft) high. The diffuser to be tested would be placed in the column and a clean water OTE would be determined over a range of airflows. The clean water procedure would usually be determined by the ASCE Clean Water

Standard (1991) which is a non-steady-state method. A steady-state method may also be used and is described in detail in the *Design Manual, Fine Pore Aeration Systems* (1989).

7.4 IN-PROCESS OXYGEN TRANSFER TESTING

The testing of aeration equipment under field conditions has been the subject of considerable research over the last 30 years (EPA, 1983; Kayser, 1969; Mueller and Boyle, 1988). In 1996, the ASCE published the *Standard Guidelines for In-Process Oxygen Transfer Testing* (ASCE, 1996) and shortly thereafter the European standard (CEN/EN, 2000) was developed which drew on much of the ASCE standard guideline. The guidelines have been developed based on over 30 years of side-by-side testing of several methods to verify reproducibility of the methods. The methods selected have proven to be the most reliable under rigorous field conditions. The technology continue to be dynamic, however, and modifications and/or new procedures will likely occur in the future.

The intent of the methods that have been developed for field conditions was to provide useful information on field performance that can be used for future design (variability in oxygen transfer, alpha values, spatial and temporal variations in oxygen demand, etc.). It provides the owner with data that can be used for operation and maintenance of aeration equipment. The procedures also offer manufacturers the opportunity to develop and improve the performance of their equipment. In some instances, engineers may use these methods for compliance guarantees. It should be emphasized, however, that performance under process conditions is affected by a large number of process variables and wastewater characteristics that are not easily controlled for a given test condition. Thus, compliance testing under field conditions can be highly subjective and uncertain.

The methods described in the ASCE In-Process Guidelines (ASCE, 1996) include a non-steady-state method, off-gas technique, and the inert gas tracer method. These methods have been well developed and provide satisfactory precision for a wide range of aeration processes. Additional provisional methods include a steady-state procedure and mass balance methods. In general, testing methods can be categorized according to whether DO is steady or nonsteady. If the influent to the test basin is diverted, these tests are referred to as batch tests and do not reflect the variability of wastewater characteristics or the actual operating conditions that might be expected. If wastewater flow to the test basin is continuous, the test more nearly represents actual operating conditions, but steady state, with respect to influent character (AOR, alpha, etc.), is difficult to achieve.

The basis of the steady-state and non-steady-state techniques is Equation 7.2. For the steady-state technique, $\frac{dC_L}{dt} = 0$, and the DO is constant in the tank, $C_L = C_R$, for a constant uptake rate, R.

$$R = \frac{(C_i - C_R)}{t_0} + K_L a_f \left(C_{\infty f}^* - C_R \right) \tag{7.6}$$

In practice, both R and C_R values are measured at a number of equal volume sampling locations, i, in the aeration tank. This technique requires using the average oxygen uptake rate and DO concentration in the tank to determine the tank oxygen transfer coefficient. Due to back dispersion and mixing in the tank, individual $K_L a_f$ values at each location are meaningless. Representative *in situ* OUR values are difficult to obtain in practice when a sample is removed from the aeration tank due to substrate or oxygen limitation (Mueller and Stensel, 1990).

$$
\left.
\begin{aligned}
& R = \frac{1}{n}\sum_{i=1}^{n} R_i, \qquad C_R = \frac{1}{n}\sum_{i=1}^{n} C_{Ri} \\[2mm]
& K_L a_f = \frac{R - \dfrac{\left(C_i - C_R\right)}{t_0}}{\left(C_{\infty f}^* - C_R\right)} \\[2mm]
& OTR_f = K_L a_f V\left(C_{\infty f}^* - C_R\right)
\end{aligned}
\right\}
\quad \text{Steady-state overall tank values} \quad (7.7)
$$

The non-steady-state equation is obtained by substituting Equation 7.6 into 7.2 thus, eliminating the constant oxygen uptake rate.

$$
\frac{dC_L}{dt} = \frac{C_R - C_L}{t_0} + K_L a_f \left(C_R - C_L\right) \tag{7.8}
$$

This equation is similar to the clean water equation except the oxygen concentration approaches the steady-state DO in the tank, C_R, not the saturation concentration. Letting $D = C_R - C_L$ and $K = K_L a_f + \dfrac{1}{t_0}$ provides the following result.

$$
\int_{D_0}^{D} \frac{dD}{D} = -K\int_{0}^{t} dt
$$

$$
\left.
\begin{aligned}
& \ln \frac{D}{D_0} = -Kt \\[2mm]
& D = D_0 e^{-Kt}
\end{aligned}
\right\} \tag{7.9}
$$

In terms of the tank DO concentration, an equation similar to Equation 7.4 is obtained allowing data analysis with the same techniques used for clean water.

$$\left.\begin{array}{l} C_L = C_R - \left(C_R - C_0\right)e^{-Kt} \\[2mm] K = K_L a_f + \dfrac{1}{t_0} \end{array}\right\} \tag{7.10}$$

In practice, both $K_L a_f$ and C_R values are again measured at a number of equal volume sampling locations, i. The average tank values are again utilized to determine the overall tank $K_L a_f$. Similar to the steady-state technique, due to back dispersion and mixing in the tank, individual $K_L a_f$ values at each location are meaningless.

$$\left.\begin{array}{l} K = \dfrac{1}{n}\sum_{i=1}^{n}K_i, \qquad C_R = \dfrac{1}{n}\sum_{i=1}^{n}C_{Ri} \\[4mm] K_L a_f = \left[K - \dfrac{1}{t_0}\right] \\[4mm] OTR_f = K_L a_f V\left(C_{\infty f}^{*} - C_R\right) \end{array}\right\} \text{Non-steady-state overall tank values} \tag{7.11}$$

Non-steady-state methods estimate an average $K_L a$ for a test section by measuring the change in DO concentration with time after a perturbation from steady-state conditions. This perturbation may be imposed on the system by changing input aeration power (up or down) or by the addition of hydrogen peroxide or high purity oxygen. The procedure requires constant OUR, DO, flow rate, and $K_L a$ over the test period, and it requires the accurate measurement of the test section DO and flow rate. It avoids the need to measure OUR and C_{∞}^{*}.

Hildreth and Mueller (1986) have shown that the above non-steady-state approach can be used in advective-dispersive systems which are not completely mixed. The K value in Equation 7.9 is defined by $K = K_L a_f + \dfrac{1}{t_0} + K_e$. The additional term, K_e, is a function of longitudinal dispersion and velocity of flow in the tank. For Ridgewood, NJ, fine pore diffusers in tanks 35.4 m (116 ft) long and 7.3 m (24 ft) wide, it varied from 0.1 to 0.3/h. In long, 91.4 m (300 ft), narrow, 9.1 m (30 ft), tanks at Whittier Narrows, CA, Mueller (1985) has shown that the batch equation where $K = K_L a_f$ could be applied near the end of the tank. For accurate results, the minimum distance, x_{min}, required downstream from a boundary in a section where OUR and $K_L a_f$ are constant was $x_{min} = 2.5\, U/K_L a_f$ where U is the forward velocity.

Non-steady-state testing is the most suitable method available for mechanical aeration systems. However, it does not provide an estimate of the accuracy of the results. During a sabbatical leave in 1980, the senior author conceived of a technique to get an estimate of how good the results were by conducting the tests twice. Each test was conducted at a different power level as shown in Figure 7.5 (Mueller, 1982; Mueller et al., 1982; Mueller and Rysinger, 1981). Changing power level can be used by itself or in conjunction with hydrogen peroxide addition to get a greater

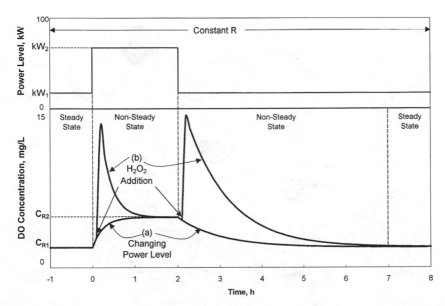

FIGURE 7.5 Dual non-steady-state analysis techniques, a) changing power levels, b) H_2O_2 addition.

spread in the non-steady-state curves. Good results can be obtained with both techniques (Mueller and Boyle, 1988).

This provides two different K_La_f and two different steady-state C_R values with one oxygen saturation value. The following equations are used with these values to calculate the *in situ* OUR and saturation concentration.

$$R = \frac{C_{R2} - C_{R1} + \dfrac{1}{t_0}\left(\dfrac{C_i - C_{R1}}{\dfrac{1}{K_La_{f1}}} - \dfrac{C_i - C_{R2}}{\dfrac{1}{K_La_{f2}}} \right)}{\dfrac{1}{K_La_{f1}} - \dfrac{1}{K_La_{f2}}} \qquad (7.12)$$

$$C_{\infty f}^* = C_{R1} + \frac{1}{K_La_{f1}}\left(R - \frac{(C_i - C_{R1})}{t_0} \right) = C_{R2} + \frac{1}{K_La_{f2}}\left(R - \frac{(C_i - C_{R2})}{t_0} \right) \qquad (7.13)$$

Close agreement of the saturation value calculated from Equation 7.13 with the clean water estimated value corrected for field conditions, Equation 2.38, indicates adequate non-steady-state results. At ratios of K_La_f values greater than 2/1, good agreement should be obtained. The oxygen uptake rate and flow must be constant during the tests, a difficult situation when K_La_f values are low requiring a long time for the tests.

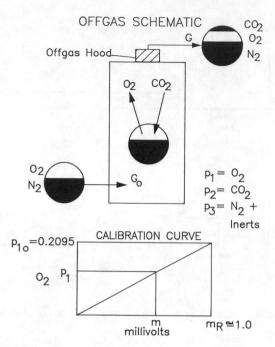

FIGURE 7.6 Off-gas analysis schematic.

The off-gas method is a gas-phase mass balance technique for directly measuring OTE of aeration devices having a diffused air component. The method requires the use of a suitable analyzer for accurately measuring the relative gas-phase oxygen content of ambient air and basin off-gas. It employs a fixed or floating collection hood for the off-gas that should cover a minimum of 2 percent of the test section area. In contrast to the non-steady-state method, off-gas methods may provide local as well as overall basin oxygen transfer data. It may also be used in zero DO systems without error. Test section DO concentration, C_{∞}^{*}, and off-gas flow rate measurements are essential if estimates of SOTR and SAE are to be obtained.

The equations governing the off-gas technique are similar to Equation 6.2 for the gas phase oxygen mass balance except only one hood location is employed as shown in Figure 7.6.

$$V_G \frac{dC_G}{dt} = \left(G_0 C_{G0} - G C_G\right) - K_L a_f V_L \left(C_{\infty f}^{*} - C_L\right) \qquad (7.14)$$

In the above equation, the subscript "0" refers to gas flow and concentration inlet to the tank, also called the reference conditions. Dividing by the mass of inlet gas at steady state provides an equation similar to Equation 2.51, except it is modified for process conditions.

$$OTE_f = \frac{G_0 C_{G0} - G C_G}{G_0 C_{G0}} = \frac{K_L a_f V_L \left(C_{\infty f}^{*} - C_L\right)}{w_o} = \frac{OTR_f}{w_o} \qquad (7.15)$$

Use of Equation 7.15 requires measurement of the inlet and outlet gas flows, a difficult task to measure accurately, especially on the inlet, which depends on accurate gas flow monitoring at the plant. This difficulty is circumvented by using the conservative nature of the mass of gas phase inerts, subscript "3", at steady state to define the influent gas flow as a function of the measured exiting gas flow through the hood.

$$G_0 C_{G30} - GC_{G3} = 0$$

$$G_0 = \frac{GC_{G3}}{C_{G30}}$$ (7.16)

Using the ideal gas law to define the concentration in the gas phase as a function of partial pressure, $C_g = \frac{pM}{RT}$, leads to the folding equation for OTE_f as a function of partial pressures.

$$OTE_f = 1 - \frac{C_g / C_{G3}}{C_{g0} / C_{G30}} = 1 - \frac{p_1 M_1 / p_3 M_3}{p_{10} M_1 / p_{30} M_3} = 1 - \frac{p_1 p_{30}}{p_{10} p_3}$$ (7.17)

In the typical off-gas measuring equipment, a desiccant is used to provide dry air, and carbon dioxide is removed by absorption onto sodium hydroxide pellets. This process leads to the measured off-gas consisting of only oxygen and inerts, allowing the inert partial pressure to be defined as follows. For dry air and no CO_2:

$$p_1 + p_3 = 1.0$$

$$p_3 = 1 - p_1$$ (7.18)

Using the mole fraction of dry air for the inlet gas as $p_{10} = 0.2095$ yields the mole fraction of inerts as $p_{30} = 0.7905$. Substituting the above with Equation 7.18 into Equation 7.17 defines OTE_f as a function of only the measured oxygen partial pressure.

$$OTE_f = 1 - \frac{0.7905 p_1}{0.2095(1 - p_1)} = 1 - \frac{3.773 p_1}{1 - p_1}$$ (7.19)

Using the millivolt DO probe readings on the inlet (reference, m_R) and exiting (off-gas, m) phases, Figure 7.6, provides the following value of p_1.

$$p_1 = 0.2095 \frac{m}{m_R}$$

The above field OTE_f is measured at the mixed liquor temperature and DO concentration at a specific hood location, i, in the tank. An average of five DO readings, alternating between off-gas and reference air, is recommended to obtain an estimate of the OTE_f variability at a location. The field results are summarized at standard conditions of 20°C and 1 atm. Knowledge of the clean water oxygen transfer efficiencies allows determination of α at each location, α_i.

$$\left. \begin{array}{c} OTE_{20i} = \alpha_i \beta SOTE_i = OTE_{fi} \left(\dfrac{\beta C^*_{\infty 20i}}{C^*_{\infty fi} - C_{Li}} \right) 1.024^{20-t_i} \\[4mm] \alpha_i = \dfrac{OTE_{20i}}{\beta SOTE_i} \end{array} \right\} \qquad (7.20)$$

For the total tank with n equal volume hood locations, the gas flow weighted average oxygen transfer efficiency and α are used.

$$\left. \begin{array}{c} G = \displaystyle\sum_{i=1}^{n} G_i \\[4mm] OTE_{20} = \alpha\beta SOTE = \dfrac{1}{G} \displaystyle\sum_{i=1}^{n} OTE_{20i} G_i \\[4mm] \alpha = \dfrac{OTE_{20}}{\beta SOTE} \end{array} \right\} \text{Off-gas overall tank values} \qquad (7.21)$$

To determine the confidence level in the OTE_i data, the standard normal distribution from the Central Limit Theorem was used at a study on the Cedar Creek plant, NY (Mueller and Saurer, 1986). Table 7.1 gives the results of the statistical analysis performed on the five OTE_{20} samples taken at each station in each test. For conciseness, a range of results is presented as opposed to individual values at each station. There is a minimum confidence level of 97.2 percent that the measured mean OTE_{20i} value is at least ± 10 percent of the true mean. A minimum confidence level of 72.9 percent exists for the mean to be within ± 5 percent of the true mean. Thus, the authors consider the off-gas technique to have a precision of ± 10 percent, about the same as the non-steady-state technique for field conditions. However, the off-gas technique provides additional information on variability of OTE_{20} and α within the tank, whereas the non-steady-state test only gives an estimate of the overall tank value.

Inert gas tracer methods may employ radioactive (Neal and Tsivoglou, 1974) or stable isotope gases such as krypton (Hovis and McKeown, 1985), noble gases, and low molecular weight hydrocarbon gases. A test section is dosed with a supersaturated level of an inert gas tracer. By monitoring the disappearance of the tracer from the liquid and applying the appropriate gas transfer equation, the value of the mass transfer coefficient of the gas is obtained. This value may be corrected for dispersion in the liquid by adding a second, conservative, nonvolatile dissolved tracer at the same time. The mass transfer coefficient of the tracer gas is related to that of oxygen by a constant, derived from theoretical and experimental investigations. Like the non-steady-state

TABLE 7.1
Variability in Off-Gas OTE Values (Mueller and Saurer, 1986)

Test	Statement	# of Samples	z Value Range	Cumulative Distribution Function Range	Minimum Confidence Level Range
1–12*	Mean OTE_{20} is ± 10% of true mean	5	2.20–16.73	0.98610–1.0000	97.2%
1–12*	Mean OTE_{20} is ± 5% of true mean	5	1.10–8.36	0.86430–1.0000	72.9%

* Statistical Analysis using Central Limit Theorem was performed on all OTE data.

Eq. 1. $z = \dfrac{\bar{x} - \mu}{s/\sqrt{n}}$
μ = unknown true mean
$\bar{x}$ = measured mean OTE_{20}
s = standard deviation of n samples taken
n = number of samples
z = standard normal distribution value for two-tail significance

Example: For Test 11, Station #1, is Mean OTE_{20} ± 10% of true mean?

$z = \dfrac{(0.1)6.97}{0.71/\sqrt{5}} = 2.20$ @ z = 2.20, cdf = 0.98610   ∴ Confidence level that mean OTE_{20}
$P[z \leq$ Eq. 1] = 0.98610   is ± 10% of true mean is 97.2%
$P[0 \leq z \leq$ Eq. 1] = .4861
$P[-(Eq. 1) \leq z \leq (Eq. 1)] = .972$

method, this method provides a measure of the overall test basin $K_L a$ and requires a constant $K_L a$ over the test period. The capital and analytical costs for this procedure are high and the technique relatively specialized (Mueller and Boyle, 1988).

At present, there is no way to assess the accuracy of the field test methods. Since there is no standard against which to make comparisons, it is only possible to compare methods with each other. The off-gas and inert tracer procedures produced estimates of process αSOTE within two to five percent of each other in parallel tests of oxidation ditches (Boyle et al., 1989). In side-by-side comparisons of six municipal and industrial waste treatment sites, the off-gas, inert tracer, and non-steady-state procedures estimated αSOTR within 10 percent of each other under conditions of relatively constant flow and OUR (Mueller and Boyle, 1988). Since these methods measure oxygen transfer in different ways, using different mechanisms, it may be presumed that they provide an accurate measurement within 10 percent under proper test conditions. The precision of these three methods also is < ± 10 %.

Currently, the steady-state method, which is the simplest to conduct, is the least precise and accurate. It is recommend only when rough estimates of transfer are required or when the method has been rigorously checked against one of the three tests above for a given facility.

7.5 QUALITY ASSURANCE FOR FINE-PORE DIFFUSERS

As described above, several instances have been reported where fine pore diffusers delivered at the construction site do not meet the specified performance. In compliance testing of aeration equipment, clean water oxygen transfer tests are normally required. If shop tests are conducted, the major concern of the engineer/owner is whether the equipment manufacturer practices quality control in the production of the diffusers. If quality control is practiced to the satisfaction of the engineer, the shop test and a field verification that proper installation has prevailed should be sufficient to ensure quality of the system. If quality assurance at the factory is not practiced or cannot be verified by the engineer, shop testing should be supplemented with verification that the diffusers shipped to the site are equivalent to those tested in the shop. Reference tests would be performed on shop-tested and field-delivered diffusers (ASCE, 2001). Statistical procedures are outlined to determine the number of diffusers required for testing and to compare the results for equivalence at some predetermined confidence level. Both OTE evaluative tests and correlative tests are described in the Guidelines (ASCE, 2001). The correlative tests include DWP and EFR methods that have demonstrated good correlation with SOTE measurements.

The concern about quality assurance is not an issue if field-scale clean water oxygen transfer tests are conducted on all basins to be placed in service. This procedure is normally not done in larger installations with multiple basins and, again, some quality assurance verification would be desirable for the remaining diffuser elements.

7.6 CHARACTERISTICS OF DIFFUSED AIR MATERIALS

Many properties can be used to characterize diffused air materials. Knowledge of these characteristics promotes better design of an aeration system for a selected set of wastewater conditions. Appropriate attention to these characteristics in the design phase may also lead to less operation and maintenance problems during the life of the system. Many of these characteristics are not routinely available for specific media. Many are most applicable and critical to porous diffusers. Several of these characteristics are used in defining quality control on media (ASCE, 2001) and may be used in specifying diffusers. These tests have also been performed to provide routine baseline data on materials to assess rates of material deterioration. The following sections briefly describe some of these characteristics. Greater detail may be found in the design manual (EPA, 1989).

7.6.1 PERMEABILITY

Initially developed in the 1900s as a simple means to specify porous diffusers, the permeability measure is such an arbitrary and inexact parameter that it is little used today. Permeability, an empirical rating that relates flux rate to pressure loss and pore size and/or pore volume, is a measure of the frictional resistance to flow in a porous medium. It is normally defined as the amount of air, at standard conditions, that will pass through 929 cm² (1.0 sq. ft) of 25 mm (1 in) thick, dry porous media at room temperature. A differential pressure of 5 cm (2 in) water gauge is used in the test. The flow rate (scfm) obtained under these conditions is referred to as the permeability (perm) rating.

This measure does not provide a true basis for comparison of porous media performance since the same permeability rating could be obtained from a diffuser with a few relatively large pores or a multitude of small pores. In addition, two diffusers with the same pore structure but different thickness would have different measured perms. Many ceramic and porous media specifications today still include permeability but until the procedure is standardized for various shapes, densities, effective area, and thickness, it will not provide a useful means of comparison. Efforts have been made to standardize permeability with the development of the specific permeability, which attempts to account for diffuser geometry (Redmon, 1985). Shortcomings still exist, however, in the method.

7.6.2 DYNAMIC WET PRESSURE

The dynamic wet pressure (DWP) is defined as the pressure differential (head loss) across the diffuser element when operating in a submerged condition expressed in cm (in) water gauge at a specified air flow rate. As a rule, the smaller the bubble size, the higher the DWP. While small bubbles may produce higher transfer efficiencies, the additional power to overcome the higher head loss may negate any potential savings.

DWP is measured in the laboratory or in the field. Figure 7.7 illustrates a typical setup for determining DWP. Air header pressure and static pressure are measured as well as the pressure just below the diffuser element. Details of the test procedure are outlined in the USEPA fine pore manual (1989). The procedure is normally more

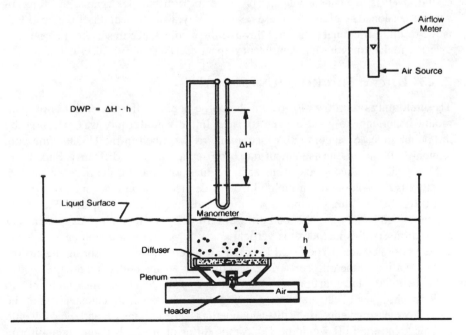

FIGURE 7.7 Apparatus for measuring dynamic wet pressure (DWP) in the laboratory.

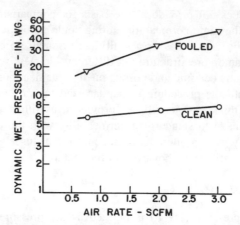

FIGURE 7.8 Impact of air flow rate and fouling on DWP of a porous diffuser.

accurate under laboratory conditions, but field installations have provided useful data on diffuser fouling and deterioration in routine plant operation. The porous media today have DWP values ranging from 8 to 100 cm (3–39 in) water gauge with typical or specified airflow rates and when new. Figure 7.8 demonstrates typical DWP vs. airflow rate for a porous diffuser. The specific value of DWP depends on the material type, surface properties, airflow rate, presence of internal or external foulant, and diffuser thickness. For new ceramic and porous plastic diffusers, most of the DWP is associated with the pressure to form bubbles against the force of surface tension. Therefore, for these devices, only a small fraction of the head loss is the result of frictional resistance through the media. Once in service, internal and external foulant may have a significant impact on DWP of a diffuser element.

7.6.3 EFFECTIVE FLUX RATIO (UNIFORMITY)

The uniformity of airflow distribution through a porous diffuser element is of paramount importance to good oxygen transfer. Initially, uniformity was measured by the bubble release vacuum (BRV) technique as described in the USEPA fine pore manual (1989). This measurement has been replaced by the Effective Flux Ratio (EFR), which measures flux of air at individual points along the diffuser surface. Air flux is the volume of air emitted from a defined area and has units of $L/s/cm^2$ ($scfm/ft^2$). Several flux parameters are used to define the EFR. Apparent Flux (AF) is determined by dividing the total diffuser airflow by the total air release area. (For dome diffusers, this includes the vertical sides; for perforated membranes, it is the entire perforated area.) The Local Flux (LF) is determined by measuring the airflow from a portion of the diffuser surface and dividing by the collection area. Effective Flux (EF) is the local airflow weighted average of the local flux measurement. An EFR is subsequently calculated by dividing the EF by the arithmetic average of the local flux measurements. If the diffusion media is uniform, the EF and the AF would be equal, and the EFR would be 1.0. As the diffusion media becomes nonuniform, the EFR increases above 1.0 because areas emitting more air are weighted more. As uniformity of air flux decreases, the ERF increases.

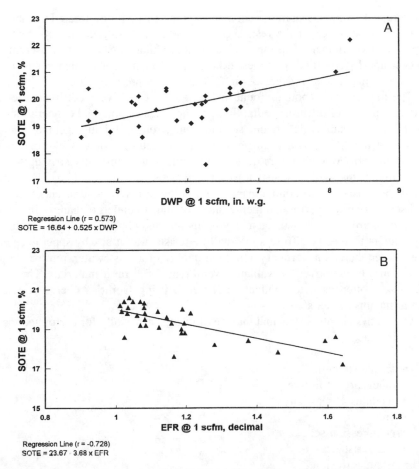

Regression Line (r = 0.573)
SOTE = 16.64 + 0.525 x DWP

Regression Line (r = -0.728)
SOTE = 23.67 - 3.68 x EFR

FIGURE 7.9 Correlation between (A) DWP and (B) effective flux ratio (EFR) with SOTE of porous diffusers in clean water at 1 scfm air flow rate.

Details of the test procedure are presented in the ASCE *Standard Guidelines for Quality Assurance of Installed Fine Pore Aeration Equipment* (2001). Both EFR and DWP are primary measurements used in evaluating quality of diffusers. A correlation between these two parameters and SOTE is proposed in these guidelines (Figure 7.9).

7.6.4 OTHER CHARACTERISTICS

A number of other physical and chemical tests may be desirable depending upon the diffuser element and the needs of the specific project. Baseline dimensions are often useful especially for membrane materials that may change shape with exposure in wastewater. Weight and specific weight are used for quality control as well as to provide baseline information on new diffusers. The structural or physical strength of ceramic or plastic media is important in assessing the potential integrity of the material under the static head of water, both during placement and during shipment and storage. Hardness is an important media characteristic for perforated membranes because it is an index of the resistance of an elastomer to deformation. Shore A durometer measurements are the most common, although Shore D measurements are occasionally specified. Changes in hardness of membranes, often occurring in wastewater, may result in decreases in OTE and back pressure.

The impact of compounds found in wastewater can have a detrimental effect on the properties of diffuser media. Some compounds of potential concern include mineral and vegetable oils, organic solvents, and strong oxidizing agents. Cleaning agents (for the diffusers) and air-phase foulant including oxidants like ozone are also of concern. Manufacturers of aeration devices are constantly striving to find new materials that will be resistant to specific agents in water and air. There are a variety of resistances to contaminants even within a given generic classification. As discussed earlier, perforated membranes continue to undergo changes in formulation to improve their resistance to environmental and physical stresses. Engineers may attempt to specify diffusers that will be resistant to attack by specific agents. Often, when there is uncertainty about the quality of a wastewater, removable test headers may be employed to evaluate several types of diffuser materials. These test headers are often used to conduct studies at existing facilities over a period of several months to years.

Other physical properties that may be of interest especially for perforated membranes include:

- tensile strength
- elongation at failure
- modulus of elasticity
- creep
- compression set
- tear resistance
- strain corrosion
- solvent extraction
- ozone resistance

7.7 NOMENCLATURE

AE_f	kg/kWh, lb/hp-h	aeration efficiency under process conditions
AOR	kg/d	actual oxygen requirements = OTR_f
C_G	mg/L	oxygen concentration in gas phase exiting aeration tank and under hood
C_{G0}	mg/L	oxygen concentration in gas phase entering aeration tank
C_{G3}	mg/L	concentration of inerts (mostly N_2) in gas phase exiting aeration tank
C_{G30}	mg/L	concentration of inerts (mostly N_2) in gas phase entering aeration tank

C_i	mg/L	influent oxygen concentration
C_L	mg/L	bulk liquid phase oxygen concentration in aeration tank
C_0	mg/L	oxygen concentration at time zero
C_R	mg/L	oxygen concentration at steady state
C_∞^*	mg/l	clean water oxygen saturation concentration at diffuser depth
$C_{\infty 20}^*$	mg/l	clean water oxygen saturation concentration at diffuser depth and 20°C
$C_{\infty f}^*$	mg/l	oxygen saturation concentration under process (field) conditions
D	mg/L	oxygen deficit based on oxygen saturation in clean water and on steady-state concentration under process conditions
D_o	mg/L	initial oxygen deficit
DWP	cm of water	dynamic wet pressure
EFR	L/s-cm^2, scfm/ft^2	effective flux ratio
F/M	lb BOD$_5$/d-lb MLSS	food to microorganism ratio
G	m$_N^3$/h, scfm	gas flow rate leaving aeration tank
G_0	m$_N^3$/h, scfm	gas flow rate entering aeration tank
G_s	m$_N^3$/h, scfm	air flow rate at standard conditions
G_{sd}	m$_N^3$/h-diff	air flow rate per diffuser at standard conditions
H	m	sidewater depth
H_s	m	diffuser submergence
K	h^{-1}	coefficient accounting for oxygen transfer, hydraulic detention time, and longitudinal dispersion in non-steady-state test
K_e	h^{-1}	coefficient accounting for longitudinal dispersion in non-steady-state test
$K_L a$	h^{-1}	oxygen transfer coefficient
$K_L a_{20}$	h^{-1}	clean water oxygen transfer coefficient at 20°C
$K_L a_f$	h^{-1}	oxygen transfer coefficient under process conditions
m	mv	oxygen probe reading in off-gas
m_R	mv	oxygen probe reading in reference gas
M_1, M_3	g/mole	molecular weight of oxygen and nitrogen, respectively
MCRT	d	mean cell residence time
n		number of sampling locations
OTE	–, %	oxygen transfer efficiency
OTE$_f$	–, %	oxygen transfer efficiency under process conditions
OTE$_{20}$	–, %	oxygen transfer efficiency under process conditions at 20°C and zero DO, overall tank value for off-gas test

OTR	kg/h, lb/h	oxygen transfer rate
OTR_f	kg/h, lb/h	oxygen transfer rate under process conditions
OUR	mg/L-h	oxygen uptake rate, R
p_1, p_{10}		partial pressure (mole fraction) of oxygen in the gas phase exiting and entering, respectively, the aeration tank
p_3, p_{30}		partial pressure (mole fraction) of inerts (mostly N_2) in the gas phase exiting and entering, respectively, the aeration tank
Q_i	m^3/h	liquid flow rate influent to aeration tank
Q_p	m^3/h	primary effluent flow rate to aeration tank
Q_r	m^3/h	return activated sludge flow rate to aeration tank
P_B	mm Hg	barometric pressure
R	mg/L-h	oxygen uptake rate, OUR
SAE	kg/kWh, lb/hp-h	standard aeration efficiency
SOTE	–, %	standard oxygen transfer efficiency
SOTR	kg/h, lb/h	standard oxygen transfer rate
SRT	d	solids retention time
t	°C	temperature
t	h	time
t_o	h	hydraulic detention time based on total flow rate into aeration tank
U	m/h	forward velocity in aeration tank
V, V_L	m^3	tank liquid volume
V_G	m^3	gas phase volume under hood
w_o	kg/h, lb/h	mass flow rate of oxygen in influent air
x_{min}	m	minimum distance downstream from a boundary where longitudinal dispersion and detention time can be ignored in non-steady-state test
α		wastewater correction factor for oxygen transfer coefficient, overall tank value for off-gas test
β		wastewater correction factor for oxygen saturation
δ		depth correction factor for oxygen saturation
μ	$N-s/m^2$	absolute viscosity
θ		temperature correction factor for oxygen transfer coefficient
τ		temperature correction factor for oxygen saturation
Ω		pressure correction factor for oxygen saturation

subscripts

| i | | sampling point or hood location |
| *1,2* | | conditions referring to power levels 1 and 2 during dual non-steady-state test |

7.8 BIBLIOGRAPHY

ASCE (1991). *Standard- Measurement of Oxygen Transfer in Clean Water,* ANSI/ASCE 2-91, ASCE, Reston, VA.

ASCE (1996). *Standard Guidelines for In-Process Oxygen Transfer Testing,* ASCE-18-96, ASCE, Reston, VA.

ASCE (2001). *Standard Guidelines for Quality Assurance of Installed Fine Pore Aeration Equipment,* ASCE, Reston, VA, in press.

Baillod, C. R. et al. (1986). "Accuracy and Precision of Plant Scale and Shop Clean Water Oxygen Transfer Tests." *Jour. Water Pollution Control Federation,* 58, 290.

Boyle, W. C. et al. (1989). "Oxygen Transfer in Clean Water and Process Water for Draft Tube Turbine Aerators in Total Barrier oxidation Ditches." *Jour. Water Pollution Control Federation,* 61, 1449.

CEN Technical Board (2000). *European Standard, Wastewater Treatment Plants-Part 15: Measurement of the Oxygen Transfer in Clean Water in Activated Sludge Aeration Tanks,* CEN/TC 165, N19.

EPA (1983). *Development of Standard Procedures for Evaluating Oxygen Transfer Devices,* EPA-600/2-83-102, Municipal Environmental Research Laboratory, Cincinnati, OH.

EPA, (1989). *Design Manual- Fine Pore Aeration Systems,* EPA/625/1-89/023, Center for Environmental Research Information, Cincinnati, OH.

Hildreth, S. B. and Mueller, J. A. (1986). "Fine Bubble Diffused Aeration: Non-Steady State Testing in Tapered Aeration Tanks." *58th Annual NYWPCA Conference.*

Hovis, J. and McKeown, J. (1985). "New Directions in Aeration Evaluation." *Seminar Workshop on Aeration System Design, Operation and Control,* EPA-600/9-85-005, 400–409.

Kayser, R. (1969). "Comparison of Aeration Efficiency Under Process Conditions." *Proc. International Conference,* Water Pollution Research, IAWPRC, Prague, 477.

Mueller, J. A. (1982). "Comparison of Dual Nonsteady State and Steady State Testing of Fine Bubble Aerators at Whittier Narrows Plant, Los Angeles." ASCE, O2 Standard Committee.

Mueller, J. A. (1985). "Comparison of Dual Nonsteady State and Steady State Testing of Fine Bubble Aerators at Whittier Narrows Plant, Los Angeles." *Seminar Workshop on Aeration System Design, Testing, Operation and Control,* EPA-600/9-85-005, 375–399.

Mueller, J. A. and Boyle, W. C. (1988). "Oxygen Transfer Under Process Conditions." WPCF, 60(3), 332–341.

Mueller, J. A., Donahue, R., and Sullivan, R. (1982). "Dual Nonsteady State Evaluation of Static Aerators Treating Pharmaceutical Waste." *37th Annual Purdue Industrial Waste Conference.*

Mueller, J. A. and Rysinger, J. J. (1981). "Diffused Aerator Testing Under Process Conditions." *36th Annual Purdue Industrial Waste Conference.*

Mueller, J. A. and Saurer, P. D. (1986). "Field Evaluation of Wyss Aeration System at Cedar Creek Plant, Nassau County, NY." Parkson Corp., New York.

Mueller, J. A. and Stensel, H. D. (1990). "Biologically Enhanced Oxygen Transfer in the Activated Sludge Process." JWPCF, 62(2), 193–203.

Neal, L. A. and Tsivoglou, E. C. (1974). "Tracer Measurement of Aeration Performance." KWPCF, 46, 247–259.

8 Aeration Systems in Natural Waters

8.1 AERATION — STREAMS AND RIVERS

In the summer of 1963, the senior author was a field engineer in Virginia evaluating an in-stream aeration system in the Jackson River (Burns et al., 1966). Two 11.2 kW (15 hp) surface aerators were located 1.8 miles downstream of the West Virginia Pulp and Paper Covington Mill, which was achieving 85 to 90 percent BOD removal in a multistage wastewater treatment plant. The objective was to increase the DO in the stream to measurable values to prevent nuisance conditions.

The location of the aerators was at the critical deficit where the rate of natural reaeration of the stream is equal to the deoxygenation rate due to BOD decay. The difference between the upstream and downstream DO measurements multiplied by stream flow provided the rate of oxygen transfer due to mechanical aeration, with no correction for deoxygenation or natural reaeration since they cancelled each other out.

The aerators were installed as floating units but were found sitting on the stream bottom after unknown sharpshooters had targeted the metal floats for practice. This event led to the installation of polyurethane filler to prevent intentional or unintentional sinking. After successful operation, DO profiles were measured upstream and downstream of the aerators along with the velocity profile and cross-sectional areas to provide flow. It was determined that the aerators were performing as desired, producing about 1.3 kg O_2/kWh (2.15 lb O_2/hph). Stream DO increased from an upstream value of zero to approximately 2 mg/L at the downstream location 61 m (200 ft) below the aerators. At the aerator stations, average stream depth was 0.52 m (1.7 ft) upstream and 0.61 m (2.0 ft) downstream, with the stream width approximately 61 m (200 ft).

Two problems, however, prevented meeting desired DO levels in the stream, both due to natural causes. The first was that the stream flow was the lowest on record during that hot dry summer with anoxic conditions prevailing above the aerators. The second was that a short distance downstream of the aerators, the stream formed a pond during these extreme low-flow conditions. Low velocity, low reaeration, and high bottom demand quickly brought the DO level back down to zero. The aerators were subsequently moved further downstream, but ponding again negated the positive effect of the aerators.

The above brief recollection highlights some of the problems that may be encountered when dealing with a natural system. However, since the above study in 1963, aeration in natural water systems has been employed in full-scale projects with beneficial results. The principles involved are the same as those in treatment plants, but applications obviously differ. This section of the book highlights several of these projects in the Chicago area with practical applications and design principles where applicable.

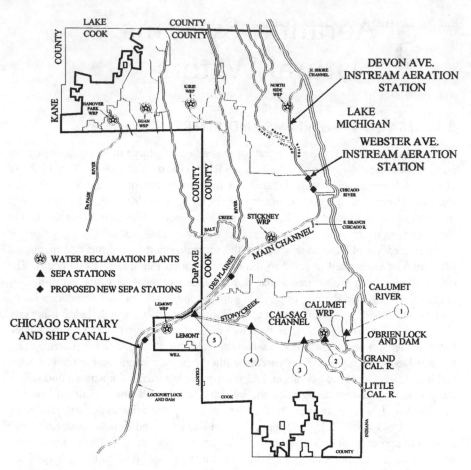

FIGURE 8.1 Chicago area waterways with present and proposed instream aeration stations (compliments of the MWRDGC, January 2001).

8.2 METROPOLITAN WATER DISTRICT OF GREATER CHICAGO: FULL-SCALE INSTREAM AERATION SYSTEMS

Early in the nineteenth century, artificial inland waterways were constructed and natural rivers in the Chicago metropolitan area were widened, straightened, and deepened by the Chicago Sanitary District, the predecessor of the Metropolitan Water Reclamation District of Greater Chicago (district). This effort provided deep draft navigation and improved drainage as water was withdrawn from Lake Michigan and discharged ultimately to the headwaters of the Illinois River. Due to low flow velocities in the waterways and increased pollutant loading over the years, historically dissolved oxygen concentrations have been low (Lanyon and Polls, 1996). Figure 8.1 shows the Chicago area waterways with the various instream aeration projects that have been constructed since 1979–1980 (Butts et al., 1996).

8.2.1 DIFFUSED AND SURFACE AERATION

As far back as 1923, the district studied the feasibility of aerating oxygen deficient water from the Chicago Sanitary and Ship Canal at the Lockport lock. Air was supplied by blowers through porous media (small wooden tanks) to increase DO with the effect of water temperature and aeration time studied.

Forty years later, the district constructed a moveable mechanical surface aeration system for testing on the north and south branches of the Chicago River and on the Sanitary and Ship Canal upstream of the Lockport powerhouse. It consisted of two Yeomans Hi-Co Wave aerators mounted on a catamaran driven by two 73 hp diesel engines. The system dimensions were 12.2 m (40 ft) long, 11.3 m (37 ft) wide, and 4 m (13 ft) high with a draft of 1.1 m (3.5 ft). The depth of the test areas varied from 3 to 6 m (10 to 20 ft) with an average of 4.6 m (15 ft). Flow rates varied from 22.7 to 159 m³/s (800 to 5600 cfs) (Lanyon and Polls, 1996).

At flow rates from 22.7 to 79.3 m³/s (800 to 2800 cfs), oxygen transfer efficiencies from 0.9 to 1.1 kg O_2/kw-h (1.5 to 1.8 lb O_2/hp-h) were obtained (Kaplovsky et al., 1964). At higher flow rates from 125 to 159 m³/s (4400 to 5600 cfs), the oxygen transfer efficiencies more than doubled to 2.4 to 2.7 kg O_2/kg-h (4 to 4.5 lb O_2/hp-h). Using this data, the instream K_LaV value was estimated with the following equation for a completely mixed section of river. Kaplovsky et al. (1964) used a plug flow approach with the same results, reporting the values as K_LA. The K_La value could not be determined from this data because no information was available on actual aeration volume, V, or interfacial area, A.

$$K_LaV = \frac{Q(C_d - C_u)}{C_\infty^* - C_d} = \frac{W}{C_\infty^* - C_d} = K_LA \tag{8.1}$$

Figure 8.2 shows that the stream flow during the tests markedly impacted the K_LaV as well as the SAE values. The authors considered a two-tier approach with the lower stream flows producing lower aeration efficiencies. Both a two-tier approach, similar to Kaplovsky et al. (1964), and a linear approach, with r² of 0.73 and 0.80, were assumed to define the observed data.

Mueller (1983) measured K_La values at various power levels in wastewater treatment plants with surface aerators where the aeration volume was known. Using an upper limit value of 5/h at high power levels with the observed K_LaV values in Figure 8.2 allowed the calculation of an "active" aeration volume, V, for the Chicago data. Figure 8.3 shows the active volume to increase with stream flow. This increase is possibly due to the physical configuration of the forebay in the Lockport lock in which the aeration units and the hydrodynamics of the system were located. One segment of the channel was about 6 m (20 ft) deep adjacent to a shallow shelf about 3 m (10 ft) deep. It is probable that at the higher flow rates, greater mixing occurred in the deeper channels thus providing a greater active aeration volume for the surface aerators. In treatment plants, depths greater than 3.7 m (12 ft) often require downdraft tubes or a bottom impeller to prevent solids from settling on the bottom.

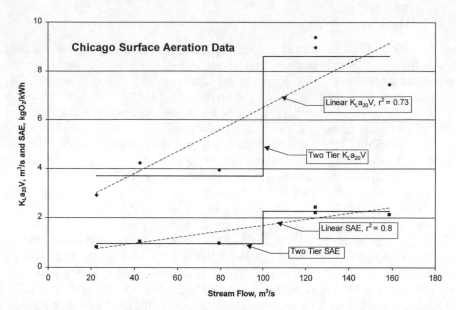

FIGURE 8.2 Impact of stream flow on $K_L a_{20} V$ and SAE for Chicago surface aeration (data from Kaplovsky et al., 1964).

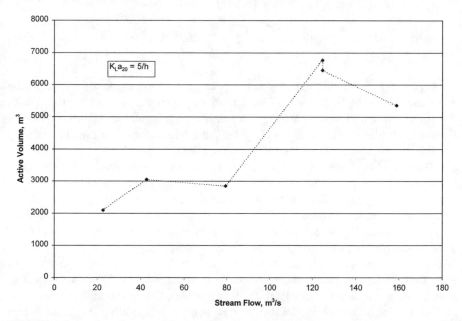

FIGURE 8.3 Impact of stream flow on active aeration volume from Chicago surface aeration analysis.

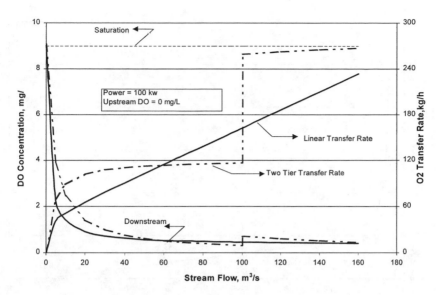

FIGURE 8.4 Impact of stream flow on downstream DO and OTR from Chicago surface aeration analysis.

Rearranging Equation 8.1 allows the calculation of the DO concentration downstream of a surface aeration zone as a function of the $K_L aV$ and stream flow rate, Q.

$$C_d = \frac{C_u + \dfrac{K_L aV}{Q} C_\infty^*}{1 + \dfrac{K_L aV}{Q}} \tag{8.2}$$

Figure 8.4 shows the impact of stream flow on the downstream DO concentration and the actual oxygen transfer rate. At higher stream flows, lower DO values occur for a given aeration power level, using both the two tier and linear transfer rates from Figure 8.2. As flow increases, more unoxygenated water is fed into the aerators resulting in a greater oxygen deficit at the aerator, a greater transfer rate, but lower downstream DO values. Cross plotting the calculated results of Figure 8.4, Figure 8.5 summarizes the impact of the required downstream DO concentration on the transfer rate. The higher the desired concentration, the lower the transfer rate. The actual Chicago tests were conducted at downstream DO concentrations <1.4 mg/L with inlet concentrations <0.5 mg/L. When DO levels of 4 or 5 mg/L are required in a stream, significantly less oxygen transfer occurs for the same power input than if DO values nearer 1 mg/L were acceptable. The impact of the stream flow on the mixing hydrodynamics in the Chicago case study makes this a more complex situation. If $K_L aV$ were constant with stream flow, a single line with a slope of $-K_L aV$ would result similar to the lower straight line portion of the two tier transfer rate.

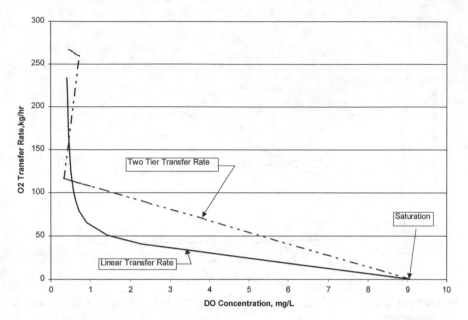

FIGURE 8.5 Effect of desired downstream DO on attainable oxygen transfer rate from Chicago surface aeration analysis.

From 1978 to 1980, two instream diffused aeration systems were constructed in the north branch of the Chicago River and in the North Shore Channel as shown in Figure 8.1. They were chosen based on economics, operational flexibility, and navigational and recreational considerations. They consisted of porous ceramic diffuser plates on the waterway bottom producing countercurrent spiral flows from parallel diffuser batteries. The first instream aeration station, constructed on the North Shore Channel at Devon Avenue during 1978–79 at a cost of $2.2 million, is equipped with four centrifugal blowers, each at 186 kW (250 hp). The second, constructed on the north branch of the Chicago River at Webster St. during 1979–80 at a cost of $2.8 million, is equipped with four rotary blowers each at 112 kW (150 hp) (Lanyon and Polls, 1996). The Devon station was designed to deliver 6000 kg O_2/d (13,300 lb O_2/d) to the waterways, and the Webster station to deliver 3600 kg O_2/d (8000 lb O_2/d).

A sketch of the layout for the Devon Avenue station for which a significant amount of data was collected is given in Figure 8.6 (Polls et al., 1982). The 0.3 m (1 ft) square diffuser plates were located in tubs 6.1 m (20 ft) into the channel with 14 plates per tub and a total of 2800 diffusers installed. The channel depth varied from 2.7 to 4.3 m (9 to 14 ft) with the width at 27.4 m (90 ft).

The data from this station, shown in Figure 8.7, are average values for each season (spring, summer, and fall) with error bars showing maximum and minimum values. Each season, the station was operated with one, two, and three blowers on line to provide a range of air flows and power draw. Both the SOTE and SAE values decreased with increasing air flow as additional blowers were brought on line, which is typical of a diffused aeration system. For fine pore diffusers, the values are lower than would be expected in a conventional aeration tank due most likely to the lower water depth

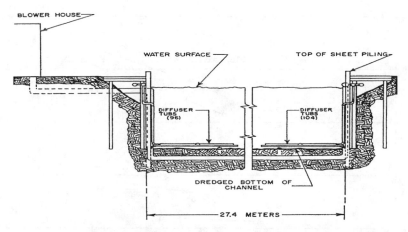

FIGURE 8.6 Schematic of the diffused aeration station in the Chicago North Shore Channel at the Devon Avenue Bridge (Polls et al., 1982, compliments of the MWRDGC, Jan. 2001).

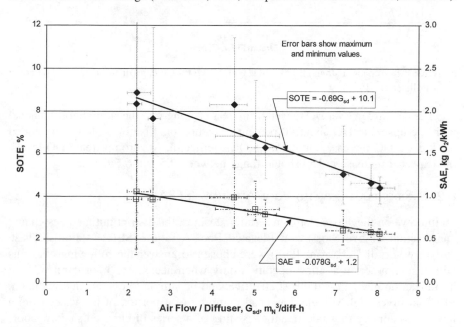

FIGURE 8.7 Diffused aeration data from the Devon Avenue Station in the Chicago North Shore Channel (adapted from Polls et al., 1982).

and the spiral roll type of aeration system instead of full floor coverage. The effect of stream flow on both parameters is given in Figure 8.8. As with the surface aeration system discussed above, increasing stream flow caused increasing SOTE values at each blower condition. The impact on SAE is much less for the one and three blower conditions. The impact of stream flow on both parameters is most likely due to increasing depth with increasing stream flow. This effect will provide higher SOTE values but draw more power. Therefore, it does not have as great an effect on SAE.

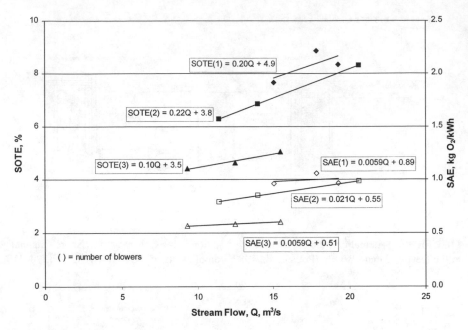

FIGURE 8.8 Impact of stream flow on SOTE and SAE for Chicago diffused aeration (adapted from Polls et al., 1982).

The increase in river DO ranged from 0.8 to 2.1 mg/L in the North Shore Channel with the upstream DO about 70 percent of saturation. Use of 3 blowers provided the highest DO increase, but it was the least efficient. At zero upstream DO levels, the expected increase in DO should range between 2.5 and 6.5 mg/L.

8.2.2 Sidestream Elevated Pool Aeration (SEPA)

To improve on the above instream aeration designs, the district initiated design and construction of five Sidestream Elevated Pool Aeration (SEPA) stations. These "urban waterfalls" (Farnan, 1998) were chosen to answer the following concerns with instream aeration: allow operation only when necessary, be off-stream, have no impact on navigation, be cost effective, and be simple to operate and maintain. SEPA involves high volume, low head pumping of a portion of the water from a low DO waterway to a sidestream consisting of a series of elevated shallow pools linked by waterfalls. Water cascades over weirs from pool to pool where it is over 95 percent saturated and discharged back to the waterway downstream of the inlet (Figure 8.9). Stations are spaced to provide incremental increases in DO to meet water quality standards. Figure 8.10 shows the expected DO levels in the Calumet Waterway System using the SEPA system.

8.2.2.1 Principles of Weir Aeration

As shown in Figure 8.11, aeration occurring over weirs involves three mechanisms. Aeration directly to the jet of water flowing over the weir is considered relatively small while aeration on the surface of the pool from the jet impact depends on the

TYPICAL URBAN WATERFALL STATION

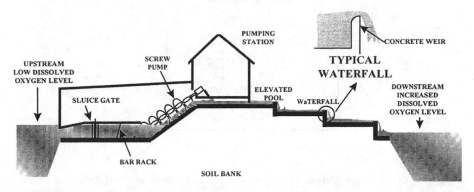

FIGURE 8.9 Typical urban waterfall station for Chicago SEPA system. (From Farnan, J. C. (1998). "Re-engineering the Design Criteria for Sidestream Elevated Pool Aeration." *Proceedings of the 1998 National Conference on Environmental Engineering*, Chicago, IL, June 1998, 62–67. With permission of ASCE.)

EXPECTED DO LEVELS IN CWS

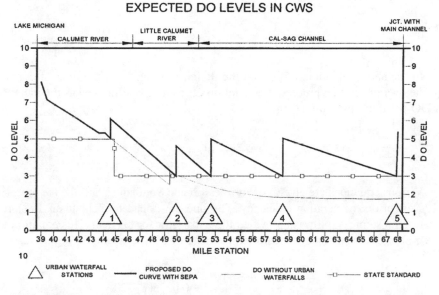

FIGURE 8.10 Expected improvement in DO levels of the Calumet waterway system due to the Chicago SEPA System . (From Farnan, J. C. (1998). "Re-engineering the Design Criteria for Sidestream Elevated Pool Aeration." *Proceedings of the 1998 National Conference on Environmental Engineering*, Chicago, IL, June 1998, 62–67. With permission of ASCE.)

intensity of surface agitation. Bubble aeration from air entrained in the jet and pool, to which the jet is discharging, is the most significant contributor to the oxygenation process (Gameson, 1957). With increasing drop height over the weir, the jet characteristics change from smooth to rough jets, then to oscillating jets, and finally to jet breakup (Wormleaton and Soufiani, 1998). Aeration efficiency increases with jet

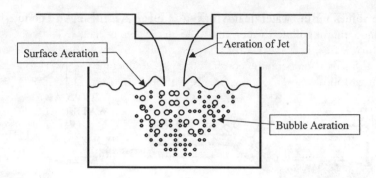

FIGURE 8.11 Weir aeration mechanisms.

height, with the rough and oscillating jets providing significant surface agitation and a large amount of closely packed bubbles entrained in the pool. Although a drop height causing jet breakup has the highest efficiency, the rate of increase with drop height is significantly lower than that of the rough and oscillating jets.The rate of aeration over the weir can be expressed by Equation 2.26. Integrating this between upstream and downstream conditions yields the following.

$$\frac{1}{r} = \frac{C_\infty^* - C_d}{C_\infty^* - C_u} = e^{-K_L a t_c} \tag{8.3}$$

In the above equation, r is called the deficit ratio and t_c is the time of contact for the overall aeration process. The aeration efficiency, E, is defined as the increase in DO per unit upstream deficit.

$$E = 1 - \frac{1}{r} = \frac{C_d - C_u}{C_\infty^* - C_u} = 1 - e^{-K_L a t_c} \tag{8.4}$$

Various equations are given by Wormleaton and Soufiani (1998) for the temperature effect on the aeration efficiency. The one developed by Tebbutt in 1977 for aeration on stepped spillways provided the least error for his data. This equation as given below is based on a standard temperature of 20°C.

$$\frac{r_T - 1}{r_{20} - 1} = \frac{E_T(1 - E_{20})}{E_{20}(1 - E_T)} = 1 + 0.0335(T - 20) \tag{8.5}$$

A number of empirical equations are available to determine the deficit ratio and aeration efficiency of weir discharges. In 1957, Gameson (1957) proposed the following equation for aeration efficiency of weirs as a function of the drop height, H_d (m) based on laboratory experimental data collected at about 10°C.

$$r = 1 + 0.5 a_w b_w H_d \tag{8.6}$$

The values for the water quality factor, a_w, were 1.25 in slightly polluted water, 1.0 in moderately polluted water, and 0.85 in sewage effluents. The coefficient b_w was 1.0 for a free weir and 1.3 for a step weir. The following year, Gameson et al. (1958) applied a temperature correction, with temperature, $T(°C)$, to the above equation as follows.

$$r = 1 + 0.34 a_w b_w H_d (1 + 0.046T) \tag{8.7}$$

In terms of aeration efficiency, Wormleaton and Soufiani (1998) presented the above with a slightly modified coefficient.

$$E_T = 1 - \left[1 + 0.361 a_w b_w H_d (1 + 0.046T)\right]^{-1} \tag{8.8}$$

The 0.361 is slightly different than the original Gameson value of 0.34 presumably due to a different temperature correction factor used to correlate the data. For the Chicago SEPA system, Macaitis (1991) indicated that an a_w value of 1.0, as suggested for moderately polluted water, and a b_w of 1.0 correlated well with their observed data.

The U.K. Water Research Laboratory (WRL) later (1973) amended Gameson's equation (Butts et al., 1999) as follows.

$$E_T = 1 - \left[1 + 0.38 a_w b_w H_d (1 + 0.046T)(1 - 0.11 H_d)\right]^{-1} \tag{8.9}$$

A value of 1.8 is suggested for the water quality factor, a_w, for clean water in the WRL equation. The b_w value ranged from 0.79 to 1.3 for field data from Butts and Evans in 1983 for sharp crested weirs (Wormleaton and Soufiani, 1998). In a full scale model for the SEPA design, Butts (1996) found b_w values to increase with increased drop height and greater number of steps. The b_w values varied from 0.9 for a single step weir at H_d of 1.5 m to 3.5 for a three step weir at H_d of 4.5 m, 1.5 m/weir. The flow rate per unit width and number of weir teeth had no significant impact on b_w.

The above equations are the simplest available, substantially only a function of drop height. Additional equations have been developed incorporating flow rate per unit width of weir, q_w, and pool depth, d_s, by Nakasone (1986). A generalized equation expressed in terms of efficiency is given as follows.

$$E_{20} = 1 - e^{-a H_d^b q_w^c d_s^d} \tag{8.10}$$

Table 8.1 gives the values of the coefficients, a, b, c, and d, for two conditions each of drop height and flow per unit width of weir. In the above equation, Nakasone defines the drop height, the difference between the upstream and downstream water levels, as $H_d = D + 1.5H_c$ where D is the weir drop height above the downstream water level and H_c is the critical water depth on the weir.

Nakasone indicates that the aeration efficiency is higher when the drop height is less than 1.2 m. When greater drops are desired, a cascade aeration system should

TABLE 8.1
Weir Coefficients in Nakasone Equation 8.10

H_d	q_w	Coefficient			
m	m³/h-m	a	b	c	d
≤ 1.2	≤ 235	0.0785	1.31	0.428	0.31
> 1.2	≤ 235	0.0861	0.816	0.428	0.31
≤ 1.2	> 235	5.39	1.31	−0.363	0.31
> 1.2	> 235	5.92	0.816	−0.363	0.31

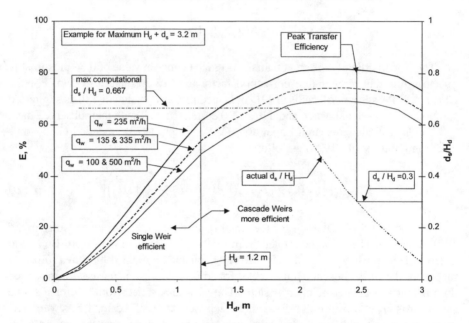

FIGURE 8.12 Impact of weir characteristics on aeration efficiency using Nakasone Equation.

be used with maximum drop heights of 1.2 m for individual weirs. The optimum discharge per unit width of weir is about 235 m³/h-m as shown in Figure 8.12. The optimum tailwater depth, d_s, in the downstream pool is $d_s = 0.3H_d$ with the maximum value of d_s in Equation 8.10 to be 0.667 H_d. The impact of both discharge per unit width and tailwater depth is less than the impact of drop height as seen in Figure 8.12 and the magnitude of the coefficients in Table 8.1. The aeration efficiency may also be increased by splitting the falling nappe into narrower individual nappes with the width of each nappe at about 1 m.

Avery and Novak (1978) developed an equation using laboratory data and flow per unit jet perimeter at the point of impact, q_p. The equation based on dimensionless weir Froude, F_w, and Reynolds, R_w, numbers is as follows, where v is kinematic viscosity and g acceleration due to gravity.

$$E_{15} = 1 - \left[1 + k_5 F_w^{1.78} R_w^{0.53}\right]^{-1}$$

$$F_w = \left(\frac{gH_d^3}{2q_p^2}\right)^{0.25}, \qquad R_w = \frac{q_p}{v} \qquad (8.11)$$

The value of k_5 was 6.27×10^{-5} for tap water and increased when salt was added to the water. The above equation only applies for a tailwater depth equal to or greater than the optimum value, the depth of maximum bubble penetration.

In the case of laboratory data on a weir with end contractions, the Nakasone approach did not fit observed data as well as the Avery and Novak equation due to a narrowing of the jet (Wormleaton and Soufiani, 1998). A study was supported by the U.S. Army Corps of Engineers (Wilhelms et al., 1993) to determine the agreement of predictive model equations with field data. The standard error of the above two models versus the field data was substantially the same, 0.166 for the Avery and Novak equation and 0.172 for the Nakasone equation.

The Chicago SEPA design involved multistage cascades. For similar drop design conditions in all n stages, Avery and Novak (1978) have shown that the deficit ratio, r_{tot}, for this type of cascade can be expressed as a function of that for a single stage, r_1. The overall aeration efficiency, E_{tot}, can also be expressed as a function of the single stage efficiency, E_1. Both r_1 and E_1 are based on the drop height from one individual stage, H_{d1}.

$$r_{tot} = r_1^n$$

$$E_{tot} = 1 - \left(1 - E_1\right)^n \qquad (8.12)$$

As indicated above, Nakasone set the breakpoint at a drop height of 1.2 m, above which staging is more efficient than a single weir. An analysis by Avery and Novak (1978) showed that a five-step cascade was more efficient than both a hydraulic jump and a single weir at the same overall head loss.

The aeration performance of labyrinth weirs, where the weir crest is not straight in planform, has been investigated by Wormleaton, Soufiani, and Tsang (1998; 2000). As the weir is indented upstream, a greater sill length results over a normal weir. The jets also collide in the drop zone causing disintegration of the solid jet and a larger surface area for aeration. The advantages of using both triangular and rectangular labyrinth weirs have been evaluated in laboratory experiments. Both the triangular and rectangular labyrinth weirs, for low drop heights < 1 m, had a significantly better aeration performance than normal weirs for the small size laboratory experiments. Overflow jets from larger weirs are less likely to collide than the jets in the above experiments, thus, labyrinth weirs may be more important in smaller installations. The authors also indicate that scaling of their aeration data to prototype size is virtually impossible, largely due to the relative invariance of bubble size.

TABLE 8.2
SEPA Design Features

Location and Capacities

Station	Mile Point	Channel Flow, m³/s	Channel Flow Treated, %	Expected Upstream DO, mg/L	Design Downstream DO, mg/L	Design Capacity, kg O₂/d
1	45.0	22.7	50	5	6.5	2950
2	50.5	32.4	7.5	4	4.3	860
3	53.0	34.0	40	3	5	5900
4	59.0	34.0	40	3	5	5900
5	68.5	34.0	48	3	5.4	7030
Total						22,640

Pump Design

Station	Number of Pumps	Type	Pump Diameter m	Speed, rpm	Lift, m	Pump Power, kW
1	4	Centrifugal Column	1.37	600	4.45	200
2	2	Screw	2.13	30	4.45	93
3	4	Screw	3.05	30	5.33	300
4	4	Screw	3.05	30	5.33	300
5	5	Screw	3.05	30	4.45	300
Total						4886
						(all pumps)

Weir Design

Station	Number of Weirs	Height/ Weir, H_{d1}, m	Total Height, H_d, m	Maximum Design Flow, m³/s
1	4	0.91	3.66	11.4
2	4	0.91	3.66	2.4
3	3	1.52	4.57	13.6
4	3	1.52	4.57	13.6
5	4	0.91	3.66	16.3

8.2.2.2 SEPA Results

The design of the five Chicago SEPA stations is given in Table 8.2. The first station utilizes a submerged axial flow, centrifugal column pump, while the remaining four use screw pumps. Station 1 is located upstream of the lock and dam system where the intake level reflects sustained high or low Lake Michigan water levels over a range of 2 m. The axial flow column pump is more efficient than the screw pumps, especially for this variation in intake elevation (Macaitis, 1991). The screw pumps provided additional aeration and had a good history of reliability for diameters less than 3.4 m.

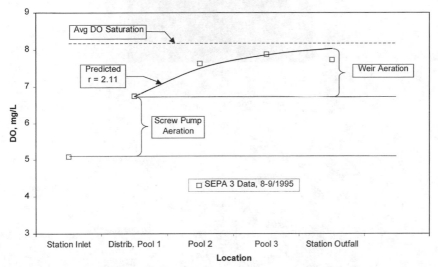

FIGURE 8.13 Comparison of predicted versus observed DOs for Chicago SEPA Station 3 using Gameson deficit ratio (photo of screw pump compliments of the MWRDGC, January 2001).

The reinforced concrete weirs are hydraulically sharp crested with weir teeth every 1.5 m to provide a ventilated nappe. The weirs of Station 1, which has a wetland feature, are designed to maintain a permanent pool and withstand ice loads. All other stations are designed with stop logs in the weirs to drain the station during the winter, thus preventing ice loads on the weirs. From the results of hydraulic studies (Kuhl, 1996) to provide maximum mixing and prevent short-circuiting, each station has a submerged weir at its discharge channel terminus. The shallow SEPA pools have crushed rock bottoms over a geotechnic membrane and soil subbase with a design velocity of 0.61 m/s to prevent sedimentation.

A two-year study has been conducted by the district to determine whether the SEPA stations are meeting their goal (Butts et al., 1999). Interstage DO data during low flow, warm weather conditions in August and September, 1995 has been given by Butts et al. (1996) for all five SEPA stations. Figures 8.13 through 8.21 compare the DO values predicted by the Gameson Equation with the observed data from an individual station as well as a photo and diagram of each station.

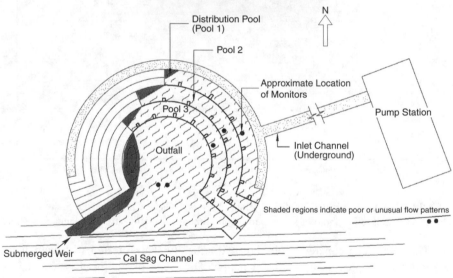

FIGURE 8.14 Plan view of geometric features of SEPA 3 showing location of continuous monitors (compliments of the MWRDGC, January 2001.)

In SEPA Station 3, Figures 8.13 and 8.14, good agreement between the observed and predicted values is obtained with both a_w and b_w taken as unity for each weir stage. This results in a deficit ratio for each stage, r/weir, of 2.11 using Equation 8.7 and aeration efficiency, E/weir, of 0.54 using Equation 8.8. Approximately 95 percent saturation was obtained for the three-weir station with the water temperature at approximately 25°C. More than half the aeration in this station occurs from the inlet screw pump due to the large amount of turbulence and agitation generated in the lifting action as seen in the pump photo. This station has a covered distribution pool so that no aquatic vegetation or photosynthetic

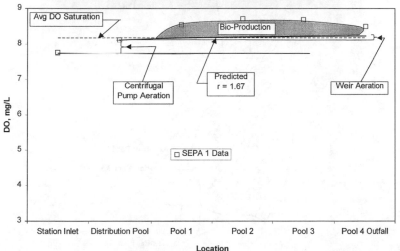

FIGURE 8.15 Comparison of predicted versus observed DOs for Chicago SEPA Station 1 using Gameson deficit ratio (plan view and photo compliments of the MWRDGC, January 2001).

oxygen production occurs prior to the first weir. Both the 1995 and 1996–97 data also indicate minimum photosynthesis occurring in the downstream weir pools although prolific aquatic vegetation is present.

In SEPA Station 1, r/weir is 1.67 and the E/weir is 0.42 for the lower weir heights compared with SEPA 3. Weir aeration is minimal compared with the centrifugal pump action and photosynthetic oxygen production as shown in Figure 8.15. This station starts at DO values about 93 percent of saturation and thus has low oxygen transfer from weir aeration. The large exposed distribution and first aeration pools favor photosynthetic oxygen production.

Figure 8.16 shows that SEPA Station 2 starts at a lower DO and has a greater contribution from screw pump aeration with relatively low weir aeration. All pools in this SEPA station are relatively small and also incur low bioproduction. Both of

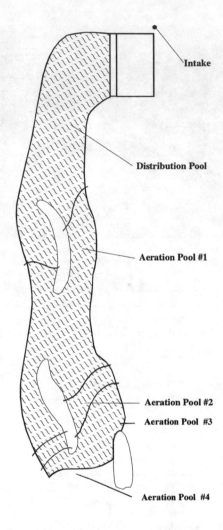

Schematic of SEPA 1

FIGURE 8.15 (continued)

the above stations attain supersaturation so that oxygen is actually removed by the downstream weirs which act as deaerators instead of aerators.

SEPA Station 4 has weir heights similar to SEPA 3 but large exposed distribution and first aeration pools similar to SEPA 1. It starts at a lower DO and therefore has more overall aeration. Bioproduction is significant in both the above pools as seen in Figure 8.17. Since aeration pool three is relatively small (Figure 8.18), little additional photosynthetic oxygen occurs in this downstream pool. In June 1997, Butts et al. (1999) showed a marked photosynthetic production after the first weir due to photosynthetic activity in the distribution pool and first aeration pool, attaining a maximum value of 147 percent saturation. This high production lasted for about 10 days after which a large drop in DO occurred. In the summer of 1996, chemical

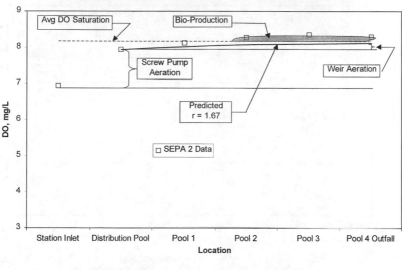

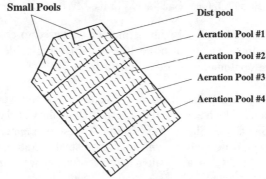

Schematic of SEPA 2

FIGURE 8.16 Comparison of predicted versus observed DOs for Chicago SEPA Station 2 using Gameson deficit ratio (plan view and photo compliments of the MWRDGC, January 2001).

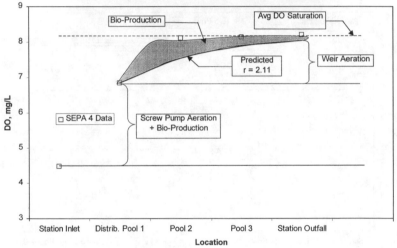

FIGURE 8.17 Comparison of predicted versus observed DOs for Chicago SEPA Station 4 using Gameson deficit ratio (photo compliments of the MWRDGC, January 2001).

treatment of the distribution pool with herbicides was used to control macrophyte growth, producing lower values than the 1997 data. During colder temperatures in October, 1996, the continuous monitoring data from SEPA Station 4 displayed no photosynthetic effect.

Figure 8.19 shows a significant amount of screw pump and weir aeration occurring in SEPA Station 5 which has a dual discharge to both the Cal-Sag Channel (CSC) and the Chicago Sanitary & Ship Canal (CSSC). This station appears to be the least affected by photosynthetic activity with relatively small aeration pools, shown in Figure 8.20. Although it was heavily silted and macrophytes were present throughout the study, no supersaturated DO concentrations were observed in the distribution pool. It has relatively small aeration pools with close weir spacing that allow minimal bioproduction, although periodically DO values slightly above saturation are obtained at individual locations.

It is obvious from the above data that the total oxygen transfer of these systems is markedly enhanced by the aeration occurring in the screw pumps. The pump

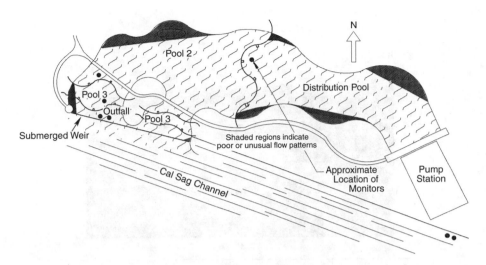

FIGURE 8.18 Plan view of geometric features of SEPA 4 showing location of continuous monitors (compliments of the MWRDGC, January 2001).

aeration is equivalent to putting an additional 0.5 to 3.1 m (1.5 to 10.3 ft) of weir aeration into the station. Based on their analysis, Butts et al. (1999) found no consistent effect of total pump height or number of pumps operational, and thus, they provide the following conservative equation to account for the screw pump aeration.

$$P_{op} = 0.5P_i + 45 \qquad (8.13)$$

P_{op} and P_i are the percent saturation at the pump outlet and intake, respectively. The increase in the pump outlet concentration reduces the number of weirs and/or the total drop height required for a SEPA station.

Taking advantage of photosynthetic oxygen production when sunlight accessible areas are available is more difficult. Once DO increases above saturation, a side channel sluiceway would probably have to be constructed to allow discharge from the upper pools directly to the receiving water to prevent deaeration from occurring over downstream weirs. The economics of this design have not been addressed. The large amount of sediment deposition occurring in large pools provides a base for attached macrophyte growth and photosynthetic oxygen production. It also causes reduced channel hydraulic capacity and pool detention times as well as increased maintenance. Future designs may include sediment traps and/or velocity control to keep sediment suspended.

8.2.2.3 Design Application

Using the Gameson Equation 8.7, with the pump Equation 8.13, the following design example shows the impact of pump aeration on the total required weir height (H_d). The temperature of the water is 25°C, and a desired effluent from the SEPA station (P_0) is 95 percent of saturation (C_{**}^* of 8.26 mg/L) at an influent value (C_i) of 2.5 mg/L.

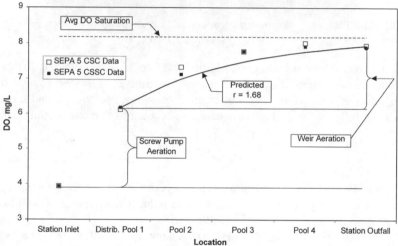

FIGURE 8.19 Comparison of predicted versus observed DOs for Chicago SEPA Station 5 using Gameson deficit ratio (photo compliments of the MWRDGC, January 2001).

P_i = 2.5/8.26 = 30.3%
P_{op} = 0.5 (30.3%) + 45 = 60.1% using Equation 8.13
C_{op} = 0.601*8.26 = 4.97 mg/l
C_o = 0.95*8.26 = 7.85 mg/L

With $C_d = C_o$ and $C_u = C_{op}$, Equation 8.3 is used to calculate r_{tot}.

$$r_{tot} = (8.26–4.97)/(8.26–7.85) = 7.97$$

The required deficit ratio for each stage (r_1) is calculated from Equation 8.12.

$$r_1 = 7.97^{(1/n)}$$

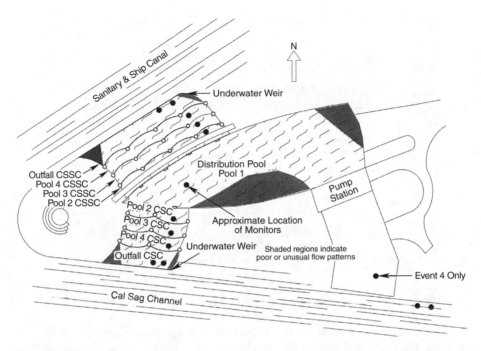

FIGURE 8.20 Plan view of geometric features of SEPA 5 showing location of continuous monitors (compliments of the MWRDGC, January 2001).

For a three-stage weir, $n = 3$

$$r_1 = 7.97^{(1/3)} = 2.00.$$

Using the Gameson Equation 8.7 with a_w and $b_w = 1$, the height of a single weir, H_{d1}, is determined.

$$H_{d1} = (2.00-1)/(0.34*(1+0.046*25) = 1.36 \text{ m}.$$

The total weir height is calculated as follows:

$$H_d = n * H_{d1} = 3 * 1.36 = 4.09 \text{ m } (13.4 \text{ ft}).$$

Figure 8.21 illustrates the impact of varying the number of weirs on total required weir height as well as the impact of pump aeration. Use of multistage weirs obviously decreases total weir height providing aeration that is more efficient as indicated by (Avery and Novak, 1978) and (Nakasone, 1986). The amount of aeration supplied by the screw pumps is also significant, supplying an equivalent weir height of 1 to 2 m similar to the measured SEPA data. No photosynthesis benefit was used in this design, although in summer months, DO would probably be equal to or greater than saturation.

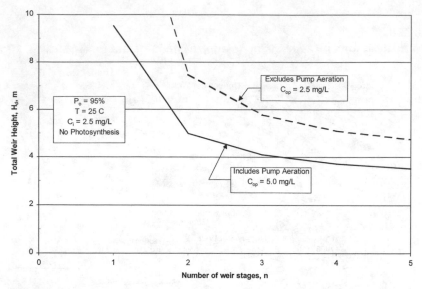

FIGURE 8.21 Effect of staging and screw pump aeration on weir design height.

Some design improvements recommended for the present SEPA system include angling the inlets and outlets and increasing their separation to prevent outflow recirculation (Farnan, 1998). By reducing approach velocities at the inlets, silting in the elevated pools should be decreased. Better access as well as vandalism proof design for these unmanned stations will allow easier maintenance and removal of unsightly vegetation and silt. The channel bottoms on future stations would be smooth, lined with fabric or cement without the present riprap, to allow easier maintenance and sediment removal by mechanical means. Screw pumps should be covered allowing periodic operation during winter to prevent bearing damage and bowing of screw shafts. All stations should also be enclosed to prevent moisture from deteriorating equipment.

8.2.2.4 Cost Analysis

A cost analysis of the instream aeration and SEPA costs is given in Table 8.3 from (Macaitis, 1991). Capital costs per kg O_2 transferred for the SEPA stations are about one half those for the instream aeration stations, while operating costs are about one third. Amortizing the capital costs over 20 years at eight percent interest and adding to the operating costs shows the SEPA unit costs are 2.5 \$/kgO_2 transferred while instream aeration is almost double at 4.9 \$/kg O_2. To prevent diffuser clogging, the instream aeration stations are operated all year, while the SEPA stations are operated about eight months per year, April through November. Use of the above aeration systems has also allowed deletion of advanced wastewater treatment projects estimated at \$300 million. The significant cost advantages and aesthetic quality of SEPA systems over advanced wastewater treatment systems make them worthy of consideration when dissolved oxygen concentrations in receiving waters are the controlling factor for water quality standards.

TABLE 8.3
Cost Analysis for Chicago Stream Aeration Systems, (Data from Macaitis, 1991)

Station	Annual O$_2$ Transfer 1000 kg/y	Capital Costs* Million $	O&M $1000/y	Annual Cost @ 8% for 20 y, $1000/y	Unit Cost $/kg O$_2$
		Instream Diffused Air Stations			
Devon	154	4.5	131	589	3.8
Webster	**93**	**5.2**	**94**	**624**	6.7
Total	247	9.7	225	1213	4.9
		Sidestream Elevated Pool Aeration Stations			
SEPA					
1	244	7.5	89	853	3.5
2	68	1.6	22	185	2.7
3	478	9.6	145	1123	2.3
4	478	9.6	145	1123	2.3
5	**575**	**10.8**	**145**	**1245**	2.2
Total	1843	39.1	546	4528	2.5

* ~1990 values. Devon station scaled up by 2.05 from 1978–79 value and Webster station scaled up by 1.86 over 1979–80 value equivalent to a 6 to 7% interest rate.

8.3 NOMENCLATURE

A	m^2	interfacial area
a$_w$		water quality factor
a,b,c,d		empirical coefficients
b$_w$		weir factor
C$_d$	mg/L	downstream DO concentration
C$_i$	mg/L	DO concentration at influent to SEPA station
C$_{op}$	%	DO concentration at pump outlet to first distribution pool of SEPA station
C$_o$	%	DO concentration at outlet of SEPA station
C$_u$	mg/L	upstream DO concentration
C_∞^*	mg/l	oxygen saturation concentration in stream
D	m	drop height from weir crest to downstream water level
d$_s$	m	water depth in pool below weir
G$_{sd}$	m$_N$^{3}/diff-h	air flow rate per diffuser at standard conditions
E		aeration efficiency, dimensionless
E$_1$		aeration efficiency, dimensionless, from one individual weir stage
E$_{tot}$		overall aeration efficiency, dimensionless, for staged weirs
F$_w$		Froude number, dimensionless
g	m/s^2	acceleration due to gravity

H_d	m	drop height, difference between water levels upstream and downstream of weir; total drop height for staged weir (cascade) system
H_{d1}	m	drop height, difference between water levels upstream and downstream of one weir in a staged weir (cascade) system
k_5		empirical coefficient
K_L	m/h	overall liquid film coefficient
$K_L a$	h^{-1}	oxygen transfer coefficient
$K_L a_{20}$	h^{-1}	clean water oxygen transfer coefficient at 20°C
n		number of weir stages
P_i	%	% oxygen saturation at pump intake of SEPA station
P_{op}	%	% oxygen saturation at pump outlet to first distribution pool of SEPA station
P_o	%	% oxygen saturation at outlet of SEPA station
Q	m³/h, m³/s	stream flow rate
q_w	m²/h	flow rate per unit weir width
q_p	m²/s	flow rate per unit jet perimeter
r		deficit ratio, dimensionless
r_1		deficit ratio, dimensionless, from one individual weir stage
r_{tot}		overall deficit ratio, dimensionless, for staged weirs
R_w		Reynolds number, dimensionless
SAE	kg/kWh, lb/hp-h	standard aeration efficiency
SEPA		sidestream elevated pool aeration
SOTE	–, %	standard oxygen transfer efficiency
T	°C	temperature
t_c	h	time of contact
V	m³	aeration volume
W	g/h, lb/h	oxygen load transferred to stream
n	m²/s	kinematic viscosity

subscripts
T   at stated temperature
15, 20   at 15 and 20°C, respectively

8.4 BIBLIOGRAPHY

Avery, S. T., and Novak, P. (1978). "Oxygen Transfer at Hydraulic Structures." *J. of the Hydraulics* Division, ASCE, 104(11), 1521–1540.
Burns, O. B., St. John, J., and O'Connor, D. J. (1966). "Pilot Mechanical Aeration Studies of the Jackson River in Covington, Virginia." *21st Annual Purdue Industrial Waste Conference*, Purdue University, Lafayette, IN, 799–811.

Butts, T. A. (1996). "The Development of Sidestream Elevated Pool Aeration Station Design Parameters Using Full-Scale Model Testing." *Rivertech 96; 1st International Conference on New/Emerging Concepts for Rivers*, Chicago, IL, 602–609.

Butts, T. A., Lanyon, R., and Polls, I. (1996). "Sidestream Elevated Pool Aeration Stations Online and Working Along the Cal-Sag Channel." *Rivertech 96; 1st International Conference on New/Emerging Concepts for Rivers*, Chicago, IL, 610–617.

Butts, T. A., Shackleford, D. B., and Bergerhouse, T. R. (1999). "Evaluation of Reaeration Efficiencies of Sidestream Elevated Pool Aeration (SEPA) Stations." *653*, Illinois State Water Survey, Chicago.

Farnan, J. C. (1998). "Re-engineering the Design Criteria for Sidestream Elevated Pool Aeration." Water Resources and the Urban Environment-98, *Proceedings of the 1998 National Conference on Environmental Engineering*, Chicago, IL, June 7–10, 1998, 62–67.

Gameson, A. L. H. (1957). "Weirs and the Aeration of Rivers." *J. Inst. of Water Engrs.*, 11(6), 477–490.

Gameson, A. L. H., VanDyke, K. G., and Ogden, C. G. (1958). "The Effect of Temperature on Aeration at Weirs." *Water and Water Engineering*, 62(November), 489–492.

Kaplovsky, A. J., Walters, W. R., and Sosewitz, B. (1964). "Artificial Aeration of Canals in Chicago." *JWPCF*, 36(4), 463–474.

Kuhl, R. A. (1996). "Hydraulic Design Criteria for Sidestream Elevated Pool Aeration (SEPA)." *Rivertech 96; 1st International Conference on New/Emerging Concepts for Rivers*, Chicago, IL, 618–624.

Lanyon, R., and Polls, I. (1996). "Artificial Aeration: Cost Effective Alternative to Advanced Wastewater Treatment." *Rivertech 96; 1st International Conference on New/Emerging Concepts for Rivers*, Chicago, IL, 625–632.

Macaitis, B. (1991). "Sidestream Elevated Pool Aeration Station Design." *Air-Water Mass Transfer: Selected Papers from the Second International Symposium on Gas Transfer at Water Surfaces*, Minneapolis, MN, 670–681.

Mueller, J. A. (1983). "Non-steady State Field Testing of Surface and Diffused Aeration Equipment." Manhattan College Environmental Engineering and Science Report to ASCE Committee on Oxygen Transfer Standards, July 1983.

Nakasone, H. (1986). "Study of Aeration at Weirs and Cascades." *J. of Environmental Engineering, ASCE*, 113(1), 64–81.

Polls, I., Washington, B., and Lue-Hing, C. (1982). "Improvements in Dissolved Oxygen Levels by Artificial Instream Aeration in Chicago Waterways." *82–16*, The Metropolitan Sanitary District of Greater Chicago, Department of Research and Development, Chicago.

Wilhelms, S. C., Gulliver, J. S., and Parkhill, K. (1993). "Reaeration at Low-Head Hydraulic Structures." *W-93-2*, US Army Engineer Waterways Experiment Station, Vicksburg, MS.

Wormleaton, P. R., and Soufiani, E. (1998). "Aeration Performance of Triangular Planform Labyrinth Weirs." *Journal of Environmental Engineering, ASCE*, 124(8), 709–719.

Wormleaton, P. R., and Tsang, C. C. (2000). "Aeration Performance of Rectangular Planform Labyrinth Weirs." *J. of Environmental Engineering*, 126(5), 456–465.

9 Operation and Maintenance

The principal objective of the design of aeration systems is to provide an effective operation with the lowest possible present worth cost, maintaining a balance between initial investment and long-term operation and maintenance (O and M) expenditures. Many long-term O and M expenditures are determined by the capabilities and constraints initially designed into the system. However, several factors under the control of the operation staff will have a significant effect on long-term O and M costs.

9.1 OPERATION

9.1.1 START-UP — DIFFUSED AIR

Prior to start-up of the aeration system, the following steps should be followed when placing an empty aeration basin into service.

- Check air piping and diffuser system and repair any loose joints, cracked piping, and other defects. Confirm that piping is free of debris such as rust or construction residue.
- Check to make sure that diffusers are installed according to manufacturer's specifications, e.g., tube diffusers are tightened and properly oriented, gaskets and O-ring seals are elastic and properly seated, the system is level, and bolts or other hardware used to apply an external sealing force are properly adjusted.
- Follow manufacturer's specifications in feeding air to the diffuser system before they are submerged. Always feed at least at the minimum recommended airflow rate per diffuser to prevent backflow of wastewater through the diffusers and into the air piping.
- Fill the aeration basin to a level of about 30 cm (12 in) above the diffusers. Observe the air distribution and check for significant leaks or maldistribution. Correct problems as needed.
- Continue to fill aeration basin while monitoring and adjusting airflow rate. Adjustment upward will be required as increase in water level will increase back pressure.
- Operate the condensation blowoffs, one at a time, until the air delivery system is free of moisture.
- Adjust flow rate of wastewater, and return sludge and airflow rates to meet desired operating conditions.

9.1.2 START-UP — MECHANICAL AERATION

- Equipment storage prior to installation and start-up may account for some operational difficulties at start-up. Most equipment can be protected up to six months for indoor storage and for four months outside. Rust and corrosion is the major culprit. Internals of gear cases and the gears themselves can become oxidized and, in some cases, the gearing can become affected due to corrosive attack of the tooth surfaces. Antifriction bearings are especially susceptible to storage damage due to moisture.
- Once installed, if delays in start-up occur, the exposure of the equipment to the elements can be even more damaging than storage. In this case, the equipment should be operated on a regular basis in accordance with manufacturer's instructions, or the equipment should be reprotected as if going into storage.
- Follow the manufacturer's specifications for start-up of all mechanical equipment. Equipment should be lubricated.
- Fill the aeration tanks prior to start-up of mechanical aerators.
- Check operation of all control equipment including variable speed drives and mechanically adjustable weirs.
- As a part of the normal start-up procedure on mechanical aeration equipment, a check is normally made for proper loading. This first power check is important for several reasons. First, a comparison of measured power load against the manufacturer's predicted power load will serve as an excellent check on proper sizing and baffling. Second, since most impellers have different power draws in the two directions of rotation, it is important that the proper direction of rotation is established at the time the motors are first phased out. Third, establishment of the steady state power level of the equipment at the time of start-up will be a useful reference to alert the operator of changes in basin liquid level or air distribution patterns. The most desirable method for initial power determination is using a recording wattmeter intended for measurement of a polyphase circuit.
- At the initial plant start-up, the plant engineer may elect to determine the vibration signature of high-speed aeration equipment (above about 600 rpm). Monitoring vibration over time will assist the operator in determining when bearings are approaching their fatigue lives.

9.1.3 SHUT-DOWN — DIFFUSED AIR

If it is necessary to shut down an aeration basin for more than two weeks, it should be drained and thoroughly cleaned. Once cleaned, the basin should be refilled to a level above the diffusers (typically, about 1 m [3 ft]) which will protect against UV light exposure and excessive temperature changes. Groundwater levels and basin buoyancy must also be considered. Airflow rates at or above manufacturer's minimum recommended levels should be maintained. Extra precautions must be considered if the basin is taken out of service during freezing conditions. In warm weather, the application of an algicide is recommended to prevent excessive algal growths.

For short-term basin dewatering for maintenance or servicing, no special servicing is required but it is advisable to perform routine inspection and housekeeping whenever possible.

9.1.4 SHUT DOWN — MECHANICAL AERATION

Use the same precautions as described above for diffused air systems relative to basin protection and inspection.

9.1.5 NORMAL OPERATION

Within the constraints placed on the suspended growth aerated system, the primary operational objective is to achieve an acceptable effluent quality while maximizing aeration efficiency. As discussed earlier, aeration efficiency is affected by several controllable parameters including

- mean cell residence time
- food-to microorganism ratio
- flow regime
- airflow rate
- dissolved oxygen concentration
- degree of diffuser fouling and deterioration
- blower efficiency
- submergence
- impeller speed
- power dissipation

The mean cell residence time, or F/M ratio, and flow regime normally constitute part of the long-term process control strategy, ranging from seasonal to many years of stable operation. As described earlier, the degree of wastewater stabilization appears to significantly affect aeration efficiencies. Plant operation that targets a high degree of wastewater stabilization, including nitrification, will likely produce a high level of OTE and SAE thereby achieving low power requirements. Seasonal changes in effluent permit requirements can result in changes in operational strategies with concomitant changes in aeration performance. Limited data suggests that flow regime may affect OTE. If the facility has capability to operate under several different regimes, it may be advantageous to experiment with them to achieve high levels of aeration efficiency. In some cases, operational stability (e.g., solids separation) may dictate flow regime, however, overriding the efficiency of the aeration process.

Diffuser airflow rate and mixed liquor DO concentration are part of the short-term, day-to-day operating strategy. As shown above, airflow rate per diffuser affects aeration system OTE for porous diffusers. Based on clean water performance data for porous diffusers, OTE will decrease by 15 to 25 percent when diffuser airflow increases from 1.6 m^3_N/h to 4.7 m^3_N/h (1.0 to 3.0 scfm) per diffuser. Little change is observed for many nonporous diffusers. Changes in airflow also affect efficiency by changing system pressure. Increasing airflow will increase the pressure drop across the flow control orifices and the diffuser element. The pressure drop across a clean porous diffuser element, as measured by DWP, is relatively small over normal airflow operating ranges. For example, the change in DWP for a ceramic disc diffuser operating at 1.6 and 4.7 m^3_N/h (1.0 to 3.0 scfm) is only 5 cm (2 in) water gauge. The pressure drop across a fixed-sized orifice for the same increase in airflow rate could be substantial, however, because the drop increases as the square of the flow rate. For a 5-mm (3/16 in) orifice, the increase in pressure drop resulting from an increase in airflow as described above is about 25 cm (10 in) water gauge.

Residual DO concentration affects OTE by changing the driving force as shown in Equation 2.52. The maximum driving force is achieved when the system is operated with a residual DO of zero. Since a positive DO residual is usually required to obtain the desired process performance, the driving force will be decreased, and OTR (OTE) will decrease below maximum, thereby requiring an increase in airflow rate. As seen earlier, as airflow increases, the value of OTE further decreases. Operation at a mixed liquor DO concentration dictated by process needs must be considered a normal cost of operation. However, operating above that required residual should be avoided because power costs will increase with no improvement in process performance. For example, operating at a residual DO of 4 mg/L instead of 2 mg/L will result in a significant increase in airflow rate and power. Assuming a 4.3 m (14 ft) submergence, a diffuser airflow rate of 1.6 m^3_N/h (1.0 scfm) for a 2.0 mg/L residual DO, and a typical relationship for airflow rate and SOTE described earlier for a porous diffuser, it would require 37 percent more air to operate at 4.0 mg/L DO instead of 2.0 mg/L. Assuming constant blower efficiency and ignoring differences in system headloss, the power consumption would be directly propor-tional to airflow. Therefore, the power consumed by operating at 4.0 mg/L instead of 2.0 mg/L would increase by 37 percent.

Operating diffusers at the lowest airflow rate possible, while not going below the manufacturer's recommended minimum rate, achieves maximum OTE and SAE. The airflow rate selected will depend upon the aeration tank oxygen demand and will vary both temporally and spatially. Tapered aeration designs are encouraged when plug flow aeration basins are employed to ensure efficient oxygen transfer throughout the system. Flexibility in design of the aeration system is important to provide sufficient oxygenation to meet all (or most) oxygen demand requirements. As a result, there will be times early in the design life when minimum recommended airflow rates will control, and excess DO concentrations may occur. Later in the design life, oxygen demand and supply may be in excellent balance. As load to the plant continues, it is possible that demands may exceed supply at points within the basin. For plug flow designs, this excess means that demands may be satisfied further downstream in the process. As long as treatment objectives are met, this method may be a satisfactory operating strategy. In fact, some operators deliberately move demand downstream in an effort to provide more efficient aeration throughout the system.

It must be emphasized that operating at low DO may result in diffuser fouling. Also, if improper orifices are employed, operation at too low an airflow rate may result in maldistribution of air, producing lower efficiencies and, possibly, resulting in fouling of diffusers that receive little or no air. The process may create an unde-sirable cycle. As some diffusers foul, the poor airflow distribution is exacerbated. For sparged turbine aerators, it is also important to ensure that sparge rings are designed to provide a uniform air-water mixture. This effect is normally accomplished by designing the ring for minimum pressure drop across the orifice holes. Manufacturers will normally specify minimum airflow rates for the sparge ring. At large turndowns when systems are operated at low airflow, uneven gassing to the turbine can result. Airflow distribution can also be a problem where multiple units are operated off a common air header much the same as might occur in diffused air headers.

Mechanical surface aerators are hydraulically dependent on liquid level in the basin since a small change in liquid level variation generally will cause a significant change in head requirements of the impeller. Different impeller designs will exhibit different sensitivities. This fact is used to control power draw and oxygen transfer rate for surface aerators. Power dissipation, measured as power per unit area or volume may also affect both transfer rate and efficiency of mechanical devices as described in detail in Chapter 5.

Plant personnel must evaluate that the mechanical aeration equipment is operating in a hydraulically stable fashion. Liquid level is important not only to control aerator power demand but also to control surge. One of the inherent physical phenomena of operating an impeller at the free liquid surface is that under a unique set of operating conditions, any contained volume of liquid can be excited into resonance. The conditions under which surge will occur relate to the tip speed of the impeller, the depth of impeller submergence, and the degree and nature of baffling. Manufacturers have determined the limits of surge for the particular impeller design being offered and can establish the point where surge may occur. Hydraulic stability may be obtained by the use of extremely long weirs such that liquid level variation between maximum and minimum conditions is low. In cases where variable levels are used for power control, proper operating controls should be established to maintain levels within equipment manufacturer' recommended range.

9.2 SYSTEM MONITORING

The aeration system must be monitored to provide data for optimizing system performance and maintenance schedules. Monitoring can lead to optimization of aeration system efficiency in several ways. First, the optimization of DO control, by which most of the power savings are achieved, relies on frequently collected DO concentration data. Second, the effects of process operational parameters including MCRT, F/M, and flow regime on SOTR can be better defined for the site-specific application. Finally, the adverse effects of diffuser fouling and/or deterioration on back pressure and OTE for fine pore diffusers can be identified so that appropriate maintenance can be initiated. Data collection frequency should be sufficient to identify normal variations and to permit recognition of long-term changes. Monitoring should include evaluation of changes in air-delivery pressures and aeration system efficiency as well as visual observations of the system.

Air-side or liquor-side fouling or diffuser element deterioration may cause changes in headloss of the diffuser. These changes may be detected in the blower discharge header or by changes in the opening of airflow control valves. Significant increases in blower pressure may be indicative of severe fouling of major portions of the aeration system. For this reason, monitoring of system pressure and airflow rate on a daily basis is recommended. Although system pressure serves to provide information on severe aeration system conditions, it is not a very sensitive indicator of increased (or decreased) diffuser headloss. Losses across the diffuser element are small relative to the pressure in the air main. Other factors, including water temperature, airflow rate, and other variable line losses, further limit the precision of this measurement. Furthermore, fouling or deterioration of only a few diffusers will typically result in redistribution of airflow with little observable change in system pressure.

A more sensitive method of monitoring diffuser headloss for porous diffusers is *in situ* DWP, measured by fixed pressure monitoring stations located throughout the system. These stations do require continual maintenance to ensure accurate and precise DWP measurements. DWP measurements can also be performed in the laboratory using diffusers taken from removable headers placed at strategic positions within the aeration basin. The advantage to this method is that the diffuser may be examined for other parameters, such as foulant, changes in physical or chemical properties, and OTE. This technique also requires careful maintenance and may impose a significant nuisance to the operator during removal and replacement of the header.

The estimate of system OTE (AE) is of great importance in evaluating the effectiveness of both operation and maintenance strategies. Rigorous methods for the evaluation of OTE (AE) are described in detail in Chapter 7. One or more method may be satisfactory for a specific site, but these methods are time-consuming and may be too costly for day-to-day monitoring. As an alternative, calculated ratios of operating data can provide good indicators of overall system performance over time. A parameter based on the ratio of the oxygen demand satisfied to the rate of oxygen supplied can be conveniently computed from operating data and used to assess aeration system efficiency. This parameter, described as the Efficiency Factor, EF, is the ratio of the oxygen demand removed (mass/time) to the mass supplied corrected by the DO driving force (EPA, 1989). Another ratio that may be used to estimate aeration system efficiency is the ratio of the oxygen demand removed per unit of electrical power consumed. This ratio includes the efficiency of the blowers and motors and air distribution system losses. A correction for DO driving force is also required.

Visual observation of the system aeration pattern can provide useful information. For diffused air systems in a grid configuration, the surface pattern should be free of localized turbulence and boiling. These maldistributions may be due to breakage of headers or diffusers, faulty joints, leaking gaskets or fouling/deterioration of diffuser elements. Coarse bubbling at the water surface may be indicative of diffuser fouling. However, it must be emphasized that a certain degree of coarse bubbling is often noted at the influent end of the aeration basin, even with new diffusers. The cause of this coarse bubbling may be due to surfactants contained in the influent wastewater. Once problems are identified by visual observation, quantitative measurements should be made to confirm the type and extent of the problem.

Mechanical aeration equipment monitoring includes evaluation of appropriate DO distribution and mixing. A DO profile can be used to assess proper oxygen dispersion. Surface mixing patterns may provide clues as to improper hydraulic mixing and surging. Impeller fouling with rags or ice can be detected by mixing patterns. Sparged turbine flooding caused by excessively high airflow is detected by observing flow patterns at the draft tube. For a downward pumping impeller, the water column should be moving downward against the sparged airflow. Monitoring for ice conditions on surface aeration equipment is an important activity in cold climates, especially during low flow periods. Auxiliary deflectors and shields are often used in severe climates to prevent icing situations from occurring.

9.3 AERATION SYSTEM CONTROL

The major objectives of aeration system control are to ensure that oxygen supply meets the dynamic spatial and temporal variations in process oxygen demand and to effectively control air delivery and oxygen transfer to minimize power consumption. The benefits of aeration control include assured integrity and uninterrupted operation of the process, increased reliability in meeting permit limits, and reduced process costs. These benefits have been discussed in some detail above. The use of manual aeration control strategies normally results in operation at a fixed airflow rate and distribution. Changes are initiated once or twice throughout the day, or perhaps, only weekly, in an effort to pace supply with demand. Since DO significantly affects process performance, airflow rates are typically set high to ensure that a positive DO is maintained during high load periods. As a result, power is wasted during extended periods of reduced loading. Today, most aeration systems are controlled by automation. Automated aeration control is the manipulation of the aeration rate by computer or controller to match the dynamic oxygen demand and maintain the desired residual or set-point DO concentration. The potential savings in aeration system energy costs achievable by automation or DO control is typically 25 to 40 percent, but can be higher (Flanagan and Bracken, 1977; Stephenson, 1985; Robertson et al., 1984; and Andersson, 1979).

An excellent reference source on the theory, design, and implementation of automatic control strategies can be found in EPA (1989). How much aeration control is required or desired and can be achieved at a plant is site specific. For new construction, the decision to incorporate aeration control is straightforward. The capital investment for even a high degree of automated control over that required for simple on-line monitoring is a small percentage of the total cost of the plant, generally one to five percent, depending on plant size. Careful attention to process and hardware flexibility is necessary to achieve maximum benefits from a well-designed aeration control system. For retrofit of manually controlled facilities, the selection of automated control must be based on achieving more effective control of the aeration system. Considerations should include minimizing operational problems and/or optimizing the aeration process to achieve energy consumption savings. The selection of the level or degree of control should be based on an incremental cost-benefit analysis.

For completely mixed systems, the conventional control scheme uses feedback from the DO sensor since oxygen demand is relatively constant and has, by definition, no spatial variation in demand. In plug flow aeration tanks, spatial variation occurs requiring a nonuniform rate of oxygen supply to accomplish a uniform DO profile. For steady-state conditions, this can be achieved by tapering diffuser density along the basin. Automated air distribution control valves can be installed to regulate airflow to each grid in an effort to maintain the set-point DO in each grid. If this is not practical, the air distribution profile can be established with manually adjusted air distribution valves, and the total airflow to the basin is automatically regulated to maintain the desired DO profile down the length of the basin. Airflow is typically controlled through the use of either analog or programmable digital controllers. The newer programmable controllers offer the advantage of facilitating the implementation of more advanced controllers and provide additional process data such as oxygen uptake rates and diffuser fouling dynamics. The primary sensors normally employed in aeration control strategies include DO monitoring equipment, airflow metering, and pressure and temperature sensors. Their accuracy and precision are critical to successful control. Field verification, calibration, and maintenance must be performed routinely to ensure proper function.

There are many different control strategies used for aeration systems and the technology is rapidly changing producing more efficient hardware and software for this application. An example of a moderate complexity strategy taken from EPA (1989) is described below for diffused aeration and is illustrated in Figure 9.1. This system is designed for a 0.23 m³/s (5.3 mgd) plant employing four parallel, plug flow basins, each containing three grids of porous diffusers. The strategy is to provide exact DO control in each basin by using individual DO set-points, controllers, airflow control valves, and air headers for each basin. In this case, it is not necessary to assume that each basin receives an identical flow or load. DO monitors may be placed in each grid, although in this example, the control is provided by a DO monitor located in the second grid of each basin. Portable probes would be used to provide manual adjustment of air distribution valves to each of the grids. Periodic adjustments may be required to achieve the most efficient DO profile.

The DO monitored in grid two of each basin provides feedback to the airflow controller for that basin. Automated valves located on the four parallel headers distribute the total blower output to the four basins. At least one of these valves is always maintained in its "most open" position to minimize the main header pressure. A pressure controller located in the main header regulates blower output by manipulating the inlet guide vanes on the centrifugal blowers. The number of on-line blowers depends on the load to the plant. Bringing them on or off-line is carried out automatically upon receiving an on/off signal from the air demand controller. The characteristic curves of the blowers are used to develop an operating map for control of the most energy efficient operating point. At this point, one of two strategies may be used to control the airflow from the blowers. One would control all on-line blowers with the same signal from the air demand controller. This strategy controls all on-line blowers at the same operating point while matching the variable airflow demand. An alternative strategy would operate one blower with the control system to respond to variable oxygen demands, and one or more of the other blowers would operate at a constant output to provide the "base supply" of air. Periodic substitution of a different blower to serve as the variable delivery source allows for load balancing and accommodates maintenance requirements.

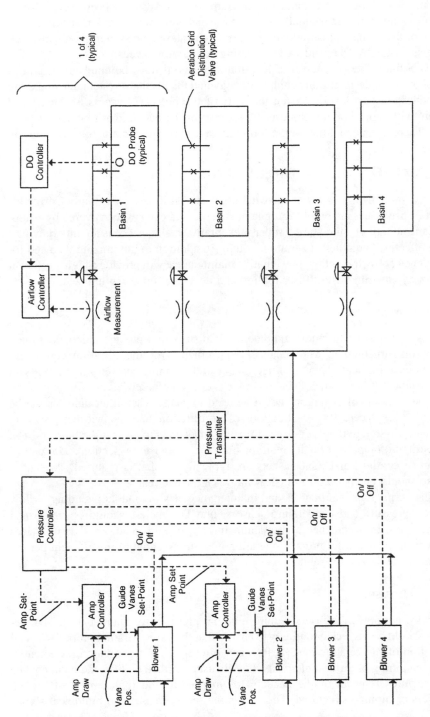

FIGURE 9.1 Moderate-complexity control schematic.

DO control for mechanical aeration equipment has typically been accomplished by DO monitoring and manual control of basin water level (submergence), aerator speed, or the number of aerators in service. Some automatic systems are being used however, whereby DO controls weir settings or aerator speeds.

As a final note on control of aeration systems, it must be emphasized that in developing and designing any control strategy and the resultant system, the operating personnel must be involved from the start of the process. Success of the control system will depend on the enthusiastic support of the people that routinely depend on it. There is no doubt that in the future most plants will adopt automated control.

9.4 MAINTENANCE — DIFFUSED AIR

This section will discuss preventative maintenance of diffused air systems. Corrective maintenance issues are highly equipment specific and can best be covered by equipment manufacturer's literature. Proper preventative maintenance is an important part of an effective and efficient aeration system. In addition to minimizing the need for emergency corrective action, preventative maintenance will provide a highly efficient system by ensuring that diffuser fouling and deterioration are minimized.

9.4.1 AIR SYSTEMS

Air systems include filtration equipment, air distribution piping, and airflow measuring instrumentation. Maintenance requirements for the filtration equipment include cleaning and changing filter media and cleaning the ionizer elements in electrostatic filtration units. The manufacturer's recommendations for maximum headloss or hours of operation should be used to gauge when filter units should be cleaned or replaced. Proper attention to air filtration maintenance can virtually eliminate air-side fouling of porous diffusers and serves to protect the blowers. The air distribution piping normally requires little maintenance. Inspection and repair of protective coatings and joint gaskets are typically all that is required. The entire system should be checked for air leaks at least once a year.

The verification, calibration, and maintenance of all monitors including airflow devices, pressure and temperature sensors, and DO meters should be performed routinely and in accordance with manufacturer's recommendations. These devices are critical to successful process operation and are essential to the efficient performance of the aeration system.

9.4.2 DIFFUSERS

Typically, nonporous diffusers require minimal preventative maintenance. The elements should be inspected routinely to ensure that they are operating properly. Visual inspection of the aeration tank surface can often provide information on potential breaks in piping or diffuser elements. For diffusers located on lifts, the maintenance only requires removal of the headers for inspection and replacement of broken diffusers or piping as required. The accumulation of greases and biological slimes on the diffuser element, causing partial plugging, is not uncommon and may require hosing or brushing on occasion to reduce back pressure in the line. If the diffusers are mounted on fixed headers, it will be necessary to dewater the aeration basin, usually annually, for inspection of the diffusers and piping. Cleaning and replacement of faulty components can take place at this time. Some manufacturers recommend air bumping for dislodging foulants as an *in situ* process noninterruptive technique. The diversity in types of nonporous diffusers requires that the operator refer to the manufacturer's recommended maintenance for best results.

Porous diffusers normally will also require the routine inspection required for nonporous diffusers but are typically more susceptible to fouling and deterioration than their counterparts. As a result, cleaning techniques are an important part of their maintenance. The next section details cleaning methods for these types of diffusers.

Cleaning Methods

A number of cleaning methods are currently used for porous diffusers. These may be generally classified as process interruptive or process noninterruptive. Process interruptive techniques require that the aeration basin be taken out of service to provide access to the diffusers. Noninterruptive methods do not require direct access to the diffusers. A further distinction in cleaning methods can be made between those that require that the diffusers be removed from the basin (*ex situ*) and those that do not (*in situ*). A list of most of the current cleaning methods is provided below.

Ex Situ
- refiring
- acid washing
- high-pressure water jetting
- alkaline washing
- detergent washing

In Situ – Process Interruptive
- acid washing
- high and low pressure water hosing
- steam cleaning
- endogenous respiration
- ultrasonic

In Situ – Process Noninterruptive
- acid injection
- air bumping

All *ex situ* methods are expensive insofar as labor and shipping costs are concerned. Very large plants may provide facilities on-site for treatment, however. Refiring which is restricted to ceramic diffuser elements requires placing the elements in a kiln and heating them in the same fashion originally used in their manufacture. The result is often the removal of most foulants from the element and restoration of the element to near-new condition. This is not always the case, however, and depends on the degree of fouling and the nature of the foulant. Jet washing, acid, and alkaline washing have all met with mixed success for ceramic diffusers. Costs are typically lower than refiring but are still high compared with *in situ* methods. When internal fouling becomes a problem, soaking of the diffuser elements in acid or alkaline solutions for an extended period (24 to 48 hrs or more) may be effective. If air-side fouling is a problem, *ex situ* methods will provide a more positive means for removal of these materials. Additional information on *ex situ* cleaning can be found in the manual of practice, FD-13 (1988).

The *in situ* process interruptive methods include hosing with either high-pressure (>415 kPa [60 psia]) or low-pressure water sprays. These methods will dislodge surface solids and biomass but are not very effective in removing in-depth foulants. Steam cleaning is about as effective as water sprays for most foulants. These methods are applicable to most porous diffusers although care must be exercised when jetting some thin film perforated membranes. Brushing or scrubbing with a stiff bristled brush often will be used in combination with jetting to improve removal of foulants. The application of 14 percent HCl (a 50 percent solution of 18 Baume inhibited muriatic acid) with a portable spray applicator to each diffuser element following hosing or steam cleaning and then rehosing the spent acid is effective in removing both organic and inorganic foulants. If the acid is allowed to penetrate the diffuser for a period of time (15 to 20 minutes) some internal foulants will also be removed in this process. Acid cleaning is restricted to ceramic and porous plastic diffusers. The diversion of wastewater flow from the aeration basin to be treated resulting in endogenous respiration of the mixed liquor and, possibly, the biomass associated with the foulant may alleviate fouling problems in some instances. To date, there has been little experience with this method.

The *in situ* process noninterruptive acid gas injection method is accomplished by injecting an aggressive gas (HCl or formic acid) into the air feed to the diffuser element. Specifically, the gas injection method includes increasing the airflow rate per diffuser to near the maximum recommended rate to insure even air distribution and to get as many pores operating as possible. The cleaning agent is then introduced into the air stream, usually until the DWP stops decreasing. Acid injection systems are most effective on Type I fouling involving inorganic acid soluble foulants, such as iron hydroxides and calcium and magnesium carbonates. The method has not been as effective against Type II or III fouling where biomass is predominant in the fouling agent. It will also not remove atmospheric dust deposited on the air-side or granular materials such as silica incorporated within Type II and III foulants. Some gas cleaning methods are proprietary processes in the U.S.

Air bumping of porous diffusers by increasing the airflow rate per diffuser to a value recommended by the manufacturer for about 15 minutes will remove some surface foulants on ceramic, porous plastic, and membrane diffusers. With perforated membrane diffusers, this "flexing" action is created by shutting off the airflow to the diffuser allowing the membrane to collapse onto the support frame. This method is followed by reintroduction of air to the units at two to three times the normal airflow rate. The highest airflow rate should never exceed the maximum recommended by the manufacturer. The bumping process is typically performed every one to four weeks for some membrane diffusers. The effectiveness of air bumping is not well documented at this time although often recommended by manufacturers.

Selection of Cleaning Methods and Frequency

It is clear that all porous diffusion aeration systems will require some form of diffuser cleaning on a periodic basis. The need for, type of, and frequency of cleaning at these installations are highly equipment and site specific. The effectiveness of cleaning methods needs to be determined by observing changes in header pressure, DWP, or measures of oxygen transfer efficiency that were described earlier. Once experience has been gained with respect to the benefits accrued by cleaning, a cost-benefit analysis can be performed to estimate cleaning frequency and method. At some plants, laboratory testing of fouled diffusers removed from test headers or from grids within a dewatered basin has provided useful information on which techniques will be most effective. The manual of practice FD-13 (WPCF, 1988) and the EPA fine pore aeration design manual (EPA, 1989) provide an excellent data base and bibliography on experiences with porous diffuser cleaning. An example of estimating cleaning frequency appears in EPA (1989) as well.

9.5 MAINTENANCE — MECHANICAL AERATION

The most significant, and generally universal, requirement for maintaining mechanical aerators is to follow the manufacturer's schedule for lubrication and other maintenance. Typically, gear reducer oil should be changed about twice a year and motor bearings greased at the same time. Those schedules may shift depending on equipment, climate, and operating conditions. For example, in areas with wide seasonal temperature changes, seasonal oil changes with oil of the proper viscosity may be necessary. Recently, motor manufacturers have introduced improved grease that permits "five-year no maintenance" operation. This guarantee may be of value to some but should not be taken as a lifetime guarantee.

As described above, monitoring of the aeration system is an important maintenance operation. Impellers should be routinely inspected and cleaned. Surface aeration equipment should be cleared of ice build-up. Routine checks of floats, cables, and other appurtenances should be performed.

9.6 NOMENCLATURE

AE	kg/kWh, lb/hp-h	aeration efficiency
DWP	cm of water	dynamic wet pressure
F/M	lb BOD$_5$/d-lb MLSS	food to microorganism ratio
MCRT	d	mean cell residence time
OTE	–, %	oxygen transfer efficiency
OTR	kg/h, lb/h	oxygen transfer rate
SAE	kg/kWh, lb/hp-h	standard aeration efficiency
SOTE	–, %	standard oxygen transfer efficiency
SOTR	kg/h, lb/h	standard oxygen transfer rate

9.7 BIBLIOGRAPHY

Andersson, L.G. (1979). *Energy Savings at Wastewater Treatment Plants,* A Report to Commissioner of European Community and Danish Council of Technology, Water Quality Institute, Denmark.

EPA (1989). *Fine Pore Aeration Systems — Design Manual,* EPA 625/1-89/023, USEPA Risk Reduction Research Labs, Cincinnati, OH.

Flanagan, M.J. and Bracken, B.D. (1985). *Design Procedures for DO Control of Activated Sludge Processes,* EPA66/2-77/032, NTIS No. PB 270960, USEPA, Cincinnati, OH.

Robertson, P. et al. (1984). *Energy Savings — Optimization of Fine Bubble Aeration,* Final Report and Replicators Guide, Water Resources Center, Stevenage Laboratories, Stevenage, UK.

Stephenson, J.P. (1985). "Practices in Activated Sludge Process Control." *Comprehensive Biotechnology: The Principles, Applications, and Regulations of Biotechnology in Industry, Agriculture, and Medicine,* Ed: M. Moo-Young, 4, 1311.

WPCF (1988). Aeration — Manual of Practice FD-13, WEF, Alexandria, VA.

Index